Principles of Analytical and Instrumental Techniques

NIPA GENX ELECTRONIC RESOURCES & SOLUTIONS P. LTD.

New Delhi-110 034

About the Author

Dr. Premasis Sukul, a retired Professor, received his M.Sc. and Ph.D. in Soil Science and Agricultural Chemistry from Indian Agricultural Research Institute, New Delhi. He has earned a wide range of research experience on pesticide analysis in different environmental compartments with a focus on environmental risk assessment. He worked as Professor and Dean in the School of Agricultural Sciences, Swami Vivekananda University, Kolkata, as an Associate Dean at Bidhan Chandra Krishi Viswavidyalaya, Burdwan campus; as Professor at Lovely Professional University, Phagwara; as Professor at Bidhan Chandra Krishi Viswavidyalaya, Mohanpur main campus, Nadia; as Senior Research Scientist at the Institute of Environmental Science (INFU), Technical University of Dortmund, Germany; as Deputy General Manager, Interfield laboratory, Cochin, and as Product Development Executive at Jai Chemicals, Faridabad. Besides, he gathered research experiences by carrying out postdoctoral studies as DAAD Scholar at GSF Institute, Munich and as Alexander von Humboldt Scholar at Kassel University and Dortmund University, Germany.

He mentored several M.Sc. and Ph.D. students as their supervisor. His numerous publications as original research papers, review papers and several book chapters in national and internationally reputable books and journals bear witness to his successful academic and research career. Recently, a textbook on 'Soil Chemistry and Plant Nutrients', authored by him, has been published by New India Publishing Agency, New Delhi.

Principles of Analytical and Instrumental Techniques

Premasis Sukul
Former Professor and Dean
School of Agricultural Sciences
Swami Vivekananda University
Barrackpore, Kolkata-700121
West Bengal, India

NIPA GENX ELECTRONIC RESOURCES & SOLUTIONS P. LTD.
New Delhi-110 034

**NIPA GENX ELECTRONIC
RESOURCES & SOLUTIONS P. LTD.**

101,103, Vikas Surya Plaza, CU Block
L.S.C.Market, Pitam Pura, New Delhi-110 034
Ph : +91 11 27341616, 27341717, 27341718
E-mail:newindiapublishingagency@gmail.com
www: www.nipabooks.com
For customer assistance, please contact
Phone: + 91-11-27 34 17 17 Fax: + 91-11- 27 34 16 16
E-Mail: feedbacks@nipabooks.com

ISBN: 978-93-91383-92-3

Composed and Designed by NIPA.

Preface

This book deals with various subject matters of analytical and instrumental techniques. Its intention is to serve as valuable text-and reference book to the undergraduates, postgraduates and the researchers in the field of agriculture, horticulture, food science, home science, forestry, biochemistry, biotechnology, agricultural chemicals and other allied fields. I believe that this book will prove to be of utility and bring clarity to the readers.

The main objective of this book is to familiarize the readers with the basics of standard solution preparation, working mechanisms of various indicators, buffer solutions, pH measurement, various titrimetric methods of analysis, separation of individual components from a mixture of substances, and their qualitative and quantitative estimation using instruments. Through this book, the readers should be able to be acquainted with the basic working principles underlying each technique along with their applications and the instruments that are invariably used in the research domain. While preparing the book, special attention was paid to not overcomplicate but provide enough background knowledge to make it useful for starters in this domain as well as experienced readers.

Each chapter of the book contains a few model questions to help the learners self-assess their grasp of the subject as well as practice frequently asked questions in various competitive examinations. Necessary references have been incorporated to motivate readers for further exploration.

Improvement is a continuous process. Therefore, I am looking forward to feedbacks and constructive comments from the readers.

Lastly, I thank my loving students and friends for their continuous encouragement while I was writing this book. My heartfelt gratitude also goes to my beloved family members: my wife Kumkum, and my children Premankur and Priyanka for their moral support.

Author

Contents

Preface .... *v*

1. Acid-base, pH, and Buffer Solution 1

1.1. Acids and bases 1

1.1.1. Arrhenius theory on acids and bases 1

1.1.2. The Brønsted and Lowry Theory 2

1.1.3. The Lewis Theory 3

1.2. Concept of pH 6

1.2.1. pH measurement 8

1.3. Buffer solution 10

1.3.1. Mechanism of buffering action 10

1.3.2. Buffer solution preparation 12

1.3.3. Henderson-Hasselbalch equation 12

1.3.4. Buffer capacity 13

1.4. Numerical problems and solutions 14

1.5. Model questions 15

1.6. Suggested readings 16

2. Volumetric Analysis 17

2.1. Introduction 17

2.2. Precautions 18

2.3. Standard solution 18

2.3.1. Expression of concentration in a standard solution 19

2.4. Dilution of standard solution 25

2.4.1. Solution preparation from impure chemicals 26

2.5. Types of titration 27

2.5.1. Acid-base titration 28

2.5.2. Redox titration 29

2.5.3. Complexometric titration 30

2.5.4. Precipitation titration 33

2.6. Model questions 34

2.7. Suggested readings 35

3. Indicators 37

3.1. Introduction 37

3.2. Acid-base indicator 37

3.2.1. Mechanism 38

3.2.2. Indicators pH range or colour change interval 43

3.2.3. Properties of a good indicator 49
3.2.4. pH Paper 49
3.3. Fluorescent indicator 50
3.4. Precipitation indicator 50
3.5. Adsorption indicator 51
3.6. Mixed indicator 52
3.7. Universal indicator 52
3.8. Solution preparation for commonly used indicators 52
3.9. Model questions 59
3.10. Suggested readings 59

4. Centrifugation 61

4.1. Introduction 61
4.2. Principle 62
4.3. Types of centrifuge 69
4.3.1. Low-speed centrifuge 70
4.3.2. Medium-speed centrifuge 70
4.3.3. High-speed centrifuge 70
4.3.4. Ultracentrifuge 71
4.3.5. Differential centrifugation 72
4.3.6. Density gradient centrifugation 73
4.4. Choice of gradient materials 76
4.5. Rotors 76
4.5.1. Swing-bucket rotor 77
4.5.2. Fixed-angle rotor 78
4.5.3. Vertical rotor 78
4.5.4. Zonal rotor 79
4.6. Care of Rotor 81
4.7. Rotor tubes 82
4.8. Model questions 83
4.9. Suggested readings 84

5. Electrophoresis 85

5.1. Introduction 85
5.2. Principle 85
5.3. Factors influencing mobility of charged molecules 86
5.3.1. Electrical field 86
5.3.2. Buffer solution 87
5.3.3. Sample under investigation 88
5.3.4. Supporting medium 88
5.4. Instrumentation and operational procedure 89
5.4.1. Instrumentation 89
5.4.2. Operational procedure 90
5.5. Types of electrophoresis 93

5.5.1. Free or moving boundary electrophoresis 93
5.5.2. Zone electrophoresis 94
5.6. Iso-electric Focusing 98
5.7. Discontinuous or Disc Electrophoresis 100
5.8. Sodium dodecyl sulfate polyacrylamide gel electrophoresis (SDS-PAGE) 101
5.8.1. Interpretation of results and estimation of molecular weights .. 102
5.9. Native-PAGE 103
5.10. Gradient gel electrophoresis 103
5.11. Pulsed-field gel electrophoresis (PFGE) 104
5.12. Two-dimensional electrophoresis (2-D electrophoresis) 104
5.13. General applications of electrophoresis 106
5.14. Model questions 107
5.15. Suggested readings 108

6. Spectroscopy 109

6.1. Introduction 109
6.2. What are electromagnetic radiation and spectrum? 111
6.3. Why do objects appear coloured? 115
6.4. Atomic spectroscopy 115
6.5. Molecular spectroscopy 116
6.6. What are colorimetry, photoelectric colorimetry, and spectrophotometry? 117
6.7. Basic theory of spectrophotometry 120
6.7.1. Lambert's law 121
6.7.2. Beer's law 122
6.7.3. Beer- Lambert law 122
6.8. Limitations of Beer's Law 123
6.9. Sources of error 124
6.10. UV-VIS spectrophotometry 126
6.10.1. Instrumentation 126
6.10.2. Standard curve 130
6.10.3. What happens when molecules are exposed to the UV-VIS spectrum? 132
6.10.4. Applications of UV-VIS spectroscopy 134
6.11. Fluorimetry 135
6.11.1. Introduction 135
6.11.2. Basic theory 137
6.11.3. Instrumentation 139
6.11.4. Intrinsic and extrinsic fluorescence 140
6.11.5. Application 140
6.12. Nephelometry and turbidimetry 141
6.12.1. Introduction 141
6.12.2. Theory 142
6.12.3. Instrumentation 144

6.12.4. Application 145
6.13. Infrared spectroscopy 146
6.13.1. Introduction 146
6.13.2. Modes of vibration 147
6.13.3. Principle 148
6.13.4. Sample preparation 150
6.13.5. Instrumentation 151
6.13.6. Application 152
6.14. Atomic absorption spectroscopy (AAS) 154
6.14.1. Introduction 154
6.14.2. Principle 155
6.14.3. Instrumentation 156
6.14.4. Background correction 160
6.14.5. Sample preparation 161
6.14.6. Interferences 162
6.14.7. Application 164
6.15. Flame emission spectroscopy (FES or Flame photometry) 164
6.15.1. Introduction 164
6.15.2. Principle 165
6.15.3. Instrumentation 165
6.15.4. Application 167
6.16. Atomic fluorescence spectroscopy (AFS) 167
6.17. Model Questions 168
6.18. Suggested readings 169

7. Mass Spectrometry 171

7.1. Introduction 171
7.2. MS applications and uses 172
7.3. MS Components 172
7.4. Ionization 173
7.5. Sources of Ionization 174
7.5.1. Electron Ionization (Electron Impact, EI) 174
7.5.2. Chemical Ionization (CI) 175
7.5.3. Field ionization (FI) / Field Desorption (FD) 176
7.5.4. Fast Atom Bombardment (FAB) 177
7.5.5. FIB (Fast Ion Bombardment)-MS 178
7.5.6. Secondary Ion Mass Spectrometry (SIMS) 179
7.5.7. Matrix Assisted Laser Desorption/Ionization (MALDI) 179
7.5.8. Atmospheric Pressure Ionization (API) 180
7.6. Mass analyser 187
7.6.1. Quadrupole analyser 188
7.6.2. Time-of-flight (TOF) analyser 189
7.6.3. Sector analyser 190
7.6.4. Ion-trap analyser 191
7.6.5. Tandem (MS-MS) mass spectrometers 193

7.7. Detector and recorder 194
7.8. Model questions 194
7.9. Suggested readings 195

8. Chromatography 197

8.1. Introduction 197
8.2. Definition 198
8.3. Principle 198
8.4. Classification of chromatographic technique 199
8.4.1. Adsorption chromatography 199
8.4.2. Partition chromatography 203
8.4.3. Miscellaneous 204
8.5. Paper chromatography 204
8.5.1. Principle 204
8.5.2. Support 205
8.5.3. Experimental procedure 206
8.6. Thin-layer chromatography 211
8.6.1. Principle 211
8.6.2. Chromatoplate preparation 212
8.6.3. Adsorbents 212
8.6.4. Solvents 213
8.6.5. Tank 213
8.6.6. Sample application 214
8.6.7. Development 214
8.6.8. Detection 214
8.6.9. Preparative-TLC and –PC 215
8.6.10. Two-dimensional chromatography 216
8.7. Column chromatography 217
8.7.1. Introduction 217
8.7.2. Column packing 217
8.7.3. Adsorbents 218
8.7.4. Solvents and elution 218
8.7.5. Sample loading 219
8.7.6. Eluent collection and detection 219
8.8. Ion-exchange chromatography 220
8.8.1. Introduction 220
8.8.2. Principle 220
8.8.3. Ion-exchanger 221
8.8.4. Ion-exchange capacity 226
8.8.5. Choice of ion-exchanger 227
8.8.6. Choice of buffer 228
8.8.7. Operation 228
8.8.8. Ion-exchange paper chromatography 229

8.9. Gel-filtration chromatography ... 229
8.9.1. Introduction ... 229
8.9.2. Principle ... 229
8.9.3. Gel-chromatographic materials ... 231
8.9.4. Applications ... 233
8.10. Affinity chromatography ... 235
8.10.1. Principle ... 235
8.10.2. Matrix ... 236
8.10.3. Ligand ... 237
8.10.4. Gel matrix activation and introduction of a spacer arm ... 238
8.10.5. Special forms of affinity chromatography ... 242
8.10.6. Operating conditions ... 243
8.11. High performance liquid chromatography ... 245
8.11.1. Introduction ... 245
8.11.2. Principle ... 246
8.11.3. Instrumentation ... 253
8.11.4. Applications ... 260
8.12. Gas chromatography ... 260
8.12.1. Introduction ... 260
8.12.2. Principle ... 261
8.12.3. Instrumentation ... 265
8.13. Model questions ... 280
8.14. Suggested readings ... 281

9. Radiotracer Technique ... 283

9.1. Introduction ... 283
9.2. Types of particles and their properties ... 284
9.2.1. Alpha particle (α-particle, $^{4}He^{2+}$) ... 285
9.2.2. Beta particle (β-particle) ... 285
9.2.3. Neutron ... 287
9.2.4. Gamma (γ) ray ... 287
9.3. Radioactive decay ... 287
9.3.1. Decay by alpha particle emission (alpha decay) ... 288
9.3.2. Decay by β-particle emission (Beta decay) ... 288
9.3.3. Decay by gamma ray emission (Gamma decay) ... 291
9.4. Interactions of particle/radiation with matter ... 292
9.4.1. Interaction of matter with α-particles ... 293
9.4.2. Interactions of matter with β-particles ... 295
9.4.3. Interactions of matter with γ-ray ... 297
9.4.4. Interactions of matter with neutron ... 301
9.5. Disintegration rate ... 302
9.6. Units used in radioactivity monitoring ... 305
9.7. Radioactivity detection ... 306

9.7.1. Gas Ionization / Geiger-Muller Counter (GM Counter) 307
9.7.2. Scintillation Counter 312
9.7.3. Semiconductive detector 319
9.8. Importance and application of radioisotopes 320
9.8.1. Neutron activation 321
9.8.2. Autoradiography 325
9.8.3. Radiometric titration 326
9.8.4. Isotope dilution analysis 327
9.8.5. Radiocarbon dating 329
9.8.6. Use of radioisotopes to study reaction mechanism 330
9.8.7. Use of radioisotope in medicine 330
9.8.8. Use of radioisotope in agriculture 331
9.9. Precautions 331
9.10. Model questions 332
9.11. Suggested readings 333

1

Acid-base, pH, and Buffer Solution

1.1. Acids and bases

Acids are chemical substances that are sour in taste and highly corrosive. The term 'Acid' was derived from a Latin word '*acidus*', meaning 'sour'. They emit sharp odour and transform litmus colour from blue to red. They lose their properties when alkalis are added to them. However, in this process, alkalis also lose their alkaline properties. This means that acids and alkalis react to each other and are neutralized to form other types of compounds, losing their own characteristic properties. Alkalis are slippery in feeling and they change the colour of litmus paper from red to blue.

1.1.1. Arrhenius theory on acids and bases

According to the theory of Svante Arrhenius (1884), acids ionize in water during its dissolution and liberate H^+ ions (protons) and a corresponding negative ion. As for instance, nitric acid (HNO_3) gives rise to H^+ and NO_3^- ions while it dissolves in water ($HNO_3 \rightarrow H^+ + NO_3^-$).

On the other hand, bases are compounds that dissociate in water and produce a positive ion (but not proton) and negative ion, specifically hydroxyl (OH^-) ions. KOH is a basic compound that ionizes in water to liberate OH^- and K^+ ($KOH \rightarrow K^+ + OH^-$)

Arrhenius theory of acids and bases holds good to prove the concept of neutralization by each other because H^+ and OH^- ions generated from acid and base, respectively react themselves to yield neutral molecule H_2O ($H^+ + OH^- \rightarrow H_2O$). According to Arrhenius theory, acid must contain H^+ ion and base must contain OH^- ion. However, this theory cannot explain why all compounds containing hydrogen do not exhibit acidity or it does not hold good for those compounds that do not directly contain OH^- ions but do exhibit characteristic properties of bases (e.g., Na_2CO_3). It cannot explain why methane is not an acid although it possesses hydrogen of oxidation number of +1. It does not explain the case of ammonia which is a weak base. Ammonia does not contain

OH^- ions but it releases OH^- ion when it reacts with water reversibly ($NH_3 + H_2O \leftrightarrow NH_4^+ + OH^-$) and nearly 99% of ammonia molecule is available in a dilute ammonia solution, leading to enrichment of the aqueous solution with OH^-.

However, considering the reversible reaction of water ionization, Arrhenius theory has been further extended to operational theory that states that any compound could be considered as acid or base, if they increase the concentration of proton or hydroxyl ions, respectively, when they are dissolved in water. It has usually been observed that the non-metallic oxides [CO_2, SO_2, SO_3, NO_2, P_4O_{10}; $CO_2 + H_2O \rightarrow H_2CO_3 \rightarrow H^+ + HCO_3^-$), hydroxides (HOCl, $HONO_2$, $O_2S(OH)_2$, $OP(OH)_3$; $HOCl \rightarrow H^+ + OCl^-$] and hydrides (HF, HCl, HBr, HCN, H_2S; $H_2S \rightarrow H^+ + HS^-$) behave like acids; while metallic oxides (K_2O, MgO, CaO; $CaO + H_2O \rightarrow Ca^{+2} + 2OH^-$), hydroxides [NaOH, KOH, $Ca(OH)_2$; $NaOH \rightarrow Na^+ + OH^-$] and hydrides (HI, NaH, CaH_2; $CaH_2 + 2H_2O \rightarrow Ca^{+2} + 2OH^- + 2H_2$) act as bases.

Now, the question may come to why non-metal hydroxide behaves like an acid, while metal hydroxide acts as a base. It may be answered by analysing the electronegativity of the individual atoms. As for instance, in case of HOCl, the electronegativity (EN) of H, O and Cl atoms are 2.20, 3.44 and 3.16, respectively. Therefore, ΔEN (1.24) between O and H is more and facilitates polarization while ΔEN is comparatively very little in between O and Cl, causing equal sharing of electrons. This results in the dissociation of HOCl in water to give rise H^+ ($HOCl \rightarrow H^+ + OCl^-$) because electrons are drawn towards oxygen in the H-O bond. But in the case of NaOH, since the electronegativity of Na atoms is 0.93, ΔEN between Na and O is more (ΔEN, 2.51) than that between O and H (ΔEN, 1.24). This, consequently, leads to unequal distribution of electrons in Na-O bonds causing an electron pull towards the O atom. This results in the dissociation of NaOH in water to liberate OH^- ion ($NaOH \rightarrow Na^+ + OH^-$). However, $Al(OH)_3$ behave as an amphoteric compound because depending on the reaction condition, it may accept OH^-, acting as an acid [$Al(OH)_3 + OH^- \rightarrow Al(OH)_4^-$] or accept H^+, acting as a base [$Al(OH)_3 + 3H^- \rightarrow Al^{+3} + 3H_2O$].

The limitations of Arrhenius theory later led to the development of two other theories on acid-base: the Brønsted and Lowry theory, and the Lewis theory.

1.1.2. The Brønsted and Lowry Theory

The Brønsted and Lowry theory was developed by the scientists, Johannes Nicolaus Brønsted and Thomas Martin Lowry, independently during 1923. According to the Brønsted and Lowry theory, any compound that can act as proton donor is grouped as acid, base is referred as proton acceptor and amphoteric compound can participate in the functioning of both acid and base together. It also states that acid contains a conjugate base and likewise, a base contains a conjugate acid. It is also important to note that any acid or base is technically a conjugate acid or conjugate base also. As for instance, butyric acid ($CH_3CH_2CH_2COOH$) is the conjugate acid of the butyrate anion ($CH_3CH_2CH_2COO^-$), a base, while butyrate is the conjugate base of butyric acid, an acid. In water, a Brønsted and Lowry acid dissociates and results in increasing H^+ ion concentration in the solution and a base dissociates by abstracting a proton from the solution to yield OH^-. Therefore, this theory is a fine-tuning of Arrhenius theory, adding an extra character to the acids and bases that they should not contain necessarily H^+ or OH^- ions. By the definition, HCl qualifies to be grouped as acid because it can donate H^+ and NH_3 is a base as it accepts H^+ in a reaction between HCl and NH_3.

HCl +	NH_3	$\longleftrightarrow$	NH_4^+ +	Cl^-
Acid	Base		Conjugate acid of base NH_3	Conjugate base of acid HCl

On application of the Brønsted and Lowry theory, water molecules behave as base because H^+ practically cannot stay within water molecules due to the pull of protons by the lone pair of electrons in the oxygen of water molecules. Therefore, in the true sense, $(H_3O)^+$ remains in aqueous solution instead of H^+ ions [$H^+ + H_2O \leftrightarrow (H_3O)^+$]. But for simplicity of expression H^+ ions are used.

Based on the extent of dissociation in water, acids and bases are classified as strong acids (HCl, HNO_3, H_2SO_4, etc.), strong bases (Gr 1 and Gr 2 metal hydroxides, such as LiOH, NaOH, KOH, etc.), weak acids (H_2S, HCOOH, CH_3COOH, etc.) and weak bases (NH_3, CH_3NH_2, etc.). Strong acids and strong bases completely dissociate in water, no undissociated acid molecule can remain in the solution. For instance 1M HCl contains 1M H^+ ions and 1M Cl^- ions. Conversely, weak acids and weak bases do not dissociate completely, but partially, and a major portion of weak acids and bases remain in their undissociated form.

1.1.3. The Lewis Theory

In this theory, the ability of a substance for transferring pairs of electrons is taken as a guide to define acid or base, without specifically considering hydrogen atoms. The Lewis theory describes the acids and bases as electron pair acceptors and electron pair donors, respectively. Gilbert N. Lewis proposed this theory in 1923. Lewis acids do not need to possess protons, but they possess outer electron shells that are capable of expansion. For instance, BF_3 acts as an acid because it can accept a pair of electrons due to its incomplete octet. Thus, Lewis acids are usually non-protonic acids. Conversely, NH_3 is characterized by a free lone pair of electrons. For this reason, NH_3 and BF_3 react within themselves to neutralize each other, forming a covalent bond in between N and B ($NH_3 + BF_3 \rightarrow H_3N\text{-}BF_3$), where BF_3 and NH_3 act as electron pair acceptor and donor, respectively. Also, the Lewis acid-base theory could be extended to consider any solute containing cationic characteristics as an acid and any solute exhibiting anionic characteristics as a base. When sulphur dioxide dissociates to SO^{+2} and SO_3^{-2}, they behave like acid and base correspondingly (**Figure 1.1**). Ag^+ acts as a Lewis acid when it reacts with Cl^- ($Ag^+ + Cl^- \rightarrow AgCl$)

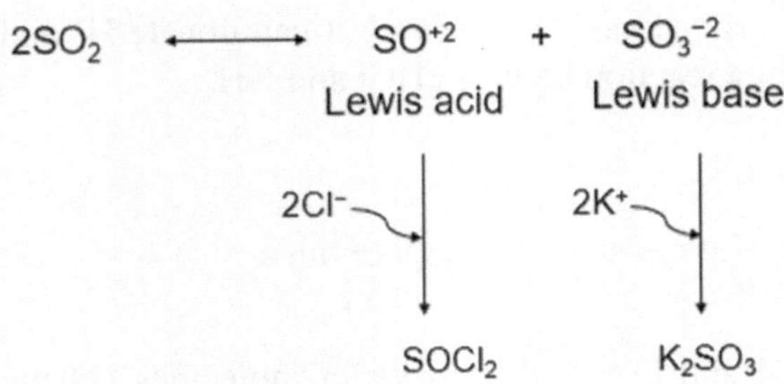

Fig. 1.1: Application of Lewis acid-base concept on ionic species of SO_2

This is important to note that acids, that are releasing protons, essentially do not function as electron acceptor, as for instance CH_3COOH. Oxidation-reduction reaction also involves transferring electrons from reducing agent to the oxidizing agent. However, unlike to the Lewis acid-base where a lone pair of electrons in the base makes a bond with the acid to form a neutral compound, it (oxidation-reduction reaction) deals with complete transfer of electrons to the oxidizing agent from the reducing agent.

Water molecules, although to a lesser extent, give rise to H^+ and OH^- ions in equal numbers by autoionization and show the property of conductivity, even in highly purified form. The conductivity goes on increasing in an aqueous solution with higher concentration of impurities within it. It has been evidenced that about one molecule of water in 10 million is ionized at any given time. For simplicity of expression, we prefer to mention the release of H^+ and OH^- ions from water molecules. But in the true sense, it is the hydronium ion (H_3O^+) instead of H^+ ion. Proton does not exist as such. It becomes solvated with one water molecule and converts to H_3O^+ through a sequential reaction as follows:

H_2O	$\leftrightarrow$	$H^+ + OH^-$	Reaction 1.1
$H_2O + H^+$	$\leftrightarrow$	$(H_3O)^+$	Reaction 1.2
$2H_2O$	$\leftrightarrow$	$(H_3O)^+ + OH^-$	Reaction 1.3

If we apply law of mass action for Reaction 1.1, we may get the following relationship at any given temperature:

$$a_{H^+} \times a_{OH^-}/a_{H_2O} = \{[H^+][OH^-]/[H_2O]\} \times (y_{H^+} \times y_{OH^-}/y_{H_2O}) = \text{Constant} \quad \text{Equation 1.1}$$

where a, y and square brackets stand for activities, activity coefficient and concentration of a specific species, respectively.

But because of lower extent of ionization, ionic concentrations of H^+ and OH^- are too little and their activity coefficients are considered as unity and for this reason, the activity of undissociated water molecules is considered as unity and its concentration remains constant.

Therefore, further we can write

$$K_w = [H+]\,[OH^-] \quad \text{Equation 1.2}$$

where K_w is known as an ionization constant or ionization product of water. Degree of ionization of water is measured by the ionization constant which varies with temperature (**Table 1.1**). K_w value increases with increase in temperature.

Table 1.1: Ionic product, pK_w and pH of water at various temperature

Temperature (°C)	K_w (ionic concentrations are expressed in moles/L)	pK_w	pH
0	0.12×10^{-14}	14.94	7.47
20	0.68×10^{-14}	14.16	7.08
25	1.0×10^{-14}	14.0	7.0
30	1.47×10^{-14}	13.84	6.92

Source: adapted from (a) Mendham, J.; Denney, R. C.; Barnes, J. D.; Thomas, M. and Sivasankar, B. (2018) Vogel's Textbook of quantitative Chemical analysis. 17th impression, Pearson Education Pvt. Ltd., India, pp: 17. (b) Sergey Bylikin (2016) Acid-base equilibria, buffers and pH (The Chemistry Tutorials Series), Kindle Edition.

In pure water at 25 °C, the concentration of H^+ and OH^- ions remain the same, each exhibiting a concentration of 1×10^{-7} moles/L of water. Therefore, pure water is considered as neutral in reaction. When electrolytes are added to pure water, it may cause reduction in H^+ ions concentration ($< 10^{-7}$ mol/L), leading to more abundance of OH^- ions in the solution due to abstraction of H^+ or release of OH^-, the solution is then considered as alkaline or basic. However, when the addition of electrolyte to pure water is responsible for further release of H^-, proton concentration becomes more than 10^{-7} mol/L and then the solution is regarded as acidic in reaction. If the added substance to water does not alter the relative concentration of H^+ or OH^- ions of pure water, the solution is regarded as 'neutral' in chemical nature.

1.2. Concept of pH

By definition, pH is negative logarithm (base 10) of hydrogen ion concentration.

$$pH = -\log [H^+], \qquad \text{Equation 1.3}$$

which may be derived into the following relationship:

$$[H^+] = 10^{-pH} \qquad \text{Equation 1.4}$$

Therefore, in pure water, considering presence of equimolar concentration of H^+ and OH^- ion concentrations (10^{-7} mol/L for each ion), we may write

$$pH = -\log[10^{-7}]$$
$$= 7 \log 10$$
$$= 7$$

For this reason, pH 7 is considered as a neutral point in a pH scale. Likewise, pOH in pure water =7.

Considering Equation 1.2 (K_w = [H+] [OH^-]), we may write the following equation, putting log in both sides:

$$\text{Log}\,(1/K_w) = \log 1/[H+] + \log 1/[OH^-] \qquad \text{Equation 1.5}$$

$$pK_w = pH + pOH \quad = 7 + 7$$
$$= 14\ (K_w = [10^{-7}]\,[10^{-7}])$$

Thus, the shift of pH may be resulted by incorporating acidic or basic substance to pure water and based on pH value, the substance is classified as 'acidic' (pH < 7) and 'basic' (pH >7). An acidic solution possesses relatively more H^+ ions and less OH^- ions than that of pure water and a basic solution contains more OH^- ions and less H^+ ions than that of pure water. Using the pH scale between 0-14, it is possible to define the extent of acidity or alkalinity between 1 mol/L of H^+ ions (pH becomes 0 because $[H^+] = 0$) and 1 mol/L of OH^- ions (pH becomes 14 because $[H^+] = K_w/[OH^-] = 10^{-14}/10^0 = 10^{-14}$).

As we find pK_w = Log ($1/K_w$), similarly we may get pK_a = Log ($1/K_a$), pK_b = Log ($1/K_b$) and pK_s = Log ($1/K_s$); where K_a, K_b and K_s are the dissociation constant of any acid , base and solubility product of any salt.

The relationship between pH and pOH is depicted in **Table 1.2**.

Table 1.2: Relationship between pH and pOH in pure water

pH	$[H^+]$ mole/L	pOH	$[OH^-]$ Mole/L
0	10^{-0}	14	10^{-14}
1	10^{-1}	13	10^{-13}
2	10^{-2}	12	10^{-12}
3	10^{-3}	11	10^{-11}
4	10^{-4}	10	10^{-10}
5	10^{-5}	9	10^{-9}
6	10^{-6}	8	10^{-8}
7 (Neutral)	10^{-7}	7 (Neutral)	10^{-7}
8	10^{-8}	6	10^{-6}
9	10^{-9}	5	10^{-5}
10	10^{-10}	4	10^{-4}
11	10^{-11}	3	10^{-3}
12	10^{-12}	2	10^{-2}
13	10^{-13}	1	10^{-1}
14	10^{-14}	0	10^{-0}

For convenience and simplicity in data generation, a pH scale (**Figure 1.2.**) is made using negative logarithm to generate positive integers. Thus, 10^{-7} is converted to $-\log[10^{-7}]$, so that finally '7' is obtained in place of 0.0000001. One-unit change in the pH value signifies a 10-fold change in the actual activity of H^+, and the activity increases as the pH value decreases. A solution of pH 5 has 10 times more hydrogen ions than a solution of pH 6. Likewise, in the alkaline range, a solution of pH 9 signifies 10 times more hydroxyl ions (or 10 times lower protons) than that of a solution of pH 8. The pH scale ranges between 0 and 14. At a high concentration of H^+ (10^{-1} M), the pH value is very low, pH = 1, while at low concentration (10^{-12} M), the pH is high, pH value becomes 12.

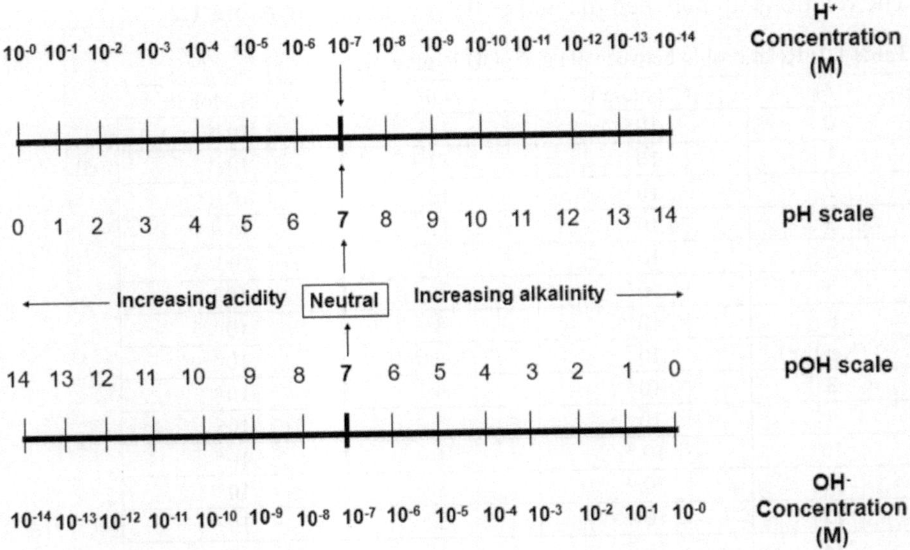

Fig. 1.2: pH scale: relationship between pH & H^+ ion concentration (moles/L of water) and between pOH and OH^- ion concentration (moles/L of water)

1.2.1. pH measurement

Colorimetric and electrometric methods are usually practised to measure pH of a solution. However, most accurate measurement of soil pH is done by pH meter, utilizing the principle of electrometric method.

Colorimetric method

Colorimetric method uses suitable dyes or acid-base indicators which usually change their colour with changing in H^+ activity or concentration. Litmus paper turns red in acid conditions and blue in alkaline conditions. Therefore, using litmus paper it is possible to examine whether the solution is acidic, neutral or alkaline. However, the extent of acidity or alkalinity cannot be assayed. It is a very simple, cheap and less time-consuming process but less precise.

pH meter

A reference electrode (known as calomel electrode) and a glass electrode (sensitive to H^+ concentration), existing as a separate entity or in a fused form as a single entity, are attached to a potentiometer for measuring electrical potential (emf) because a potential difference (voltage difference) is developed between two electrodes, which is proportional to pH. The glass membrane in the glass electrode facilitates the permeability of hydrogen ions excluding the other ions, thus creating a potential across the glass membrane due to the presence of H^+. The glass electrode consists of a thin H^+ sensing glass containing dilute HCl (0.1M) where an Ag-AgCl wire with a constant potential is fixed. Outer surface of the glass membrane comes in contact with the test solution under the backdrop of the inner surface in contact to a constant hydrogen ion concentration. In most cases, the reference and the glass electrode are fused together in one tube. The calomel electrode is a half-cell, containing mercury and calomel (mercury(I) chloride) covered with KCL solution of definite concentration that may be saturated or 1M or 0.1M. The potential is dependent on H^+ concentration of the test solution. Eventually, the potential difference formed between the electrodes pair is measured with a voltmeter. The relation between emf and pH is governed by the Nernst equation:

$$E = E^0 - (RT/nF) \times \log [H^+] \qquad \text{Equation 1.6}$$

Where,

E = emf produced by electrode system,

Eo = a constant dependent on the electrodes used,

R = gas constant,

T = absolute temperature,

N = number of electrons involved in equilibrium, and

F = Faraday constant.

Since E^0 value differs from electrode to electrode, standardization of the electrode is essential using buffer solutions (solutions with known H^+ concentration). It is also important to adjust the temperature of the test solution and buffer solution with a temperature control knob to eliminate the error due to temperature difference as the Nernst equation says about the dependency of generated potential by temperature.

Initially the electrodes are rinsed with distilled water and wiped with tissue paper. After setting the temperature knob to the room temperature, the pH meter is calibrated, using two buffer solutions (4 and 9). Usually, one buffer solution should have a pH below the pH of the test solution and another buffer solution above the pH of the test solution. After such standardization of the pH meter, the test solution is measured for its pH. After completion of pH measurement, the electrode should be washed properly with distilled water, wiped with tissue paper and to avoid drying immersed in suitable buffer solution as per the manufacturer's guidelines.

1.3. Buffer solution

Buffer Solution is an aqueous solution containing a weak acid and the conjugate base of that weak acid in a form of salt, or a weak base and the conjugate acid of that weak base in a form of salt, as for instance a mixture solution of acetic acid and sodium acetate or a mixture solution of ammonia and ammonium hydroxide. Similarly, mixture solution of pyridine (C_5H_5N) which is a weak base and its salt, pyridinium chloride where pyridinium cation ($C_5H_5NH^+$) acts as its conjugate acid, is another common example of buffer solution. A buffer solution exhibits minimum change in its pH, when it is diluted or solutions of acid/base are added to it in small quantity because the combined presence of the weak acid and a salt containing its conjugate base (or weak base and a salt containing its conjugate acid) resists a change in pH by maintaining a chemical equilibrium in the reactions they involve and hence, helps to maintain a relatively stable solution pH. The composition of a buffer solution is supposed to be such that it facilitates the removal of any extra amount of hydrogen ions or hydroxide ions, if they are added to the solution in smaller amount. This unique characteristic of buffer solution helps to carry out several biological and chemical reactions that are supposed to be conducted at a specific and relatively stable pH. Whether one should use a buffer solution consisting of a weak acid and its conjugate base (acidic buffer solution exhibiting pH <7.0) or a weak base and its conjugate acid (alkaline buffer showing pH>7.0), it depends on the desired and working pH of the reaction. According to the Henderson-Hasselbalch equation, the pH of the buffer solution can easily be changed by manipulating the ratio of acid to salt. However, the Henderson–Hasselbalch equation is not applicable to strong acids and strong bases.

1.3.1. Mechanism of buffering action

The main principle of mode of action of any type of buffer solution is to remove H^+ or OH^- ions which are added to the solution in small amounts. Attempts are made here to understand the reaction mechanism, using examples of acetate buffers which contain acetic acid and its salt sodium acetate. Acetic acid, being a weak acid, dissociates feebly to give rise to a limited number of acetate ions and protons. Therefore, concentration of undissociated acetic acid remains more at equilibrium or it may be said that reaction proceeds to the left side.

$CH_3COOH \rightleftharpoons CH_3COO^- + H^+$ Reaction 1.4

[Practically the reaction proceeds as $CH_3COOH + H_2O \rightleftharpoons CH_3COO^- + H_3O^+$]

When sodium acetate is incorporated into the system, it dissociates faster, and acetate concentration is increased in the system.

$CH_3COONa \rightleftharpoons CH_3COO^- + Na^+$ Reaction 1.5

As a result, according to Le Chatelier's Principle, there will be a further shift in Reaction 1.4 from right hand side to left side. Under this condition, if acid (as for instance, one drop of HNO_3) is added in small amount to the solution, the buffer solution shows a tendency to remove the extra added H^+ because extra amount of added protons reacts to acetate ions of the system to form acetic acid and since acetic acid is week in dissociation, extra added protons disappear, and the solution pH is maintained.

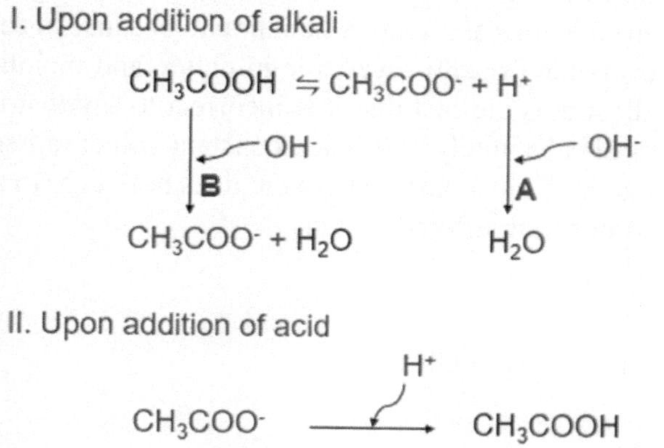

I. Upon addition of alkali

$CH_3COOH \rightleftharpoons CH_3COO^- + H^+$

B: $CH_3COOH + OH^- \rightarrow CH_3COO^- + H_2O$

A: $H^+ + OH^- \rightarrow H_2O$

II. Upon addition of acid

$CH_3COO^- \xrightarrow{H^+} CH_3COOH$

Fig. 1.3: Resistance to change in pH in an acetate-sodium acetate solution system upon addition of alkali (I) and acid (II)

Conversely, when OH^- ions (as for instance one drop of NaOH solution) are added to the buffer solution (obviously in small quantities), they interfere with the equilibrium of Reaction 1.4 by reacting with H^+ to produce H_2O (**Figure 1.3 A**). However, during that time the equilibrium of Reaction 1.4 shifts from left to right side because there is a continuous removal of protons by neutralization with added OH^-. Again, OH^- directly reacts with acetic acid giving rise to acetate ions and water molecules (**Figure 1.3 B**). As a result, extra added OH^- is consumed in the buffer solution without changing the solution pH.

Alkaline buffer solution also acts in a similar way; as for instance, the mixture solution of NH_3 and NH_4Cl. In this case NH_3 in water dissociates to form NH_4^+ and OH^- and NH_4Cl dissociates to NH_4^+ and Cl^- ions. As a result, in the system there is an abundance of undissociated NH_3 concentration as it is a weak base and also a higher amount of NH_4^+ ions, coming mostly from NH_4Cl and partly from NH_3. During formation of NH_4^+ ions from NH_3 in water, OH^- ions are generated to make the solution alkaline. The reaction takes place as follows:

$NH_3 + H_2O \rightleftharpoons NH_4^+ + OH^-$ Reaction 1.6

$NH_4Cl \rightleftharpoons NH_4^+ + Cl^-$ Reaction 1.7

When acids are added to such a system, protons are consumed by OH^- ions (Reaction 1.6) to form neutral water and at the same time they are consumed by NH_3, yielding more of NH_4^+ [**Figure 1.4 (I)**]. When extra OH^- ions are added to this system, ammonium ions then react with OH^- ions to form an undissociated form of NH_3 and water. Thus, extra amounts of H^+ or OH^- added to the system are neutralized to some extent. However, it should be kept in mind that in both acetic acid-sodium acetate and ammonia-ammonium chloride system, majority of the added protons or hydroxide ions are removed but not all because in each case the reactions are of reversible or semi-reversible in nature and maintain an equilibrium state. This illustrates the fact that it is incorrect to say that the buffer solution maintains its pH absolutely 100% at a constant value; rather it is better to say that the change in pH in a buffered system does not occur to the same extent as it is expected in an unbuffered system.

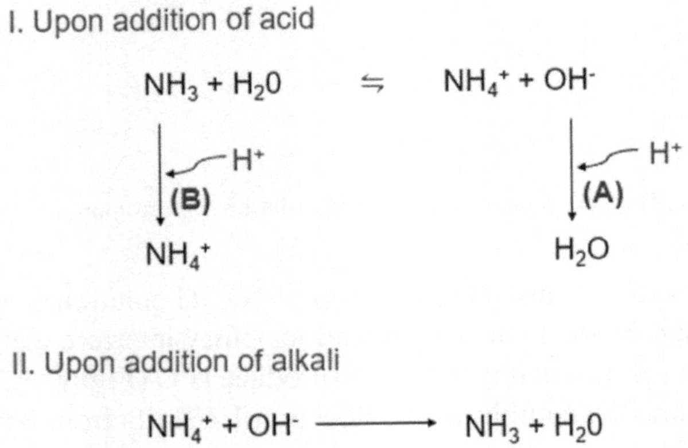

I. Upon addition of acid

$NH_3 + H_20 \leftrightharpoons NH_4^+ + OH^-$

(B) H^+: $NH_3 \rightarrow NH_4^+$

(A) H^+: $OH^- \rightarrow H_2O$

II. Upon addition of alkali

$NH_4^+ + OH^- \longrightarrow NH_3 + H_20$

Fig. 1.4: Resistance to change in pH in an ammonia-ammonium chloride solution system upon addition of acid (I) and alkali (II)

1.3.2. Buffer solution preparation

A buffer solution of desired pH can be prepared easily with the help of Henderson-Hasselbalch equation by regulating the ratio of salt and acid or salt and base. However, for this, pK_a (acid dissociation constant) of the acid and pK_b (base dissociation constant) of the base should be known at a given condition.

1.3.3. Henderson-Hasselbalch equation

In case of a buffer solution which is composed of a weak acid (HA) and its salt with a strong base; HA in water ionizes partially as follows:

$HA + H_2O \rightleftharpoons H^+ (H_3O^+) + A^-$ (conjugate base) Reaction 1.8

$K_a = [H^+][A^-]/HA$ (K_a = acid dissociation constant)

$-\log K_a = -\log[H^+] - \log\{[A^-]/[HA]\}$

$pK_a = pH - \log\{[base]/[Acid]\}$

pH (of acid buffer) = $pK_a + \log\{[base]/[Acid]\}$ Equation 1.7

Equation 1.7 is known as Henderson-Hasselbalch equation. Actually, Lawrence Joseph Henderson in 1908 worked on a buffer solution using carbonic acid and depicted an equation. However, afterwards, Karl Albert Hasselbalch put a logarithmic expression on it and the final equation is termed as Henderson-Hasselbalch equation. This equation is very important to calculate buffer solution pH.

For an alkaline buffer solution, the equation becomes as follows:

pOH (of alkaline buffer) = $pK_b + \log\{[base]/[acid]\}$ Equation 1.8

pH (of alkaline buffer) = pKa – $\log\{[base]/[acid]\}$ Equation 1.9

1.3.4. Buffer capacity

Buffer capacity (β), which is also referred to as 'buffer value', 'buffer index' or 'buffer efficiency', tells about the number of moles of strong acid or base you can add in a litre of buffer solution to obtain an increase or decrease of one pH unit. In other words, it is defined as the moles of a strong acid or strong base necessary to be added to change the pH of 1L solution by 1, divided by the pH change. It is a unitless number. Thus, the buffering capacity of a solution explains its efficiency to resist the change in pH during addition of strong acid or base. This can be measured by titrating the buffer with an acid or a base and is reflected by the slope of the titration curve.Van Slyke in 1922 first explored buffer capacity in an equation form which is described as follows:

$\beta = \Delta n / \Delta pH$ Equation 1.10

Where,

β = buffer capacity

Δn = a finite change in gram equivalent of strong base or acid to 1L of buffer solution, and

ΔpH = a finite change in pH

Therefore, from the Equation 1.10, it is understood that buffer capacity becomes 1, when 1 gram equivalent addition of strong acid or base to 1L buffer solution leads to a change of 1 unit of pH value. Thus, a lower change in pH of a buffer solution signifies its higher buffering capacity.

1.4. Numerical problems and solutions

Problem1:

What is the pH of a solution containing 3.5 $\times 10^{-6}$ g ions of H^+/L?

Solution1:

pH= $-\log[H^+] = -\log(3.5 \times 10^{-6}) = -\log 3.5 - \log 10^{-6} = -\log 3.5 + 6\log 10 = -0.54 + 6.0$

$= 5.46$

Problem 2:

What is the $[H^+]$ at solution pH 3.67?

Solution 2:

We can write $[H^+] = 10^{-pH}$ (as pH $= -\log[H^+]$) $= 10^{-3.67}$

$= 2.1 \times 10^{-4}$ g ions of H^+/L

Problem 3:

According to Henderson – Hasselbalch equation, what could be the ratio of base to acid in an acid buffer solution, when (a) pH = pK_a, (b) pH = pK_a + 1 and (c) pH = pK_a + 2?

Solution 3:

(a) If the ratio becomes 1:1, pH becomes equal to pK_a [pH = pK_a + log{[base]/[Acid]}, pH = pK_a + log{[1]/[1]}, pH = pK_a + log1, pH = pK_a + 0, pH = pK_a]

(b) If the ratio becomes 10:1, pH becomes equal to pK_a + 1 [pH = pK_a + log{[base]/[Acid]}, pH = pK_a + log{[10]/[1]}, pH = pK_a + log10]

(c) If the ratio becomes 100:1, pH becomes equal to pK_a + 2 [pH = pK_a + log{[base]/[Acid]}, pH = pK_a + log{[100]/[1]}, pH = pK_a + log10^2, pH = pK_a + 2log10, pH = pK_a + 2]

Problem 4:

A buffer solution contains 1 mL of 1M acetic acid and 100 mL of 1M sodium acetate. What will be the buffer pH if the pKa value of acetic acid is 4.76?

Solution 4:

pH = pKa + log {[salt]/[Acid]} = 4.76 + log{[100 mL x 1M]/[1 mL x 1M]}

= 4.76 + log (100/1) = 4.76 + 2log10

= 4.76 + 2 = 6.76

Problem 5:

Calculate $[OH^-]$ in a solution where $[H^+]$ is 4.33×10^{-4}.

Solution 5:

$[OH^-] = K_w/[H^+] = (1.0 \times 10^{-14})/(4.33 \times 10^{-4}) = 2.31 \times 10^{-11}$

1.5. Model questions

i. What are acids and bases? Explain various theories to define acids and bases. What happens when acids and bases react? How are acids and bases measured?

ii. On which concept the Arhenius theory on acids and bases does hold good? Write down the weakness of the theory.

iii. Using the Brønsted and Lowry theory, explain the reason of acidity in H_2SO_4.

iv. Why do non-metal hydroxides behave like acids and metal hydroxides as bases? Explain it with an example.

v. 'Lewis acids are usually non-protonic acids' – justify the statement.

vi. Discuss the Lewis Theory to define an acid or a base.

vii. What is the relationship between pH of water and temperature?

viii. Why does pure water show neutral pH at 25 °C? The pH of a solution is zero. What does it mean?

ix. Why does a solution of pH 4 show a pOH value of 10?

x. Discuss briefly the principle involved in measuring pH by pH meter. Why is pH scale developed at a negative logarithmic scale? What is the H^+ ion concentration in a solution at pH 6 and why?

xi. Define buffer solution. What are the constituents in a buffer solution? What are the uses of the buffer solution? Explain buffering capacity.

xii. How does a buffer solution work? Explain it with examples.

xiii. In a buffer solution, when acid or base is added in small amounts, the pH of the solution does not change remarkably. Why? Explain, taking examples of both acidic and basic buffer solutions.

xiv. Calculate the pH of a buffer solution containing 0.1M ethanoic acid and 0.2M sodium acetate. K_a for acetic acid is 1.74 x 10^{-5} mol dm^{-3}.

xv. Calculate the pH of a buffer solution containing 0.2 mol dm^{-3} ammonia and 0.2 mol dm^{-3} ammonium chloride. K_a for the ammonium ion (NH_4^+) is 5.62 x 10^{-10} mol dm^{-3}.

1.6. Suggested readings

(i) Petrucci, R. H. (2007) General Chemistry: Principles and Modern Applications. Upper Saddle River, N. J., Pearson / Prentice Hall.

(ii) Chang, R and Goldsby, K. (2014) General Chemistry: The Essential Concepts. 7th Edition, McGraw Hill.

(iii) Bylikin, S. (2016) Acid-base equilibria, buffers and pH (The Chemistry Tutorials Series), Kindle Edition.

2

Volumetric Analysis

2.1. Introduction

Volumetric analysis, also known as titrimetric analysis, is an analytical method of estimating the concentration of a known solute (analyte) in a solution by measuring the volume of a second known solution of known concentration (standard solution), that reacts completely with the known solute of unknown concentration. The prerequisite of titration is that the two different analytes, present in solution form, must involve in a fast chemical reaction. The standard solution is more commonly known as *'titrant'* and the solution whose concentration needs to be determined is frequently called *'titrand'*. In general, the process, also known as *'titration'*, involves the slow addition of titrant from a burette to the titrand in a conical flask till the reaction between the two solutes in solution form is completed. When the reaction occurs in between an acid and a base, the titration is known as an acid-base titration; and if the nature of the reaction is of oxidation and reduction, it is referred to as a redox titration and likewise precipitation, adsorption, complexation type of reactions may also involve in titration. The completion of the reaction is detected usually with the help of an indicator that sharply changes the colour at or near the equivalence point. The indicator is initially (before beginning the titration) added to the titrand solution. The point at which the colour change takes place is considered as the end point of the titration or the equivalence point because at this point the solutes from titrand and titrant react completely and stoichiometrically. For this, the equivalence point or end point of titration is also known as *'stoichiometric end point'*. However, a 'titration error' is inevitable due to the difference between a theoretical endpoint and the experimental end point. The choice of indicator plays a significant role in minimizing the titration error. The reasons for changing colour are explained separately in Chapter 3, Section 3.2.1. In such a process, the quantity of standard solution required for completion of the reaction is monitored from the burette readings. Because of the known concentration of the titrant, one can calculate the number of moles of titrant required to complete the reaction, using the titrant volume. As the equation of the reaction is known, the number of moles of the titrand solute is determined easily.

2.2. Precautions

One should be very much careful in titration reaction in two aspects: One relates to the reading out of the apparent position of the standard solution meniscus in the burette. The meniscus position should be seen keeping the eye level parallel to the meniscus level. Otherwise, looking below or above the meniscus level leads to parallax and the reading becomes low or high. The other one relates to the detection of end points. If the endpoint is not detected properly, a possibility of under-estimation or over-estimation remains. To avoid such error, the titrant should be added carefully at a slow rate, most preferably dropwise. As the colour changing end point is approached, one can see it momentarily but fades out quickly upon stirring the solution. Finally, the colour persists, and the end point is achieved. Usually, fading occurs more slowly as the endpoint is approached. Proper control on the stopcock of the burette is of utmost importance in this regard.

2.3. Standard solution

A solution of known concentration is referred as standard solution. However, based on the quality and mode of preparation, we may have primary standard and secondary standard. '*Primary standard*' solution can be prepared by dissolving an accurately weighed known quantity of an analyte into a known volume of solvent, preferably using a volumetric flask. The prerequisites of a primary standard reagent must be of its availability in pure form (>99.9% purity), its easy solubility in a suitable solvent and its stability in pure form as well as in solution form. The reagent should not be hygroscopic so that it should not absorb moisture or react with oxygen or carbon dioxide while weighing. It should be dry enough during weighing in an analytical balance; otherwise, accurate weighing of the analyte is not possible, leading to an erratic and inconsistent concentration of the solution. *'Secondary standard'* solutions refer to those solutions which are made up of the reagents other than the primary reagents. Like primary standard solution, it cannot be made by direct weighing the solute in a certain specific amount and dissolving it in a definite volume of solvents. The secondary reagents do not contain the properties that are essentially required for primary standards. The most common characteristics of the secondary standard reagents are that they are hygroscopic and not obtained in pure form. As for instance, when NaOH is weighed, it starts absorbing moisture due to their hygroscopicity and also it absorbs CO_2 from the atmosphere. So, an accurate weighing is not possible. In such cases, a solution is prepared approximating the desired concentration and the solution is then standardized by titrating it against a primary standard solution to obtain its accurate concentration with high accuracy. Thus, if stoichiometry is utilized to determine the concentration of a titrant, it is called a secondary standard solution [*Source:IUPAC. Compendium of Chemical Terminology, 2nd ed. (the "Gold Book"). Compiled by A. D. McNaught and A. Wilkinson. Blackwell Scientific Publications, Oxford (1997). Online version (2019) created by S. J. Chalk. ISBN 0-9678550-9-8.* https://doi.org/10.1351/goldbook]. However, there may be a build up of errors, if sufficient care is not taken to glassware calibration, technique, stability of solutions, etc. Oxalic acid, benzoic acid, anhydrous Na_2CO_3, NaCl, KCl, etc. are few examples of primary standard, while NaOH, KOH, HCl, H_2SO_4, $KMnO_4$, etc. are secondary standards.

2.3.1. Expression of concentration in a standard solution

There are many different ways to express the concentration of solutions, such as molarity, molality, normality, formality, volume percentage, weight percentage and part per million (ppm).

2.3.1.1. Mole

Before going through the different concentration expression, one should possess a clear knowledge of 'Mole' that corresponds to the mass of a substance containing 6.02 x 10^{23} particles (atoms, molecules, ions, or electrons) of the substance and is expressed as a symbol 'mol'. One mol of a substance is the quantity equivalent to the substance's atomic or molecular mass (atomic or molecular weight). Thus, 1 mol carbon possesses a mass of 12 g and contains 6.02 x 10^{23} of carbon atoms. 1 mol of water molecules have a mass of 18 g because water possesses a molecular mass of 18 and 1 mol of glucose (molecular weight 180) is of 180 g. We express the quantities of reactants and products in a chemical reaction in terms of mole. If a substance 'A' reacts with substance 'B' to give rise products of 'C' and 'D', as per the reaction equation: $A + 2B \rightarrow 3C + D$, we presume that 1 mole of A reacts with 2 moles B to yield products of 3 moles of C and 1 mole of D. The concept of mole is applicable to any elementary units that may be of an ion, a radical, an electron, an atom, and a molecule.

2.3.1.2. Molality (m)

Molality, also known as molal concentration, is defined as the total moles of a solute contained in a kilogram of a solvent. Thus, it may be expressed by the following formula with a unit of 'moles/kg':

Molality (m) = Number of moles of solute / Mass of solvent (kg) Equation 2.1

Thus, if a solution contains 5 moles of solute / kg of solvent, the solution is considered as 5 molal or 5m solution.

2.3.1.2.1. Calculation of molality from a given mass

Problem 1:

Calculate the molality of a solution where 2 g NaCl is dissolved in 150 ml of water.

Solution 1:

First, mole of NaCl in 2 g NaCl is to be calculated.

Moles of NaCl = Mass in g / Molecular weight

= 2/58.4

= 0.03 mole

Since density of water is 1 g/ml, 150 ml water is equivalent to 150 g or 0.150 kg water.

Therefore, Molality = Moles of NaCl/Mass of water in kg

= 0.03/0.150 m

= 0.2 m

2.3.1.2.2. Calculating mass of a solute from given molality

Problem 2:

Calculate the mass of NaCl per kilogram of water in a 10 *m* aqueous solution.

Solution 2:

We know, Molality = Moles of solute/ mass of solvent in kg

So, 10 = Moles of NaCl/1

Moles of NaCl = 10

Again, we know that Moles of solute = Mass in g/molar mass

Or, 10 = Mass in g/58.4

Therefore, Mass of NaCl = 10 x 58.4

= 584 g

Thus, 584 g NaCl is to be added to 1 kg water to make a 10 m NaCl solution.

2.3.1.3. Molarity (M)

Molarity, also known as molar concentration, is defined as the total moles of a solute contained in a litre of a solvent. Thus, it may be expressed by the following formula with a unit of 'moles/L':

Molarity (M) = Number of moles of solute/volume of solvent (L) Equation 2.2.

So, we may write

1M (molar solution) = 1 mol/ L

= 1 mmol/mL

= 1 μmol/μL

1mM (millimolar solution) = 1 mmol/L

= 1 μmol/mL

Thus, if a solution contains 5 moles of solute/L of solvent, the solution is considered as 5 molar or 5M solution. Therefore, the 5M NaCl solution contains 5x58.46 g of NaCl in I L distilled or deionized water.

However, often the solutes are not present in the pure form or it is found in liquid state, such as HCl, H_2SO_4, etc. During the preparation of a standard solution of such reagents, it is necessary to prepare their solution, considering purity and density. Once the secondary solution is prepared, it should be titrated with a primary standard solution to know the strength of the secondary standard solution more accurately.

Problem 1:

Prepare 1M HCl solution from a concentrated HCl solution, containing 37.1% purity and 1.2 g/cm^3 density.

Solution 1:

To prepare 1M HCl solution, 36.46 g (molecular weight or gram formula weight of HCl) should be taken and mixed with deionized water in 1 L volumetric flask. However, as HCl is in a liquid state, the density should be considered to determine the volume of HCl which is equivalent to 36.46 g. The given density of HCl is 1.2 g/cm^3. Therefore, 30.38 mL (36.46/1.2) corresponds to 36.46 g of HCl. The purity is 37.1%. So, it can be written further:

37.1 mL pure HCl is present in 100 mL of crude HCl.

30.38 mL pure HCl is present in (100x30.38/ 37.1) or 81.9 mL crude HCl.

Therefore, 81.9 mL of HCl possessing 1.2 g/cm^3 density and 37.1% purity should be mixed properly with deionized or distilled water in a 1L volumetric flask. However, this solution should be titrated against the standard solution of sodium carbonate (primary standard) using methyl orange as indicator.

The volume of crude HCl to be taken for preparation of 1M HCl may be directly calculated from the following relationship:

Volume of crude HCl = $M_{required} \times V_{required} \times$ molecular weight / (% purity × density) Equation 2.3

[where, $M_{required}$ = required molarity, $V_{required}$ = final solution volume to be prepared]

= 1M x 1L x 36.46 / (0.371 x 1.2)

= 81.9 mL

Since molality considers the mass of the solute and solvent, it has got an edge over the molarity concept because mass of solute and solvent never change with a fluctuation of temperature and pressure, while volume of solvent changes. Thus, the molarity of a solution may change with fluctuation of temperature and pressure, and molality is considered a better measurement than the molarity.

2.3.1.3.1. Molarity calculation

Problem 1:

Calculate the molarity of a solution that contains 0.5 moles of NaOH in 1.5 lit of water.

Solution 1:

Molarity (M) = Number of moles of solute / volume of solvent (L)

= 0.5/1.5

= 0.33 M

Problem 2:

Calculate the molarity of NaOH solution containing 5 g NaOH dissolved in 750 ml water.

Solution 2:

NaOH molecular weight = 40

Total moles of NaOH in 5 g = Mass in g/Molecular weight

= 5/40 = 0.125 mole

Therefore,

Molarity of the solution = Number of moles of solute / volume of solvent (L)

= 0.125 mole/0.750 L

= 0.17M

2.3.1.4. Normality (N) and its calculation

Normality is the expression of concentration of standard solution where we express the solute amount as gram equivalent weight in one litre of solvent, while in molarity we use gram molecular weight of solute in one litre of solvent. However, the term 'equivalent' is expressed differently in acid-base and redox (oxidation-reduction) chemistry. In acid-base chemistry, an equivalent means the mass of a chemical that donates or accepts one mole of protons.

More elaborately, the equivalent weight of an element is the mass which combines with or displaces 1.008 gram of hydrogen or 8.0 grams of oxygen or 35.5 grams of chlorine. The concept of equivalent weight considers the solute's valence (number of hydrogen ions that can be displaced). Therefore, 1N NaOH contains 1 gram-equivalent NaOH [Gram-molecular weight/valence = (23 + 16 + 1)/1 = 40 g] in 1 litre volume of water. There are solutes in which 1N and 1M solutions are made up of the same quantity of the solute, such as NaOH. Likewise, 1N and 1M NaCl solution contain same amount of solute, 58.4 g of NaCl because gram-molecular weight (58.44) and gram-equivalent weight (58.44/1) for NaCl are same in 1 L distilled water. However, there are solutes that differ in their molar and normal solution composition, such as Na_2CO_3. To prepare 1M Na_2CO_3, 106 g (molecular weight of Na_2CO_3, 2 x 23 + 12 + 3 x 16 = 106) of Na_2CO_3 are added to 1 L distilled water. But to prepare 1N Na_2CO_3 solution, 53 g of Na_2CO_3 is added to 1 L distilled water because gram equivalent weight of Na_2CO_3 is 53 g (gram molecular weight/valence, 106/2). Similarly, 1M and 1N H_2SO_4 differ in their solution composition. Gram-molecular weight of H_2SO_4 is 98 (2 x 1 + 32 + 4 x 16 = 98). So, 98 g of H_2SO_4 are added to 1 L distilled water to prepare 1M H_2SO_4 solution. But 49 g of H_2SO_4 should be added to 1 L distilled water to prepare 1N H_2SO_4 solution because gram-equivalent weight of H_2SO_4 is 49 that is obtained dividing its gram-molecular weight (98) by its valence (2). So, it may be written that the molarity of this 1N solution of H_2SO_4 would be 0.5 (M = g /gram-molecular weight per litre or 49g/98g = 0.5). H_2SO_4, being diprotic acid, one-half mole of sulphuric acid provides one mole of protons.

Therefore, from this relationship, a formula as below may be developed:

$$W = N \times V \times M_w/v_a \qquad \text{Equation 2.4}$$

Where,

W = Weight (W) in grams,

N = desired normality,

V = volume needed in litre,

M_w = Gram-molecular weight, and

v_a = valence

Problem 1:

Calculate the amount of NaCl required to prepare 2 L of 2N NaCl solution.

Solution 1:

If we apply above mentioned formula (Equation 2.4),

W = 2 x 2 x 58.44/1.0 =233.76 g

This means that one has to add 233.76 g of NaCl in 2L volume of water to prepare 2N NaCl solution.

In redox chemistry, an equivalent refers to the mass of a chemical that donates or accepts one mole of electrons. For this reason, the number of electrons a chemical donates or accepts from its half-reaction is determined. For example, one mole of aluminium (III) reacts with three moles of electrons to give one mole of aluminium metal ($Al^{+3} + 3e^{-} \rightarrow Al$). Therefore, 1N solution of aluminium (III) is prepared by mixing one-third gram formula weight or molecular weight of an aluminium (III) compound with distilled water in 1L volumetric flask up to its mark.

In redox reaction, the number of electrons involved in a reaction or the change in oxidation number of an element in the oxidant or reductant, or the total change in the oxidation number (in the presence of more than one atom of the reactive element) are considered in determining equivalent weight of an analyte. In such cases, the equivalent weight is calculated as follows:

Equivalent weight = Molecular weight/Number of electrons gained or lost
Equation 2.5

Or we can write,

Equivalent weight = Molecular weight/Change in oxidation number of significant element
Equation 2.6

e.g., $MnO_4^- + 8H^+ + 5e^- \rightleftharpoons Mn_2^+ + 4H_2O$

Mn in $KMnO_4$ exists in +7 state. In acidic medium, this Mn +7 goes to Mn+2 state and hence there is a net gain of 5 electrons.

In this case, equivalent weight of $KMnO_4$ = Molecular weight of $KMnO_4$/5

= 158/5

= 31.6 g

2.3.1.5. Percent solutions

Mass percent

It refers to the number of grams of solute per 100 g of solution. When 20 g potassium sulphate is administered in 80 g water, the mass percent of the solution is 20 or it may be called as 20% by mass solution.

Mass percent = mass of solute/mass of solution

= 20 g/(20 g + 80 g) × 100%

= 20%

Volume percent

It means the number of millilitres of solute per total amount of 100 mL of solution. For instance, if one has to prepare a 5% by volume solution of ammonia in water, 5 ml ammonia is mixed with deionized water, taking volume of water in such a way that total solution volume becomes 100mL.

2.4. Dilution of standard solution

The following equation is very helpful while diluting a mother standard solution. However, for that you need to know the concentration (C_m) of the stock solution, the required volume (V_d) and concentration (C_d) of the newly diluted solution. In the true sense, you are finding out the volume (V_m) of the stock solution to be diluted to a desired volume (V_d).

$$C_m \text{ x } V_m = C_d \text{ x } V_d \qquad \text{Equation 2.7}$$

For instance, suppose there is a need to prepare a 250 mL of 0.25 M NaCl solution from a 5M NaCl stock solution. Calculate as follows, using Equation 1.5.:

$$C_m \text{ x } V_m = C_d \text{ x } V_d$$

$$5 \text{ x } V_m = 0.25 \text{ x } 250$$

$$V_m = 0.25 \text{ x } 250/5$$

$$= 12.5 \text{ mL}$$

Therefore, a volume of 12.5 mL of 5M NaCl should be dispensed in a 250 mL volumetric flask and finally, the volume is made up to 250 ml with deionized water. However, one should take reasonable care while making up the desired volume. Initially the volumetric flask should be partially filled with deionized water, calculated volume of the mother stock solution is then added, and the container is swirled properly to mix, followed by the further addition of deionized water to the mark. The container is stoppered properly and is inverted several times to mix.

2.4.1. Solution preparation from impure chemicals

2.4.1.1. Chemical in solid state

Initially, the number of moles (n) of pure chemical needed to prepare the solution is to be ascertained by multiplying the volume (v) and molarity (M) of the resulting solution. Later, the mass (m) of pure chemical required to prepare the solution is calculated by multiplying the number of moles of pure chemical and the gram formula weight of the chemical. Finally, the mass of the impure chemical is calculated by dividing the mass of pure chemical needed by the percent purity of the given product.

Problem 1:

What is to do if you have to prepare 0.2 M solution of 1 L from 75% pure sodium hydroxide? The molar mass of NaOH is 40 g/mol.

Solution 1:

i. Calculate the number of moles (n) of pure chemical needed to prepare the solution.

   n = v x M

   = 1 L x 0.2 M

   = 0.2

ii. The mass of pure chemical required to prepare the solution is obtained as follows:

   m = = n x Molar mass

   = 0.2 x 40

   = 8.0 g

iii. Mass of impure chemical required to prepare the solution is obtained as follows:

   Mass of impure chemical = the mass of pure chemical needed / percent purity

   = 8/0.75

   = 10.7 g

Therefore, 10.7 g of 75% pure sodium hydroxide are required to prepare 0.2 M solution of 1 L.

2.4.1.2. Chemical in liquid state

If the chemical exists in a liquid state, density of the liquid chemical is considered, and the volume of chemical is obtained by dividing the mass of impure chemical by its density.

Thus, volume of impure chemical (mL) = mass of impure chemical (g) / density of impure chemical (g/mL)

Now, considering the purity of the product; the final volume of impure chemical needed should be as follows:

Final volume of impure chemical required = Solution molarity × solution volume × molar mass of the chemical / (percent purity × density)

Problem 1:

What is to do if you have to prepare 0.2 M HCl solution of 1 L from 37% pure hydrochloric acid? The molecular weight of hydrochloric acid is 36.46 g/mol and its density is 1.2 g/mL.

Solution 1:

Volume of impure chemical (37% pure HCl) required = Solution molarity × solution volume × molar mass of the chemical / (percent purity × density)

$= 0.2 \times 1 \times 36.46 / (0.37 \times 1.2)$

$= 16.4$ mL

Therefore, 16.38 mL of 37% pure HCl are required to prepare 0.2 M solution of 1 L.

2.5. Types of titration

In two ways, concentration of an analyte can be measured using the technique of titration: direct and indirect titration. *Direct titration* involves the direct estimation of titrand concentration by adding a standard titrant to the titrand and the process is continued until the equivalence point is reached. However, in *back titration (or indirect titration)*, excess but known amount of standard titrant is mixed with the titrand, so that the titrand is consumed completely by the reaction with the titrant and the excess amount of titrant, which remains unreacted, are allowed to be titrated against a second standard reagent. Back titration is typically applied in acid-base titrations when (i) the acid or base, more commonly base, is an insoluble salt ($CaCO_3$); (ii) the endpoint in direct titration is difficult to detect (weak acid and weak base titration) and (iii) the rate of reaction is slow.

2.5.1. Acid-base titration

In this titration method either a base from the burette is added to acid in the conical flask to estimate the concentration of acid (acidimetry) or acid from the burette is added to a base in the conical flask to determine the concentration of the alkali (alkalimetry) in the presence of an indicator. The titration is stopped when the endpoint is detected by observing the colour change due to the presence of an indicator. Salt and water are produced by the reaction of acid and alkali. Whether the salt solution is neutral, acidic or alkaline at the end point; it depends on the strength of the reacting acid and base and this influences the choice of the indicator. When both acid and base are strong, a neutral salt solution ($HCl + NaOH \rightarrow NaCl + H_2O$) is formed at the end point. But when strong acid and weak base is titrated, salt solution become acidic ($NH_4OH + HCl \rightarrow NH_4Cl + H_2O$); while the reaction between weak acid and strong base leads to the formation of alkaline salt solution ($CH_3COOH + NaOH \rightarrow CH_3COONa + H_2O$) at the end point. A graph, known as titration curve, may be constructed by plotting the volume of titrant (x-axis) and the pH of the reaction solution (Y-axis). At the end point, a steep slope is obtained in the curve, meaning a rapid change in the solution pH. This titration curve, more specifically the steep slope in the curve, is used as a guide to select the indicator for such an acid-base titration, so that the indicator's pH range must overlap this steep change in pH at the end point of the titration (refer Section 3.2.1.).

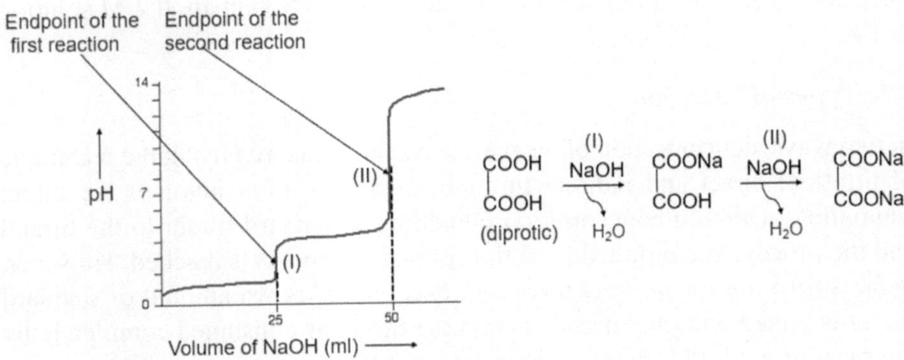

Fig. 2.1: Titration curve of oxalic acid (diprotic) and sodium hydroxide, showing two neutralization points

However, the steep slope may be obtained in one or more than one occasion in the same titration curve, depending on the nature of the reaction. As for instance, when NaOH and HCl are reacted, one equivalence point will be reached because HCl is a monoprotic acid (release one H^+), while oxalic acid is a diprotic acid and releases two H^+ ions when it is reacted with NaOH (**Figure 2.1**). In the latter case, the reaction will be completed within two phases and for this reason, two equivalence points will be obtained.

2.5.2. Redox titration

In oxidation-reduction titration method, a reducing substance is titrated with a standard solution of an oxidizing agent or an oxidizing substance is titrated with the standard solution of the reducing agent. This type of titration is possible due to the redox reaction between the analyte or titrand and the titrant, using a redox indicator, such as titration of iodine and thiosulphate in presence of a starch indicator (and/or a potentiometer). In this reaction iodine is reduced to iodide and when all iodine is consumed, the blue colour disappears. $K_2Cr_2O_7$ is an oxidizing agent and undergoes the following reaction: $Cr_2O_7^{-2}$ (orange) + $14H^+ + 6e^- \rightarrow 2Cr^{+3}$ (green) + $4H_2O$. At the endpoint, ΔE becomes zero.

The principle of this titrimetric method is based upon the change of the oxidation number or electrons transfer between the reactants. Oxidation process involves the loss of electrons (increases oxidation number) while the reduction process involves the gain of electrons (decreases oxidation number), that is, oxidizing agents undergo reduction and reducing agents undergo oxidation.

Oxidant + n e^- $\rightarrow$ Reductant

Reductant $\rightarrow$ Oxidant + n e^-

A redox titration curve may be constructed using the change in electrochemical potential (Y-axis) as a function of the titrant volume (X-axis). The electrochemical potential is related to the concentrations of titrant and titrand by the Nernst equation as follows:

$$E = E^\circ + (0.0591/n) \times \log([Ox]/[Red]) \text{ at } 25^\circ C \qquad \text{Equation 2.8}$$

Where, E° represents standard electrode potential and equals to E at unit molar concentrations of oxidant [Ox] and reductant [Red].

Similar to acid-base indicators which identify the sudden change in pH, redox indicator detects the sudden change in redox potential at the equivalence point. The redox indicators may be of three types: internal (e.g., diphenylamine, N-phenyl anthranilic acid, etc.), external [$K_3Fe(CN)_6$ is used as an external indicator in the titration of iron with $K_2Cr_2O_7$] and self-indicating ($KMnO_4$) reagent.

2.5.3 Complexometric titration

A complexometric titration (also known as chelatometric titration) involves the mixing of a complexing agent from a burette to an analyte solution in a conical flask. The end point of the titration is monitored by using metal ion indicators or spectrometric or electrometric methods. The principle of such titration depends on the capability of a complexing agent to form a stable complex with metal ions with known stoichiometry, as for instance, EDTA (multidentate ligand) forms a complex with Ca^{+2} in a ratio of 1:1. Metal ions and ligands act as the Lewis acids (accepts electron pairs) and Lewis bases (donate Lewis pairs), respectively. The stoichiometric complex remains in soluble form and undissociated which is an added advantage of this titration technique. Besides, it does not involve any possibilities of co-precipitation as it happens in precipitation titration and moreover, reaction selectivity may be employed in this technique as the complexing agent coordinates with specific metal ions.

Several metals, more specifically the transition elements, are evidenced to form complex compounds (coordination compounds) with ligands (binding or complexing agent), where the metal atoms or positively charged metal ions are bound to a number of anions or neutral molecules containing lone pairs of electrons by forming coordinate bonds. The ligands, that are acting as simple ions (Cl^-), small molecules (H_2O, NH_3, etc.), larger molecules (EDTA) or very large macromolecules containing higher molecular weight (proteins), may result in the formation of different number of coordinate bonds. Ligand, binding the metal with only one coordinate bond, is known as unidentate-ligand. Likewise, ligand may react with a metal atom or ion as bidentate, tridentate, tetradentate, forming two, three, four coordinate bonds, respectively and so on; as for instance, EDTA is termed as hexadentate ligand as it can bind a metal with six coordinate bonds. Hexadentate nature of EDTA is derived from its six potential sites: a pair of unshared electrons located on each of the two nitrogen atoms and each of the four ionised carboxyl groups, which are capable of complexing with a metal ion (**Figure 2.2**).

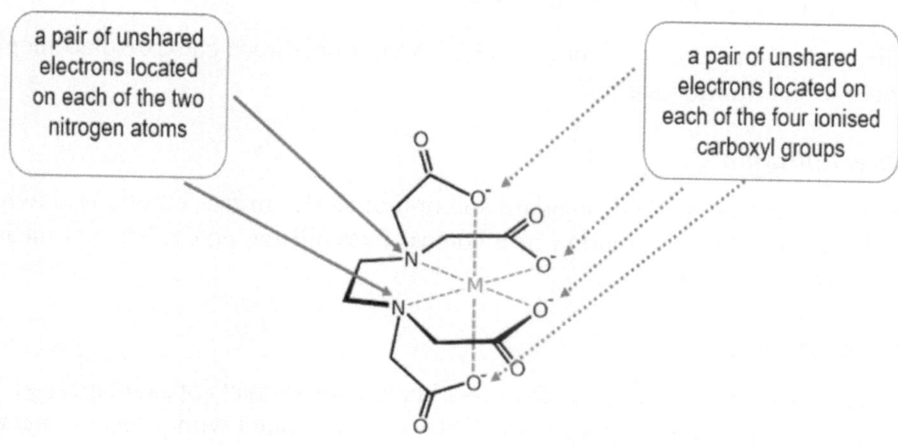

M = Metal ion

Fig. 2.2: EDTA molecule complexing with a metal ion at its six potential sites

In this field, another term 'chelate' is more frequently used. When a ring structure is developed by a chemical interaction between a metal atom or ion and a polydentate ligand through formation of more than one bond, the formed heterocyclic ring structure is called chelate and the ligand is referred to as chelating agent. A chelate is produced when a metal ion coordinates with two or more donor groups of a single ligand to form a five- or six-membered heterocyclic ring. The stability of the chelates increases with their denticity. It is usually a frequently asked question that why the chelation causes the metal ions to be more soluble in water. This is simply due to the presence of acidic or basic hydrophilic groups (COOH, OH, NH_2, etc.). The chelating agents, containing both acidic and basic groups, show their solubility in a wide range of pH. After formation of metal chelates, the metal does not participate in any ionic reactions. For this reason, the chelating agents forming soluble metal chelates are also referred as sequestering agents because they take out the metal from the solution. Thus, polyvalent metals, which cannot be extracted from aqueous solutions with organic solvents, may easily be squeezed out from aqueous solutions with the help of water-soluble chelating agents. However, all chelating agents do not form water soluble complexes. The metal chelates that are devoid of the hydrophilic groups prefer organic solvents for their solubility. Dimethylglyoxime, salicylaldoxime, etc. are few examples of ligands that form chelates which are soluble in organic solvents but insoluble in water.

In the complexation reaction, binuclear (containing two metal ions) or polynuclear (containing more than two metal ions) complex formation is also common. This occurs when metal ion concentration is very high.

2.5.3.1. EDTA titration techniques

There are mainly three techniques in EDTA titration: direct, back, displacement and indirect titration method.

Direct titration

In direct titration method, standard solution of EDTA from a burette is slowly added to a metal ion solution in a conical flask till the equivalence point is reached.

Back titration

Metal ion solution is mixed with excess but known amounts of standard EDTA solution. The unreacted portion of EDTA is then titrated with a second metal ion standard solution.

Displacement titration

In displacement titration, the analyte metal ion is added to a metal - EDTA complex which is already formed with other metals. The analyte metal ion displaces the second metal from the already existing complex and the metal is subsequently titrated with standard EDTA solution. The condition of this titration states that the stability constant of the analyte with EDTA must be higher than that of the starting metal-EDTA complex. A typical common example of displacement titration is the determination of Hg^{+2} as the analyte, using Mg-EDTA. This technique is usually applied to those metal ions that do not have a good indicator.

Indirect titration

Sometimes as a titrant, EDTA is used to estimate and quantify the amount of anions in solution. For instance, sulphates are precipitated as barium sulphate which is then separated, washed and boiled in excess but known amounts of EDTA, so that Ba^{+2} is complexed with EDTA. In the next step, the unreacted portion of EDTA is back titrated with a standard metal solution. This helps to determine Ba^{+2} and consequently, SO_4^{-2} content. However, it should be remembered that Ba-EDTA stability constant should be considerably high.

The complexometric titration, using EDTA as a complexing agent, is most commonly applied to determine hardness of water which is due to the abundance of Ca and Mg salts. Total Ca and Mg in water are titrated with standard EDTA solution in the presence of EBT indicator.

2.5.3.2. Metal indicator

A metal indicator may be used in complexometric titration using EDTA as a complexing agent. Initially the indicator and metal produce a coloured complex. When EDTA is added to this system, metals are released from the initial indicator-metal complex and form complexes with EDTA. As a result, by such a gradual process, when indicator molecules are converted completely to their free form, the solution colour is changed, and the equivalence point is obtained. Therefore, the precondition of an indicator selection in complexometric titration is that the formation of initial indicator metal complex should undergo through a weak binding, so that EDTA can easily remove metal ion from the metal-indicator complex; i.e. the indicator should bind metal ions weakly than EDTA does. Metal ion indicators behave as acid-base indicators because the exhibition of colour is pH-dependent. Organic dyes, such as Fast Sulphon Black, Eriochrome Black T (EBT), etc. are common metal indicators used in a metal ion-EDTA titration. It behaves as a weak acid, and its acid form and its counterpart of conjugate base possess different colours [H_2-Ind^- (Red) + $H_2O \rightarrow$ H-Ind^{-2} (Blue) + H_3O^+].

2.5.3.3. Masking and demasking agent

When an analyte metal ion, which is to be estimated, is present in a solution mixed with other metal ions, these other metal ions are then chemically masked in such a way that they cannot participate in reaction with the indicator and EDTA. As for instance, cyanide ions can form very stable cyanide complexes with Hg(II), Fe (II), Ni, Zn, Ag and Cd; but they do not produce any stable complex with Ca^{+2}, Mg^{+2}, Pb^{+2} and Mn^{+2}. This behaviour of cyanides facilitate the titration of Ca^{+2}, Mg^{+2}, Pb^{+2} and Mn^{+2} separately with EDTA standard solution in the presence of Hg(II), Ni, Zn, Ag, Zn and Cd that are masked with sodium cyanide. Titanium(IV), iron(III), and aluminium can be masked with triethanolamine; mercury with iodide ions. Conversely, demasking is opposite to masking. It is a process of converting a masked substance to regain its reacting ability. The cyanide complexes of zinc and cadmium may be demasked by treatment with formaldehyde-acetic acid solution, chloral hydrates, etc.; so that the metals form complexes with EDTA.

2.5.4. Precipitation titration

The objective of precipitation titration involves the formation of a precipitate due to the reaction between titrant and analyte. Once the whole amount of the analyte is consumed by the reaction, the excess amount of titrant due to its extra addition (even a fraction of a drop of titrant solution) reacts with the indicator, leading to a colour change and indicating the completion of the reaction.

The precipitation reaction can only be utilized for titrimetric analysis, if the reaction is very fast, stoichiometry of the reaction is known and obviously, there should not be any other secondary reaction to be occurred because the added precipitant should be equivalent to the precipitated substance to be formed in the precipitation titration. Estimation of chloride ion content in a solution is commonly done by precipitation titration, using silver nitrate as a titrant, where the reaction takes place as follows:

$$AgNO_3 + NaCl = NaNO_3 + AgCl\downarrow$$

However, there are three main methods: Mohr, Volhard and Fajans method that differ within themselves depending on the type of application and the mode of action of the indicator to be used. These methods are separately dealt in Section 3.4 and 3.5. Among the three methods, the Volhard method is widely used because of its easy detection of the endpoint.

In the Mohr method, it is to be remembered that chromate ion, which is used as an indicator, precipitates with silver ions but the precipitate thus formed is relatively more soluble in aqueous solution than that of silver chloride. Thus, during the first phase of titration silver chloride is formed but when all chlorides are consumed by the reaction, further addition of one excess drop of silver nitrate solution leads to the reaction between silver ion and chromate indicator, resulting in a reddish precipitate. Most importantly, a neutral medium is maintained in the Mohr method; otherwise, chromate is converted to dichromate in acidic condition and silver ions may react with hydroxyl ions to form silver hydroxide in alkaline medium.

Precipitation titration contains few limitations, such as possibility of co-precipitation, difficulties in end point detection and moreover, the most important limitation is that this titration technique is applied primarily on halides (Cl^-, Br^- and I^-).

2.6. Model questions

i. What is a standard solution? What are the differences between primary and secondary standard solutions? Illustrate with examples.

ii. What are molal, molar and normal solutions? Illustrate with examples.

iii. How are 0.1N H_2SO_4 and 0.1N NaOH prepared and standardised?

iv. How is 0.2M solution of acetic acid prepared, considering the molecular weight of acetic acid as 60.05 g/mol and its specific gravity as 1.05 g/mL?

v. How is it possible to prepare 500 ml of 0.5N NaOH from a stock solution containing strength of 2N NaOH?

vi. Write down the principle behind the complexometric titration. What are the different types of EDTA titration technique? What are masking and demasking agents used in EDTA titration? Illustrate them with examples.

vii. Write down the principles of acid-base, redox and precipitation titration. Illustrate with reactions in each case.

viii. Standard solution of sodium hydroxide must be protected from atmospheric gases ----- Why? Sodium hydroxide solution containers should not be glass stoppered ------ Why?

ix. How do you prepare 200 mL of 0.5 N $KMnO_4$ solutions? The molar mass of $KMnO_4$ is 158 g/mol.

x. How do you prepare 500 mL of 0.5 N NaCl solutions? The molar mass of NaCl is 58.44 g/mol.

2.7. Suggested readings

(i) IUPAC (1997) Compendium of Chemical Terminology, (2nd Edition, the "Gold Book"). Compiled by McNaught, A.D.; Wilkinson, A. Blackwell Scientific Publications, Oxford. https://doi.org/10.1351/goldbook

(ii) Bassett, J.; Denney, R. C.; Jeffery, G. H., Mendham, J. (1978) Vogel's Textbook of Quantitative Inorganic Analysis Including Elementary Instrumental Analysis. Revised by 4th Edition. The English Language Book Society and Longman.

(iii) McPherson, P. (2014) Practical Volumetric Analysis. Royal Society of Chemistry.

(iv) Patnaik, P. (2004) Dean's Analytical Chemistry Handbook (2nd Edition), McGraw-Hill.

3

Indicators

3.1. Introduction

Indicators are the natural or synthetic chemical compounds that are used to detect the equivalence point in a volumetric analysis where unknown compound in a solution form is quantitatively estimated with the help of a known standard solution. The equivalence point, also known as the stoichiometric end point, is the point or stage of a titration reaction where the equivalent amount of the reacting substances are mixed together, and this point is identified by monitoring the rate of change in pH, conductivity, flow of current, adsorption, fluorescence, changing of solution colour, occurrence or disappearance of turbidity, flocculation or deflocculation, etc. However, indicators which help to change the colour of the titrating solution are most common. However, there are instances that titrating solutions, such as $KMnO_4$ may also act as an indicator. In a redox reaction, $KMnO_4$ in acid solution is reduced to a colourless product. When the stoichiometric end point is reached, an excess additional drop of $KMnO_4$ shows its pink colour indicating the end point of the titration.

3.2. Acid-base indicator

An acid-base indicator which is also referred as neutralisation indicator may be a weak acid or weak base that dissociates in water and forms the weak acid and its conjugate base or the weak base and its conjugate acid, respectively. The species and its conjugates exhibit different colours. Therefore, in other way, it may be said that colour changes in a solution by the acid-base indicators occur due to the deprotonation or protonation of the indicator themselves over a pH range.

Various dyes or indicators are primarily used to detect a pH of a solution because these indicators change their colour with changing in H^+ or OH^- ion activities in solution. Thus, acid-base indicators are chemicals that help to tell about the chemical nature of a solution whether it is acidic, neutral, or alkaline. For this reason, they are known as pH indicators. However, these indicators are also commonly used in an acid-base titration to identify the endpoint. Since the colour of different indicators change at various levels of H^+ ions in the solution, they are used to identify varying pHs of the solutions. The indicators must give a stable colour change to detect the pH level of a solution. Although the method is not so precise and accurate, one can use the method in the field to obtain a preliminary idea on soil pH, as it is simple and less expensive. Several indicators along with their functioning at a specific range of pH are enlisted in Table 3.1.

3.2.1. Mechanism

Indicators, in general, contain several alternating (conjugated) carbon-carbon double bonds and single bonds that absorb wavelengths from visible light to bring the molecule to a coloured state. Indicators on dissolution in water, which are weak acids or weak bases in their chemical nature, give rise to the formation of ions. This leads to the changes in the electronic arrangement in the molecule and thus, facilitates absorption of different wavelengths of light. Consequently, the colour of the solution will be different from the earlier one. In general, it has been evidenced in all weak acids and bases that if a weak acid is added to a solution which is exhibiting pH lower than the acid's pKa, the weak acid becomes in a protonated stage. Deprotonation of the weak acid molecule is quite obvious when the solution pH is more than the acid's pKa. Most importantly, when the weak acids and bases are concerned as indicators, they exhibit colour changing behaviour during their stay as acidic or basic form due to their attainment of different energy levels for proton attachment or detachment.

At the equivalence point, the ionization constant of the weak acid or the weak base form of the indicator and their concentrations in the solution help to assess the pH of the solution because H^+ ion concentration, influenced by the nature of the titrant and titrand, actually dictate to define the equivalence point.

If a weak acid (H-ind) dissociates in water, the chemical reaction at equilibrium is illustrated in **Figure 3.1** as follows:

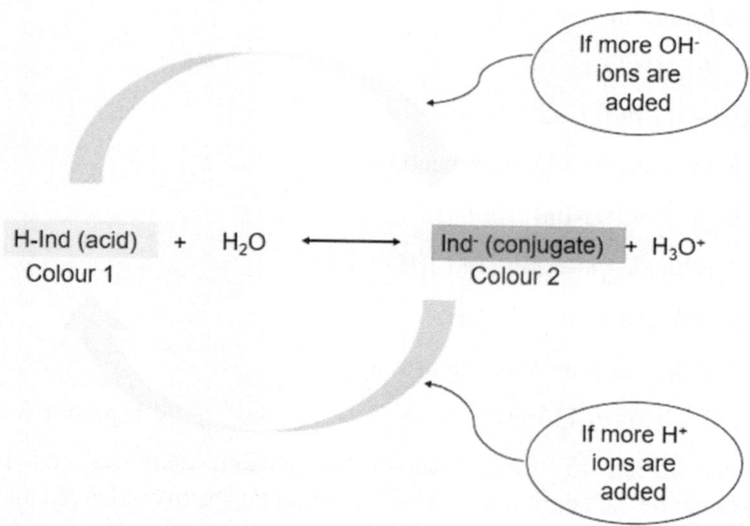

Fig. 3.1: Behaviour of weak acid indicators at equilibrium

According to Le Chatelier's Principle, the equilibrium shifts toward left and the colour of the acid form is displayed at lower pH; while at higher pH, the equilibrium shifts toward the right side to compensate the depleted hydronium ions due to addition of hydroxyl ions and the colour of the conjugate is exhibited.

The equilibrium constant for the reaction becomes

$K_{Ind} = [H_3O^+][Ind^-] / [H\text{-}Ind]$ Equation 3.1

Where, K_{Ind} is the indicator dissociation constant.

The concentrations of the acid form and basic form turn to be equal at the point when colour changing takes place. At this point, half of the indicator is in acid form and the other half is its conjugate base and the pH of the solution at this point is called the pK_{Ind}. Thus, the end point for the indicator depends entirely on the pK_{ind}.

Therefore, from Equation 3.1 we get

$K_{Ind} = [(H_3O)^+]$, as $[H\text{-}Ind] = [Ind^-]$

Or, pK_{Ind} = pH (of the solution) Equation 3.2

Rearranging the Equation 3.1, we may write

$[Ind^-]/[H\text{-}Ind] = K_{Ind}/[H_3O^+]$

Or, $[(H_3O)^+]/K_{Ind} = [H\text{-}Ind]/[Ind^-]$

Putting log in both sides, the equation becomes

$Log([(H_3O)^+]/K_{Ind}) = log([H\text{-}Ind]/[Ind^-])$

Or, $log([(H_3O)^+]) - log(K_{Ind}) = -log([Ind^-][H\text{-}Ind])$

Or, $-pH + pK_{Ind} = -log([Ind^-]/[H\text{-}Ind])$

[pK_{Ind} is known as apparent indicator constant]

Or, $pH = pK_{Ind} + log([Ind^-]/[H\text{-}Ind])$ Equation 3.3

Equation 3.3 is similar to the Henderson-Hasselbalch equation and is meaningful to describe the equilibrium of indicators. It has been evidenced that when $[H\text{-}Ind]/[Ind^-]$ goes beyond 10, acid colour predominates. Oppositely, when $[Ind^-]/[H\text{-}Ind]$ attains a value beyond 10, the alkaline colour prevails. Thus, we get the following two equations (Equation 3.4 & 3.5).

when $[H\text{-}Ind]/[Ind^-] > 10$:

$pH = pK_{Ind} + log([Ind^-]/[H\text{-}Ind])$

Or, $pH = pK_{Ind} - 1$ Equation 3.4

and, when $[Ind^-]/[H\text{-}Ind] > 10$:

Or, $pH = pK_{Ind} + 1$ Equation 3.5

Considering the Equation 3.4 & 3.5, colour change interval may be postulated as follows (Equation 3.6):

$pH = pK_{Ind} \pm 1$ Equation 3.6

Therefore, it is assumed that the pH range or colour change interval is concerned about approximately two pH units. Since the colour change is dependent on the concentration of acidic and basic form of indicator, it does not become abrupt but is a gradual process. One point is reached when the concentration of acidic and basic form of indicator becomes equal and then, Equation 3.3 may be written as follows (Equation 3.7):

$pH = pK_{Ind} + log([Ind^-]/[H\text{-}Ind])$ (Equation 3.3)

Or, $pH = pK_{Ind} + log(1)$

Or, $pH = pK_{Ind}$ Equation 3.7

At this pH of the solution, the colour change stays in the midway of its changing and this solution pH, where the pK_{Ind} equals to it, is considered as critical pH, below or above of which determines the dominance of the acidic form or basic form of indicator, respectively and accordingly, the colour will be displayed in the solution.

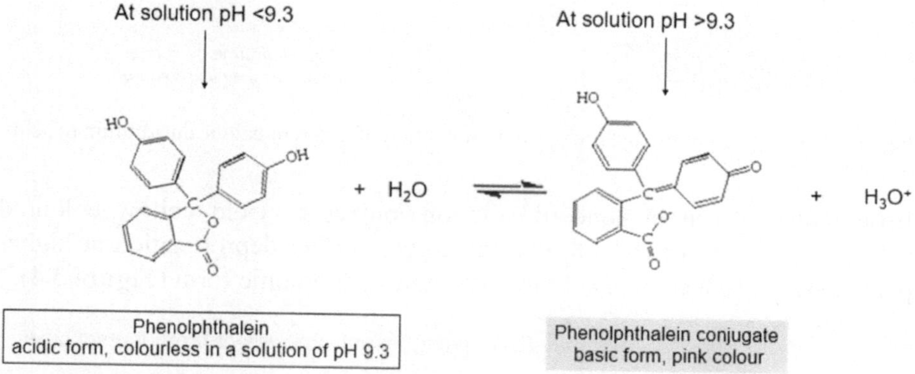

At solution pH <9.3

At solution pH >9.3

+ H_2O ⇌ + H_3O^+

Phenolphthalein acidic form, colourless in a solution of pH 9.3

Phenolphthalein conjugate basic form, pink colour

Fig. 3.2: Dissociation of phenolphthalein and production of different colour on addition of acid or alkali

As for example, phenolphthalein is a colourless molecule in acidic and neutral pH. In acidic solutions which are rich in proton concentration, the molecule does not dissociate. Eventually, the colour of the original indicator molecule remains the same. However, under alkaline conditions, H^+ is abstracted from the molecule and the ionized phenolphthalein gives rise to pink colour. Phenolphthalein has a *pKa* of about 9.3. Therefore, acid and base form of phenolphthalein exist in the equal amount at the solution pH of 9.3. While solution pH remains below 9.3, phenolphthalein remains in protonated form and shows the colour of its acid form which is colourless. But addition of phenolphthalein in the solution of pH >9.5 shows a colour change from no-colour to pink colour because base form (deprotonated form), possessing pink colour, predominates at this pH due to dissociation (**Figure 3.2**).

Similarly, methyl orange dissociates in water and exhibits red colour at a pH <3.7 and yellow colour at a pH >3.7 on addition of acid or alkali, respectively (**Figure 3.3**). Loss of one hydrogen ion from the -N-N- bridge between the rings is evidenced in the basic form of methyl orange. The positively charged terminal nitrogen is neutralized by the electrons that were earlier utilized during hydrogen binding. Thus, a negatively charged methyl orange basic conjugate is formed, which shows yellow colour.

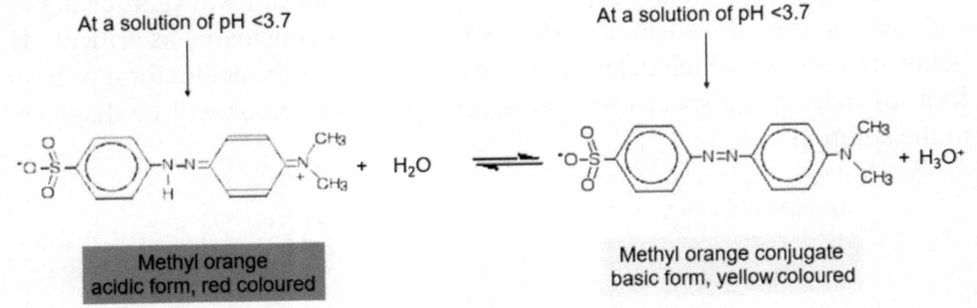

At a solution of pH <3.7

At a solution of pH <3.7

Methyl orange acidic form, red coloured

Methyl orange conjugate basic form, yellow coloured

Fig. 3.3: Dissociation of methyl orange and production of different colour on addition of acid or alkali

Bromocresol green in aqueous solution ionizes to yield yellow coloured mono-ionic form at pH <3.8, that undergoes further deprotonation at higher pH (>5.4) to produce blue coloured resonating di-anionic form (**Figure 3.4**).

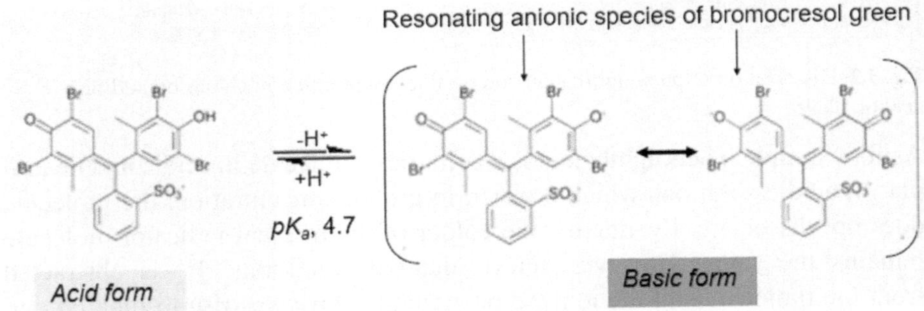

Resonating anionic species of bromocresol green

pK_a, 4.7

Acid form

Basic form

Fig. 3.4: Dissociation of bromocresol green and production of different colour on addition of acid or alkali

Similarly, bromocresol purple indicator possesses two forms namely cyclic sulfonic ester (yellow colour) and sulfonate structure (purple colour) at pH <5.2 and pH >6.8, respectively (**Figure 3.5**).

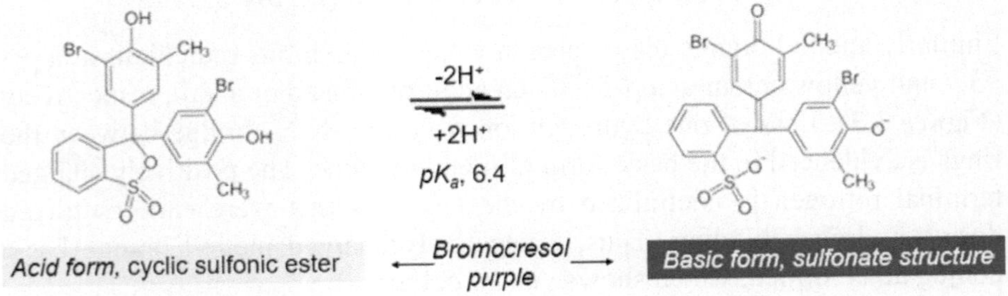

pK_a, 6.4

Acid form, cyclic sulfonic ester ← Bromocresol purple → Basic form, sulfonate structure

Fig. 3.5: Dissociation of bromocresol purple and production of different colour on addition of acid or alkali

There are some indicators that dissociate more than once, associated with difference in colour change; as for example, thymol blue (*pKa*, 1.5) dissociates once in the pH range of 1.2-2.8 along with a colour conversion from red (acid form) to yellow (base form), while the indicator (*pKa*, 8.9) goes for its second change at a pH range of 8.0-9.6 where colour changes from yellow to blue (base form) [**Figure 3.6**].

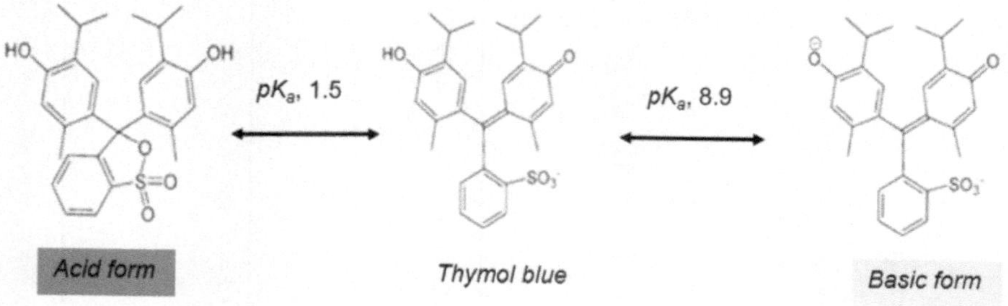

Fig. 3.6: Dissociation of thymol blue and production of different colour on addition of acid or alkali

3.2.2. Indicators pH range or colour change interval

It is also to be remembered that an indicator changes colour over a range of hydrogen ion concentrations. It does not promote changing of colour sharply at specific hydrogen ion concentration (given by their pK_{ind}). The colour change occurs over a narrow range of pH. This range is termed '*colour change interval*' which is expressed as a pH range, confining usually within about two pH units (Equation 3.6). In the pH scale, indicators vary within themselves about the position of the *colour change interval.* Each indicator possesses a specific *colour change interval.* A list of these pH ranges for a few commonly used indicators are given in **Table 3.1**. At pK_{ind}, the indicator remains in such a state that half of the amount of the indicator remains in its acid form and the other half portion of the indicator is shared by its basic form. Colour change takes place gradually over a narrow solution pH range by shifting of the equilibrium toward left or right by addition of proton or hydroxyl ions and one point will reach when a permanent colour is visible. For instance, methyl orange shows its dominant red colour under acidic conditions. Under this situation, if alkali is mixed slowly drop by drop with the solution; indicator's equilibrium starts shifting toward the formation of basic conjugate form with an orange tint. Further addition of alkali causes the changing of orange colour to permanent visible yellow colour. The whole event takes place within a pH range of 3.2-4.4 with a pK_{ind} of 3.7.

Table 3.1: Commonly used indicators in acid and alkaline solutions with their colour change intervals

Indicators	Colour		*'Colour change interval'* or pH range	*pKa*
	Acid form	Base form		
Methyl Violet: $C_{24}H_{28}N_3Cl$,N-(4-(bis(4-(dimethylamino)phenyl)methylene)cyclohexa-2,5-dien-1-ylidene)methanaminium chloride) 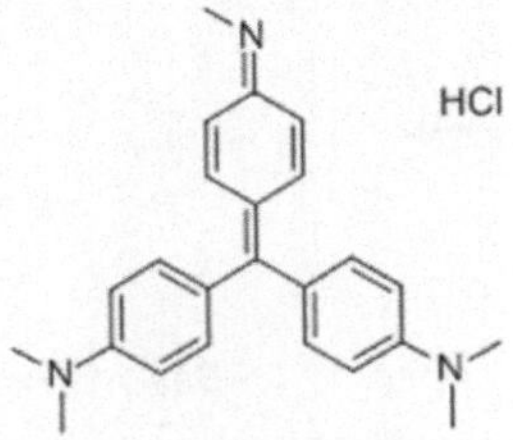	Yellow	Blue	0-1.6	0.8
Penta methoxy Red: $C_{24}H_{26}O_6$,Bis(2,4-dimethoxyphenyl)(2-methoxyphenyl)methanol 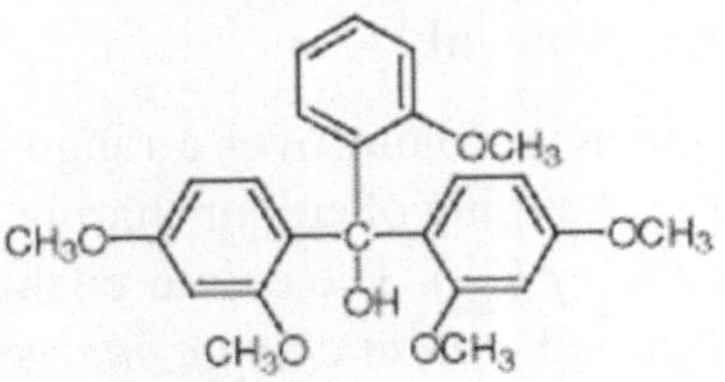	red violet	No colour	1.2-2.3	1.9
Thymol Blue: $C_{27}H_{30}O_5S$, α-hydroxy-α,α-bis(5-hydroxy carvacryl)- o-toluenesulfonic acid γ-sultone; thymol sulfonephthalein 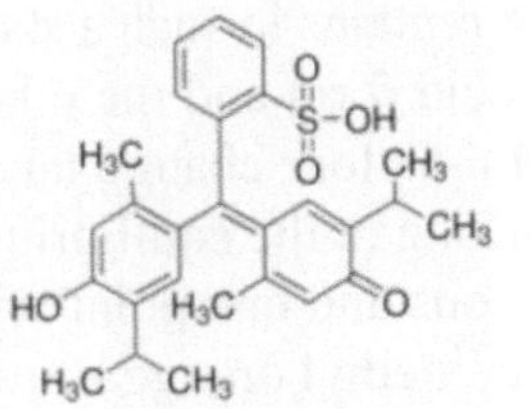	Red	Yellow	1.2-2.8	1.5
Dimethyl Yellow/Methyl Yellow: $C_{14}H_{15}N_3$ or $C_6H_5N{=}NC_6H_4N(CH_3)_2$,4-(Dimethylamino)-azobenzene 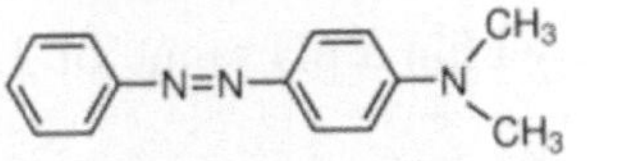	Red	Yellow	2.9-4.0	3.3

Indicators	Colour		'Colour change interval' or pH range	pKa
	Acid form	Base form		
Methyl Orange: $C_{14}H_{14}N_3NaO_3S$, Sodium 4-[(4dimethylamino)phenylazo] benzenesulfonate 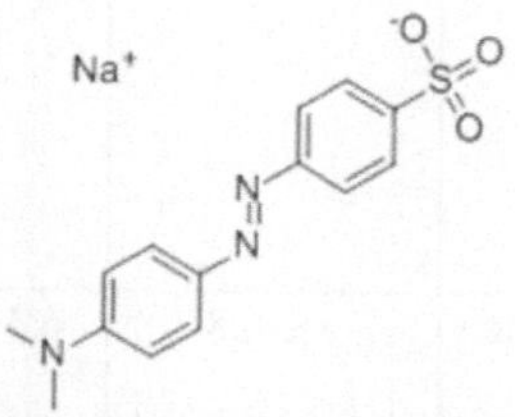	Red	Yellow	3.2-4.4	3.7
DNP: $C_6H_4N_2O_5$ or $C_6H_3(OH)(NO_2)_2$, 2,4-Dinitrophenol 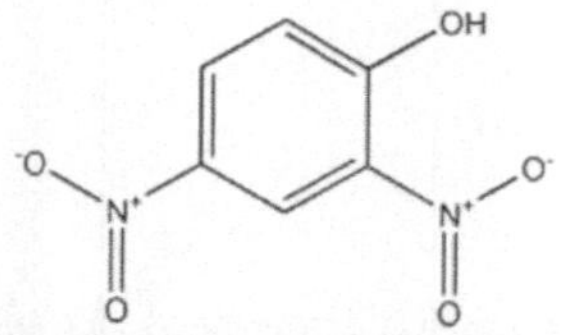	No colour	yellow	2.4-4.0	4.1
Tetrabromphenol blue: $C_{19}H_6Br_8O_5S$, 4,5,6,7-Tetrabromo-3,3-bis(3,5-dibromo-4-hydroxyphenyl)-3H-benzo[c][1,2]oxathiole 1,1-dioxide 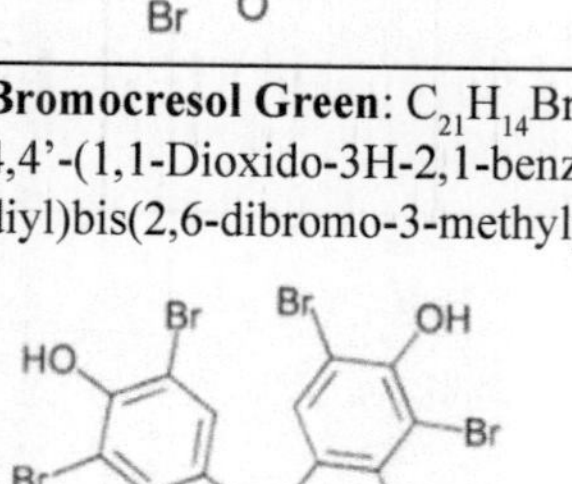	Yellow	Blue	3.0-4.6	4.1
Bromocresol Green: $C_{21}H_{14}Br_4O_5S$, 4,4'-(1,1-Dioxido-3H-2,1-benzoxathiole-3,3-diyl)bis(2,6-dibromo-3-methylphenol) 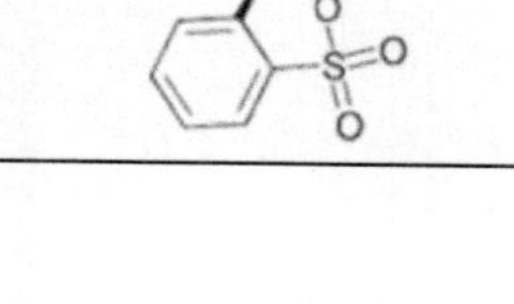	Yellow	Blue	3.8-5.4	4.7

Indicators	Colour		'Colour change interval' or pH range	pKa
	Acid form	Base form		
Methyl Red: $C_{15}H_{15}N_3O_2$,2-{[4-(Dimethylamino)phenyl]diazenyl}benzoic acid 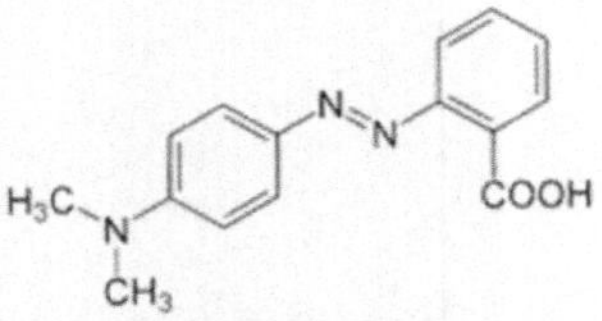	Red	Yellow	4.8-6.0	5.1
Bromocresol Purple: $C_{21}H_{16}Br_2O_5S$, 4,4'-(1,1-Dioxido-3H-2,1-benzoxathiole-3,3-diyl)-bis(2-bromo-6-methylphenol) 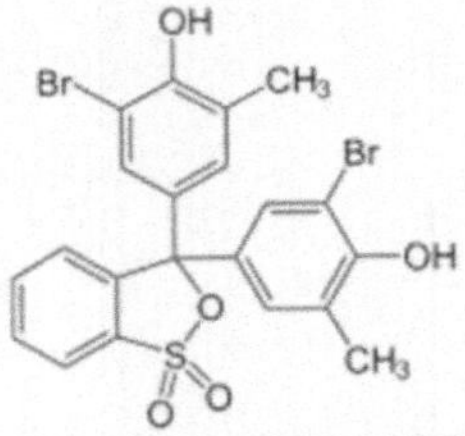	Yellow	Purple	5.2-6.8	6.4
PNP: $C_6H_5NO_3$, 4-Nitrophenol or *p*-Nitrophenol 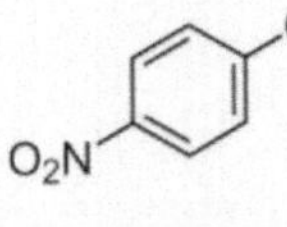	No colour	yellow	5.0-7.0	7.1
Litmus: 7-hydroxyphenoxazone (chromophore in various litmus components) 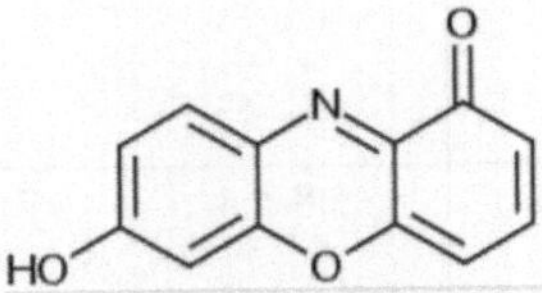	Red	Blue	5.0-8.0	6.5

Indicators	Colour		'Colour change interval' or pH range	pKa
	Acid form	Base form		
Bromothymol Blue: $C_{27}H_{28}Br_2O_5S$, 3,3'-Dibromothymol sulfon phthalein or 4,4'-(1,1-Dioxido-3H-2,1-benzoxathiole-3,3-diyl)bis(2-bromo-6-isopropyl-3-methylphenol) 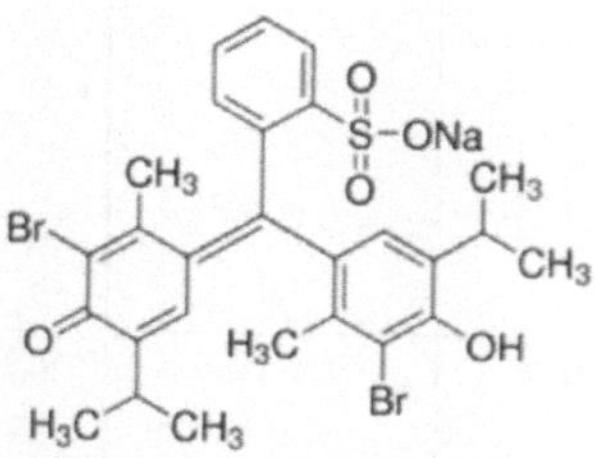	Yellow	Blue	6.0-7.6	7.0
Phenol Red: $C_{19}H_{14}O_5S$, 4,4'-(1,1-Dioxido-3H-2,1-benzoxathiole-3,3-diyl) diphenol 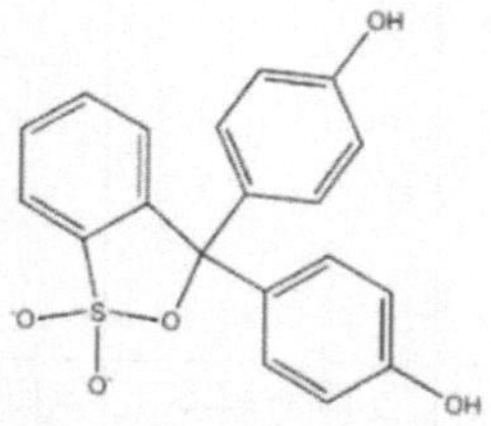	Yellow	Red	6.8-8.4	7.9
Naphtholphthalein: $C_{28}H_{18}O_4$, 1-Naphtholphthalein 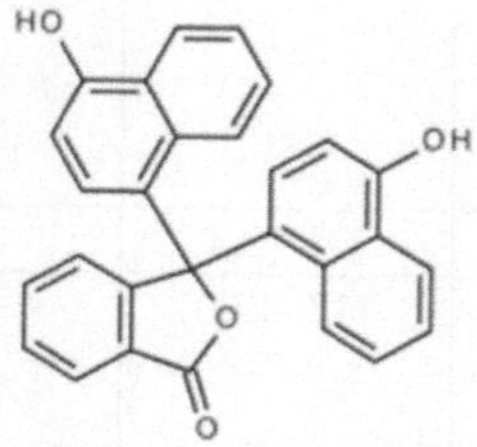	Yellow	Blue	7.3-8.7	8.4
Thymol Blue (Second change): $C_{27}H_{30}O_5S$, α-hydroxy-α,α-bis(5-hydroxy carvacryl)-o-toluenesulfonic acid γ-sultone; thymol sulfone phthalein 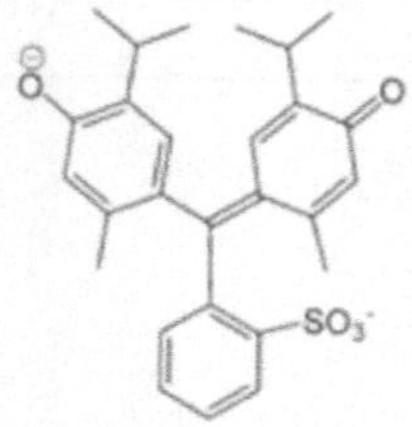	Yellow	Blue	8.0-9.6	8.9

Indicators	Colour		'*Colour change interval*' or pH range	*pKa*
	Acid form	Base form		
Phenolphthalein: $C_{20}H_{14}O_4$, 3,3-Bis(4-hydroxyphenyl)isobenzofuran-1(3H)-one 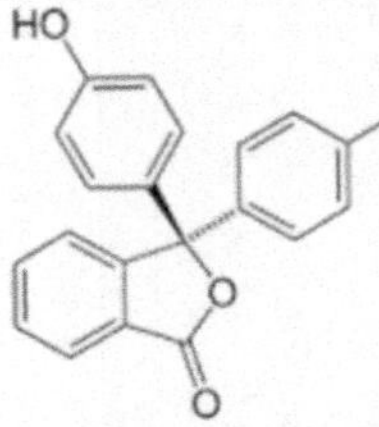	No colour	Pink	8.2-10.0	9.3
Thymolphthalein: $C_{28}H_{30}O_4$, 3,3-bis(4-hydroxy-2-methyl-5-propan-2-yl phenyl)-2-benzofuran-1-one 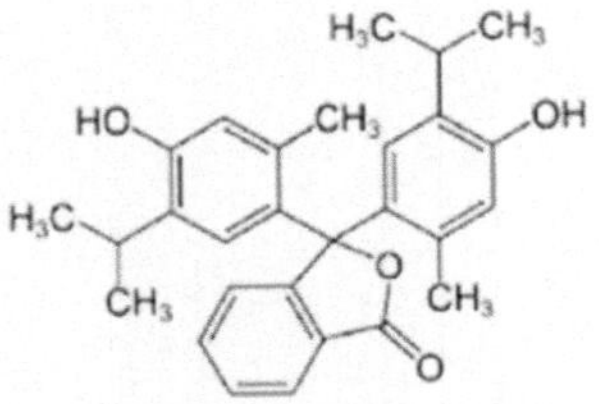	No colour	Blue	9.4-10.6	9.9
Alizarine Yellow R: $C_{13}H_8N_3NaO_5$ (Na salt), 5-[(p-Nitrophenyl)azo]salicylic acid sodium salt 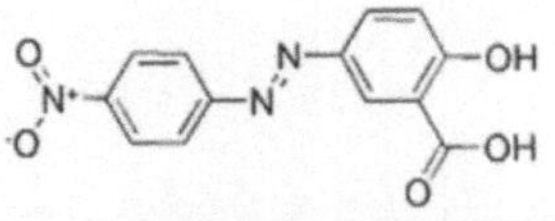	Yellow	Red	10.1-12.0	11.0
Universal indicator	Red	Violet	1.2-12.4	
Tropaeolin O: $NaC_{18}H_{14}SN_3O_3$, 4-[(2,4-Dihydroxyphenyl)azo]benzenesulfonic acid monosodium salt 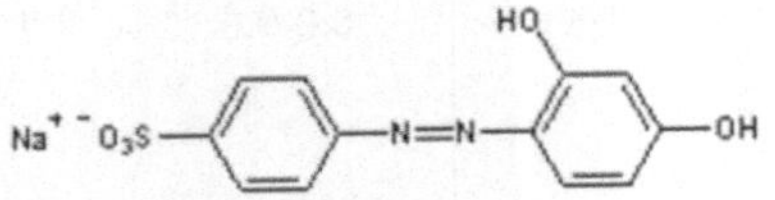	yellow	Orange	11.0-13.0	
Poirier's blue	Blue	Violet-pink	11.0-13.0	

3.2.3. Properties of a good indicator

A good indicator should possess the following properties:

(i) The colour change must be easily detected.

(ii) The colour change must be rapid.

(iii) The colour change should take place over a narrow pH range.

(iv) The indicator molecule must not react with the substances being titrated. Being weak acids or weak bases, the acid-base indicators are weaker than the titrants and do not interact with them before the completion of the stoichiometric reaction.

(iv) The indicator should have a pK_{ind} value that is close to the expected pH at the equivalence point. This reduces the errors.

3.2.4. pH Paper

pH papers are paper strips or plastic strips impregnated with combinations of chemical indicators. They promote colour changing at different pH values. A solution pH may be approximately ascertained by immersing a paper strip into the solution followed by comparing the colour to the standards provided. Frequently, these paper strips are referred to as litmus paper. When red litmus, a weak diprotic acid, is exposed to a basic compound, the H^+ ions are consumed by the added base. As a result, the conjugate base, possessing blue colour, is formed. For this reason, the wet red litmus paper turns blue in alkaline solution.

Litmus is actually a mixture of 10-15 different dyes that are obtained from lichens (*Roccella montagnei, Roccella tinctoria, Dendrographa leucophaea, Lecanora tartarea*, and many others). All these component dyes contain a common chromophore, 7-hydroxyphenoxazone. Erythrolitmin, azolitmin, spaniolitmin, orcein, leucoorcein, leucazolitmin are few common components in litmus. These are water-soluble in nature and they are adsorbed onto filter papers to manufacture pH paper or pH indicator for testing acidity or alkalinity. Blue litmus paper turns red under acidic conditions and red litmus paper turns blue under basic conditions. Litmus is also available in solution form. The colour change interval of litmus ranges between pH 5.0–8.0 with pK_{ind} of 6.5. Wet litmus papers are also utilized to test the influence of water-soluble gas on acidity or basicity. For example, NH_3 gas, being alkaline in nature, converts the colour of litmus paper from red to blue. Chlorine gas makes the blue litmus white. However, exposure of litmus to chlorine results in the bleaching of the dye by hypochlorite ions, leading to an irreversible reaction. Due to the irreversibility of the reaction, the litmus paper is not used as an indicator in such cases.

3.3. Fluorescent indicator

Fluorescence refers to a phenomenon where a molecule absorbs light to form a short-lived excited state and re-emit light at a longer wavelength. The fluorescent indicators whose fluorescent properties in solution are influenced by a change in hydrogen-ion concentration, oxidation potential, or metal-ion concentration can be used for endpoint determination. The main principle of fluorescence indicators in acid-base titrations involves giving rise of fluorescence in either its acidic or basic form or vice versa. However, basic fluorescence is most common. For instance, eosin (pH range, 0.0-3.0; green fluorescence), dichlorofluorescein (pH range, 4.0-6.6; blue fluorescence), coumaric acid (pH range, 7.2-9.0; green fluorescence), salicylic acid (pH range, 2.0-4.0; violet fluorescence), etc are all base fluorescence which means that their acid form does not generate any fluorescence, but their basic form normally produces fluorescence. The fluorescence helps to detect the endpoint of any titration reaction, where conventional acid-base indicators cannot function properly due to solution turbidity or highly coloured solution.

BCECF [2′,7′-bis-(2-carboxyethyl)-5-(and-6)-carboxyfluorescein] is a commonly used fluorescent pH sensor to detect the changes in cytosolic pH, possessing *pKa* of 6.97 which is close to physiological pH. It has been evidenced that the indicator is poorly fluorescent at low pH but the fluorescence becomes strong with increasing pH. The excitation spectrum of the BCECF faces a little shift with pH change, while the wavelength of the emission maximum remains the same. The pH is determined ratiometrically by the relative fluorescent intensities at 535 nm when the dye is excited at 439 nm and 505 nm.

3.4. Precipitation indicator

In this case, after completion of the titration, a coloured precipitation is obtained to identify the end point. Mohr and Volhard methods for determination of chloride concentration are the classical examples where precipitation indicators are used.

Mohr's method

Chloride ion concentration of a solution is determined by titration with silver nitrate which is slowly added from a burette to a chloride solution in presence of an indicator potassium chromate, so that silver chloride precipitates.

$Ag^{+}(aq) + Cl^{-}(aq) \rightarrow AgCl \downarrow$ (solid precipitate, white) Reaction 3.1

When all chloride ions are consumed due to the precipitation of silver chloride, addition of an extra drop of silver nitrate leads to the reaction between the silver ions and the chromate ions from the indicator potassium chromate. Such reaction results in the formation of a red-brown precipitate of silver chromate.

$2\ Ag^{+}(aq) + CrO_4^{-2}\ (aq) \rightarrow Ag_2CrO_4 \downarrow$ (solid precipitate, red brown)

Reaction 3.2

The Mohr's method is primarily operated at a pH range between 6.5 and 10. If the solutions are acidic, the Volhard method is suitably used.

Volhard method

In the Volhard method excess silver nitrate is added to chloride solution in presence of nitric acid to acidify solution, ammonium ferric sulphate and nitrobenzene. This facilitates the precipitation of all chlorides as silver chloride. Nitric acid is added to make the solution acidic, so that precipitation of metal hydroxides like ferric hydroxide may be avoided. Nitrobenzene is used to prevent further dissolution of silver chloride during the titration reaction with potassium thiocyanate because silver thiocyanate's solubility is slightly lower than solubility of silver chloride.

$AgCl(s) + SCN^{-} \rightarrow AgSCN(s) + Cl^{-}$ Reaction 3.3

Nitrobenzene makes a coating on the silver chloride precipitations. In the next step, excess silver is back titrated against potassium thiocyanate. Endpoint is detected by a distinct wine-red complex which is formed by the reaction between excess thiocyanate and Fe^{+3}.

$Ag^{+} + SCN^{-} \rightarrow AgSCN(s) \downarrow$ (white colour) Reaction 3.4

$Fe^{+3} + SCN^{-} \rightarrow Fe(CNS)_3$ (red colour) Reaction 3.5

3.5. Adsorption indicator

This is a type of indicator used in the titration reactions that involve adsorption of the indicator on the already existing precipitated molecules.

Fajans method

During addition of silver nitrate solution to the chloride solution, silver chloride is formed and precipitated. Before reaching the endpoint of the titration, chloride ions remain in excess. So, chloride ions are adsorbed on the AgCl surface, imparting a negative charge to the precipitate. Once the endpoint is reached, excess addition of one drop of silver nitrate causes excess abundance of silver ions (no excess Cl^{-} at that stage) that are adsorbed on the silver chloride precipitate surface, giving rise to a positive charge. At that point, anionic dichlorofluorescein (Fl^{-}) dyes are adsorbed on the particles and dark red silver fluoresceinate is formed on the surface of the precipitate. Thus the exhibition of its reddish colour is regarded as the end point of the titration.

$Ag^+ + Fl^- \rightarrow AgFl$ (silver fluoresceinate, dark red) Reaction 3.6

3.6. Mixed indictor

A mixed indicator, which is a mixture of indicator substances or a mixture of indicator and an organic dye, provides a sharp endpoint over a narrow pH range in an acid -base titration. It is difficult to obtain a sharper endpoint usually with an ordinary acid-base indicator due to their functioning over about two units of pH. In such a case, a mixed indicator may be used to detect the sharp endpoint, making it more discernible to the eye. A mixture of 3 parts of 0.1% aqueous sodium salt of thymol blue and 1 part of 0.1% of aqueous salt of cresol red, which is used for the titration of CO_3^{-2} to HCO_3^{-1} stage, shows a colour change from yellow to violet at pH 8.3. Sometimes the mixed indicator is also referred to as a screened indicator, such as screened methyl orange where xylene cyanol FF (2.8 g) and methyl orange (2 g) are added to ethanol (50%, 1L). This causes a colour change from green (alkaline side) to magenta (acidic side) via grey (pH 3.8).

3.7. Universal indicator

Universal indicator is one type of mixed indicator which comprises several indicators or dyes that cause colour changes gradually over a range of pH and used for approximation of pH measurement. This may be in a paper form (indicator paper) or in solution form. This is generally not meant for acid-base titration.

3.8. Solution preparation for commonly used indicators

A list of indicators with their method of preparation in the solution form is provided in **Table 3.2**.

Table 3.2: Methods of solution preparation of few important indicators

Name of the indicator	Solution preparation
pH Indicators	
Bromocresol Green, (0.4 g/l)	Bromocresol green (100 mg) is mixed up with sodium hydroxide solution (0.05 N, 2.9 ml) and rectified spirit (5 ml) and warmed until dissolved. Once it is dissolved, further addition of rectified spirit (50 ml) along with a volume make-up to 250 ml with distilled water is made.
Bromocresol Purple, (0.4 g/l)	Bromocresol purple (100 mg) is added to rectified spirit (5 ml) and warmed until dissolved. Rectified spirit (50 ml) and sodium hydroxide solution (0.05 N, 3.7 ml) are further added to it, followed by a volume make-up to 250 ml distilled water.
Bromophenol Blue (3′,3″,5′,5″-tetrabromophenolsulfonphthalein, 0.4 g/l.	Bromophenol blue (100 mg) is warmed with sodium hydroxide solution (0.05 N, 3.0 ml) and rectified spirit (50 ml) and finally the volume is made up to 250 ml with distilled water.
Bromothymol Blue, 0.4 g/l.	Bromothymol blue (100 mg) is warmed with sodium hydroxide solution (0 05 N, 3.2 ml) and rectified spirit (50 ml) and finally the volume is made up to 250 ml with distilled water.
m-Cresol Purple (4-[3-(4-hydroxy-2-methylphenyl)-1,1-dioxo-2,1λ6-benzoxathiol-3-yl]-3-methylphenol), 4 g/l	m-cresol purple (100 mg) is dissolved in rectified spirit (50 ml) and sodium hydroxide solution (0.05 N, 5.2 ml) and the final volume is made up to 250 ml with distilled water.
Cresol Red (4-[3-(4-hydroxy-3-methylphenyl)-1,1-dioxo-2,1λ6-benzoxathiol-3-yl]-2-methylphenol), 0.2 g/l.	Cresol red (50 mg) is dissolved by warming in sodium hydroxide solution (0.05 N, 2.65 ml) and rectified spirit (5 ml), followed by further addition of rectified spirit (50 ml) and rest portion with distilled water, so that the final volume becomes 250 ml.
Dimethyl Yellow (4-dimethylaminoazobenzene), 2 g/l.	Dimethyl yellow (500 mg) is directly solubilized in 250 ml of rectified spirit.
2,4-Dinitrophenol, 1 g/l.	2,4-dinitrophenol (250 mg) is directly dissolved in 250 ml of water.
Methyl Orange, 0.4 g/l	Methyl orange (100 mg) is solubilized in 50 ml of rectified spirit and the volume is made up to 250 ml with distilled water.
Methyl Red, 0.1 g/l.	Finely powdered methyl red (25 mg) is mixed with NaOH solution (0.05 N, 0.95 ml) and rectified spirit (5 ml), followed by warming and further addition of rectified spirit (125 ml) and distilled water to make the final volume of the solution to 250 ml.

Name of the indicator	Solution preparation
Naphtholphthalein, 5 g/l.	1-naphtholphthalein (1.25 g) is dissolved in 250 ml of rectified spirit.
3-Nitrophenol, 1 g/l.	3-nitrophenol (250 mg) is solubilized directly in 250 ml of water.
4-Nitrophenol, 1 g/l.	4-nitrophenol (250 mg) is solubilized directly in 250 ml of water.
Phenolphthalein, 10 g/l.	Phenolphthalein (2.5 g) is dissolved in 250 ml of rectified spirit.
Phenol Red, 0.2 g/l.	Phenol red (50 mg) is mixed with NaOH solution (0.05 N, 2.85 ml) and rectified spirit (5 ml), and is warmed. After that, further additions of rectified spirit (50 ml) and distilled water (195ml) are made to obtain a final volume of 250 ml.
Thymol Blue, 0.4 g/l.	Thymol blue (100 mg) is warmed with NaOH solution (0.05 N, 4.3 ml) and rectified spirit (5 ml), followed by further addition of rectified spirit (50 ml) and final volume make-up to 250 ml with distilled water.
Thymolphthalein, 2 g/l.	Thymolphthalein (0.5 g) is dissolved in rectified spirit (60 ml) and is further diluted to 250 ml with distilled water.
Brilliant Green ([4-[4-(diethylamino) phenyl]-phenylmethylidene] cyclohexa-2,5-dien-1-ylidene]-diethylazanium;hydrogen sulfate), 1% aqueous solution. Besides its use as a pH indicator, it is also used for staining plant cytoplasm.	1 g of brilliant green is dissolved in 50 mL of deionised water, and the solution is diluted to 100 mL.
Chlorophenol Red (3′,3′-Dichlorophenolsulfonaphthalein), 0.04% aqueous	23.5 mL of 0.01M sodium hydroxide is added to 226.5 mL of deionised water. Later, 0.1 g of chlorophenol red is dissolved in this NaOH solution.
Congo Red Indicator [disodium 4-amino-3-[4-[4-(1-amino-4-sulfonato-naphthalen-2-yl)diazenylphenyl]phenyl] diazenyl-naphthalene-1-sulfonate], 0.1% aqueous solution	0.1 g of Congo red is dissolved in 50 mL of deionised water, and then the solution is further diluted to 100 mL.
Indigo Carmine (5,5′-indigodisulfonic acid sodium salt), 0.25% aqueous-alcoholic solution	0.25 g of Indigo Carmine is dissolved in 80 mL of 50% ethyl alcohol solution, followed by further dilution to 100 mL with 50% ethyl alcohol solution. Due to its poor self-life, it is always suggested to prepare the solution fresh before use.

Name of the indicator	Solution preparation
Litmus, 0.5% aqueous solution	0.5 g of litmus is dissolved in 80 mL of boiling deionised water. After cooling to room temperature, the solution is further diluted to 100 mL with deionised water.
Metal indicators	
Calcein, (2-[7'-[bis(carboxymethyl) amino]methyl]-3',6'-dihydroxy-3-oxospiro[2-benzofuran-1,9'-xanthene]-2'-yl]methyl-(carboxymethyl)amino] acetic acid) Calcein is a fluorescein-based metallofluorescent indicator. It produces a green fluorescence and an orange-yellow colour with Ca ions in alkaline solution. In absence of free Ca ions, the colour becomes yellowish red without any fluorescence.	Calcein (1 g) with potassium chloride (100 g) was ground to a fine powder and mixed properly.
Calcon (3-hydroxy-4-[(2-hydroxy-4-sulfonaphthalen-1-yl)diazenyl] naphthalene-2-carboxylic acid), 1 g/l. It gives a purple-red colour with Ca ions in alkaline solution. In absence of free metal ions, the solution becomes blue.	Calcon (250 mg) is dissolved in 250 ml of water.
Catechol Violet, 1,2-Benzenediol, 4,4'-(1,1-dioxido-3H-2,1-benzoxathiol-3-ylidene) bis-Pyrocatechinsulfonephthalein, 1 g/l Under acidic conditions, with Bi it gives a blue colour and red with Th ions. The solution is yellow in absence of these metal ions	Catechol violet (250 mg) is added to 250 ml of water.
Methylthymol Blue (3,3′-Bis[N,N-di(carboxymethyl)-aminomethyl] thymolsulfonphthalein, Sodium Salt). With this indicator, a blue colour is obtained in the presence of Ca and Mg ions under alkaline conditions. The solution becomes grey in absence of these metallic ions.	Methylthymol blue (1 g) is ground uniformly with potassium nitrate (100 g) to a fine powder.
Murexide (ammonium purpurate). An orange colour with Cu, yellow colour with Ni & Co (pH 10 to 11) and red colour with Ca (pH 11) in alkaline solution are obtained by this indicator. The solution gives a purple colour in the absence of free metal ions.	Murexide (1 g) is triturated with anhydrous K_2SO_4 (100 g) to a fine powder.

Name of the indicator	Solution preparation
Sodium Alizarin Sulphonate (Sodium 1,2-dihydroxy-9, 10-anthraquinone-3-sulphonate), 1 g/l. In presence of Th and Al ions, it produces a bluish-red colour at acidic condition (about pH 4.0) and in their absence, the solution is yellow.	Sodium alizarin sulphonate (250 mg) is dissolved in 250 ml of water.
Xylenol Orange [3,3'-bis { N,N-di (carboxymethyl) aminomethyl}-o-cresolsulphonphthalein)],1 g/l. With Hg, Pb and Zn ions under acidic conditions, xylenol orange exhibits a reddish purple colour. However, it produces a yellow colour in the absence of these metal ions.	Xylenol orange 250 mg is mixed with 250 ml of water and is filtered, if necessary.
Eriochrome Black T (EBT), Sodium 1-[1-Hydroxynaphthylazo]-6-nitro-2-naphthol-4-sulfonate, 1% alcoholic solution: This is used for EDTA titrations.	1 g of Eriochrome Black T is dissolved in 80 mL of 95% ethyl alcohol, and the solution is further diluted to 100 mL with 95% ethyl alcohol.
Indicators for non-aqueous titrations	
Crystal Violet (hexamethyl pararosaniline hydrochloride), 5 g/l. This indicator changes from violet (basic) to yellowish green (acidic) via bluish-green (neutral)	Crystal violet (1.25 g) is dissolved in 250 ml of glacial acetic acid.
1-Naphtholbenzein (a)1-Naphtholbenzein indicator solution (in acetic acid), 2 g/l. In non-aqueous media, it changes from green (basic) through orange (neutral) to yellow (acidic). (b) 1-Naphtholbenzein indicator solution (in propan-2-ol), 10 g/l. In non-aqueous media, it makes a colour change from green (basic) to yellow orange (acidic).	(a) 1-naphtholbenzein (500 mg) is added to glacial acetic acid (250 ml). (b) 1-naphtholbenzein (2.5 g) is solubilized in propan-2-ol (250 ml).
Quinaldine red [2-(4-dimethylaminostyryl) quinoline ethiodide],1 g/l.	Quinaldine red (250 mg) is solubilized in methanol (250 ml).
Oxidation and reduction indicators	
Barium Diphenylamine-4-Sulphonate, 7.5 g/l At a transition emf of +0.84V, it transforms from colourless to violet.	Barium diphenylamine-4-sulphonate (750 mg) is dissolved in water (100 ml)

Name of the indicator	Solution preparation
1, 10-Phenanthroline, 17 g/l Phenanthroline with ferrous ions changes from red to blue at a transition emf of + 1.08V.	1,10-phenanthroline (1.49 g) is mixed with 70 ml of water containing ferrous sulphate (0.7 g). Finally, the volume of the solution is made up to 100 ml.
Fehling's Solution A **Fehling's Solution B** Mixture of Fehling's Solution A & B: This solution is used for testing reducing sugars and aldehydes.	34.6 g of copper (II) sulfate pentahydrate is dissolved in 500 mL deionised water. 125 g of potassium hydroxide and 173 g of potassium sodium tartrate tetrahydrate are dissolved in 500 mL of deionised water. Fehling's Solution A and B are combined at a ratio of 1:1 just before use.
Adsorption indicators	
Amaranth [trisodium 1-(4-sulpho-1-naphthylazo)-2-naphthol-3, 6-disulphonate], 2 g/l. During the titration of iodine and iodides with potassium iodate, the colour changes from orange red to yellow.	Amaranth (500 mg) is mixed directly with water (250 ml).
Dichlorofluorescein [2,7-dichlorofluorescein], 1 g/l. With this indicator, the precipitate of silver chloride, during titrating a chloride solution with silver nitrate, becomes orange at pH 5-8. However, after completion of the precipitation, the precipitate turns to red colour.	Dissolve 0.25 g of dichlorofluorescein in 250 ml of rectified spirit.
Phenosafranine (3,7-diamino-5-phenylphenazonium chloride), 2.5 g/l. With this indicator, the precipitate of silver chloride or bromide, during titrating a chloride or bromide solution with silver nitrate, becomes red in acid solution. However, after completion of the precipitation, the precipitate turns to blue colour.	Phenosafranine (250 mg) is solubilized in distilled water (100 ml).
Miscellaneous indicators	
Ammonium Ferric Sulphate, 100 g/l It produces a deep red colour with thiocyanate ions under acidic condition.	Ammonium ferric sulphate (25 g) is solubilized in water (250 ml). In absence of transparency of the solution, few drops of nitric acid (5N) may be added.

Name of the indicator	Solution preparation
4-dimethylaminobenzylidene rhodanine, 0.2 g/l. Making a complex with silver ions, it causes a colour change from very pale yellow to red.	4-dimethylaminobenzylidene rhodanine (50 mg) is dissolved in acetone (250 ml).
Methylene Blue (Tetramethylthionine chloride dihydrate), 0.15 g/l,	Methylene blue (38 mg) is dissolved in water (250 ml).
Potassium Chromate, 50 g/l.	Potassium chromate (12.5 g) is dissolved in water (250 ml).
Starch, 5 g/l. It forms a blue colour with free iodine in the presence of a soluble iodide.	Initially, soluble starch (0.5 g) is mixed with 5 ml of water. The solution is then stirred in boiling water (95 ml). The solution is boiled further for few minutes, followed by cooling.
Coomassie Brilliant Blue G-250 (triphenylmethane dye) It is used for staining proteins in polyacrylamide and agarose gels in electrophoresis.	0.1 g of Coomassie brilliant blue G-250 is dissolved in 25 mL methyl alcohol. 40 mL deionised water is added followed by 5 mL acetic acid. Finally, the mixture is diluted to 100 mL with DI water.
Clayton Yellow, Disodium 2,2′-[(1E)-triaz-1-ene-1,3-diyldi(4,1-phenylene)] bis(6-methyl-1,3-benzothiazole-7-sulfonate), 1% aqueous solution. This is used as a pH indicator and also as a fluorescent dye for microscopy.	1 g of Clayton yellow is dissolved in 50 mL of deionised water, followed by its dilution to 100 mL.
Eosin Y Stain, 2-(2,4,5,7-tetrabromo-6-oxido-3-oxo-3H-xanthen-9-yl)benzoate [in its deprotonated form], 0.5% aqueous solution It is used as a cytoplasmic stain.	0.5 g of eosin Y is dissolved in approximately 80 mL deionised water and the final volume is made up to 100 mL. A few drops of chloroform are added to the solution as a preservative.
Ninhydrin (2,2-Dihydroxyindane-1,3-dione) This is used for testing proteins	2.5 g of ninhydrin is added to 50 mL of n-butyl alcohol in a 600 mL beaker. The solution is gently heated, using a magnetic stirrer/hot plate in a fume hood until the entire solid is dissolved. The resultant solution is further diluted to 500 mL with n-butyl alcohol.
Molisch Reagent This is used for testing aldehydes, sugars, and carbohydrates.	5 g 1-naphthol is dissolved in 100 mL of 95% ethyl alcohol.
Rhodamine B, [9-(2-Carboxyphenyl)-6-(diethylamino)-N,N-diethyl-3H-xanthen-3-iminium chloride], 1% aqueous solution It is used as a biological stain.	1 g of rhodamine B is added to 50 mL deionised water, followed by further dilution to 100 mL.

Source: adapted and modified from IS:2263-1979 (First Revision, Reaffirmed 1998) Methods of preparation of indicator solutions. Bureau of Indian Standards.

3.9. Model questions

i. Why are acid-base indicators also known as pH indicators?

ii. How do acid-base indicators work? Explain with proper examples.

iii. What is pK_{ind}? How is it derived? What does it signify when pK_{ind} equals the value of solution pH?

iv. How does phenolphthalein work? Illustrate the mechanism using its chemical structure.

v. What happens to methyl orange when acid or alkali is added? Describe its functional mechanism using the chemical structure of methyl orange.

vi. Write down the functional mechanism of the following indicators:

(a) Bromocresol purple

(b) Bromocressol green and

(c) Thymol Blue

vii. Name a few fluorescent indicators. Describe the functional mechanism of fluorescent indicators, using an example. Why is BCECF used to detect cytosolic pH changes?

viii. Describe the principle of Mohr's and Volhard methods during quantitative determination of chloride concentrations in solution. What is the purpose of using nitrobenzene in the Volhard method?

ix. Mention the principle of Fajans method where an adsorption indicator is used.

x. How does xylenol orange work with metal ions, such as Pb, Hg and Zn?

3.10. Suggested readings

(i) Kolthoff, I. M. (2010) Indicators. World Public Library Association.

(ii) IS:2263-1979 (First Revision, Reaffirmed 1998) Methods of preparation of indicator solutions. Bureau of Indian Standards.

(iii) Bassett, J.; Denney, R.C.; Jeffery, G. H.; Mendham, J. (1978) Vogel's Textbook of Quantitative Inorganic Analysis Including Elementary Instrumental Analysis (4th Edition). The English Language Book Society and Longman.

4

Centrifugation

4.1. Introduction

Centrifugation is a technique that promotes accelerated settling of particles in a solid-liquid mixture by the application of centrifugal force. The device that accelerates the separation with the help of centrifugal force is known as 'Centrifuge'. The centrifuge consists of a fixed base or frame and a rotating part in which the mixture is placed and is then spun at high speed. Normally if a solid-liquid immiscible mixture is left undisturbed, many particles, which are comparatively larger in size, settle at the bottom of a container due to gravity (1 x *g*) and the colloidal particles, extremely small in size, are not separated at all or they settle at a length of time that seems to be impractical. Two particles of different masses settle at different rates in response to gravity. But it is possible to get all the particles settled at reasonable time by subjecting the solid-liquid mixture to high centrifugal force. The separation which proceeds very slowly by gravity can be accelerated greatly in a centrifuge by introducing centrifugal force to increase the settling rate depending on the rotation speed of the rotor. With this technique it is not only possible to separate solid and liquid from a mixture but also it is possible to separate particles of different masses in a tube at different rates. Centrifugal force, that causes the particles to move radially away from the axis of rotation, is used to enhance the settling rate in a 'Centrifuge'. The centrifugal force is created by rotating the suspension at a certain speed, often designated as revolutions per minute (rpm). The spinning motion forces the solid to settle at the bottom of the centrifuge tube. The subjected force on the particles is measured as x *g*, gravity and called as Relative Centrifugal Force (RCF). An RCF of 1000 x *g* specifies that the centrifugal force applied on a particular suspension is 1000 times greater than the Earth's gravitational force. The particle's settling velocity in centrifugation is a function of the particle's shape and size, centrifugal acceleration, the volume fraction of solids present, the density difference between the particle and the liquid, and the viscosity. A common example of using centrifugal force in our daily life is washing clothes in a washing machine. During drying, the centrifugal force is created in the spin cycle and it facilitates faster drying by squeezing water out of the fabric. In dairies skimmed milk is obtained by separating cream from milk with the help of centrifugation.

Centrifugation is now widely used in the field of biochemistry, cellular and molecular biology, and in medicine for sedimentation of cells and viruses, separation of subcellular organelles, and isolation of macromolecules, such as DNA, RNA, proteins, lipids, etc. Many industrial processes use this technique for solid-liquid phase separation in pharmaceutical, chemical, textile, and wastewater treatment processes. In effluent treatment plants (ETP), raw sewage that comes from homes, hospitals, industries, etc. are centrifuged to separate solid materials and produce clear water or effluent.

4.2. Principle

In a centrifuge during sedimentation the applied centrifugal field (G) is directed radially outward and is determined by the Equation 4.1.

$$G = \omega^2 \times r \qquad \text{Equation 4.1}$$

Where,

ω = angular velocity of the rotor in radians/sec, and

r = radial distance of the particle or molecule from the axis of rotation in cm.

Taking into consideration revolutions per minute (rpm) and the value of one revolution of the rotor as 2π radians, the angular velocity (ω) could be expressed in radians/sec (Equation 4.2).

$$\omega = \frac{2\pi \times \text{rpm}}{60} \qquad \text{Equation 4.2}$$

By substituting the value of ω in Equation 4.1, the centrifugal field (G) becomes

$$G = \frac{(2\pi \times \text{rpm})^2}{(60)^2} \times r$$

$$= \frac{4\pi^2 \times rpm^2 \times r}{3600} \qquad \text{Equation 4.3}$$

The centrifugal field is usually compared to the earth's gravitational field (g = 981 cm/sec^2) and expressed in multiple of g. It is reported as Relative Centrifugal Field (RCF) in g units. Therefore, RCF becomes

$$\text{RCF} = \frac{4\pi^2 \times rpm^2 \times r}{3600 \times 981}$$

$$= 1.12 \times 10^{-5} \times \text{rpm}^2 \times r \qquad \text{Equation 4.4}$$

RCF is dependent upon the speed of the rotor (rpm), the radius of the rotation of the rotor (r), and also the rotor design (fixed-angle or swinging-bucket). It is possible to keep the factors like rotor design and rotor speed constant, but r varies from the top to the bottom of a centrifuge tube. The difficulty with using this formula (Equation 4.4) is to establish the value for r. Three r-values are provided by the manufacturer of the centrifuge: r_{max}, r_{min} and r_{av} which correspond to the distances from the centre of rotation to the bottom, top and middle of the sample tube, respectively. Hence, it is usually a common practice to take r_{av} (average radius) to calculate RCF. It is necessary to keep in mind that RCF is a constant that is independent of the apparatus used.

It is interesting to note that in many research publications rotational speed (rpm) during centrifugation is quoted instead of the amount of acceleration to be applied in the sample. This is not proper because centrifuges may come in different shapes and sizes and the rotors vary with their dimensions. Two rotors with different diameter running with the same rotational speed result in different accelerations. Therefore, the universal unit of centrifugation to define its acceleration should be the centrifugal force (x g). This is a transferable unit among different centrifuges.

When a particle or molecule with an effective mass (m) on rotation about a central axis is subjected to a centrifugal field, a centrifugal force (F_c) is generated (Equation 4.5).

$$F_c = m \times \omega^2 \times r \qquad \text{Equation 4.5}$$

From the Equation 4.5 it is clear that the centrifugal force becomes more; facilitating rapid sedimentation in case of larger particles (higher value of m), faster the centrifugation (more angular velocity or more rpm) and higher the distance of the migrating particles from the central axis of rotation (more r-value). Effective mass of the particle is defined as the actual mass corrected by a correction factor, which is nothing but a buoyancy factor due to the weight of the solvent displaced (Equation 4.6).

$$\mathrm{m} = m_0 - m_0 \times \upsilon \times \rho \qquad \text{Equation 4.6}$$

Where,

m_0 = actual mass of the particle,

υ = partial specific volume of the molecule or cc of the solution volume increase occurring by addition of 1 g particle solute in a large excess of solvent, and

ρ = density of the solvent.

Substituting *m* (obtained from Equation 4.6) in Equation 4.5, the centrifugal force becomes

$$F_c = (m_0 - m_0 \times \upsilon \times \rho) \times \omega^2 \times r$$

$$= m_0(1 - \upsilon \times \rho) \times \omega^2 \times r \qquad \text{Equation 4.7}$$

Therefore, it is evident that higher partial specific volume of the particle and higher density of the solution result in decreased sedimentation rate.

The buoyancy factor contains the medium density ρ (g/ml) and the specific volume of the particle υ (ml/g). The specific volume may be considered as $1/\rho_p$, where ρ_p is the effective density of the particle. So buoyancy factor (1- υ × ρ) becomes $(1-\frac{\rho}{\rho_p})$. Therefore, if the medium density and the effective particle density are same ($\rho_p = \rho$); the particle no longer sediments because net force on the particle becomes zero.

However, a frictional force (F_r, Equation 4.8) is also generated with the movement of the particle in the liquid medium and this resistance force opposes the motion.

$$F_r = f \times v \qquad \text{Equation 4.8}$$

Where,

f = frictional coefficient of the particle in the solvent, and

v = sedimentation velocity.

Assuming the particle to be rigid, unhydrated and spherical (radius r_p) which is moving through a medium of viscosity η, according to the Stoke's law the frictional coefficient (*f*) can be expressed as Equation 4.9.

$$f = 6\,\pi \times \eta \times r_p \qquad \text{Equation 4.9}$$

The frictional coefficient of a particle thus depends upon its shape, size and hydration, and of the viscosity of the medium. For hydrated particles the actual radius of the particle should be replaced by the effective or Stoke's radius.

From Equation 4.8, it is explored that higher sedimentation velocity of the particle causes higher F_r value. However, a point comes when these two opposing forces (F_c and F_r) are balanced, and a constant velocity is reached. Therefore, by combining Equation 4.7 and 4.8, Equation 4.10 can be deduced.

$$f \times v = m_0(1 - \upsilon \times \rho) \times \omega^2 \times r$$

$$v = \frac{m_0(1 - \upsilon \times \rho) \times \omega^2 \times r}{f} \qquad \text{Equation 4.10}$$

It is clear that v depends upon both the rotor speed and *r*. By dividing both sides of the Equation 4.10 by $\omega^2 \times r$, an expression could be achieved (Equation 4.11) which is independent of these. This is called the Svedberg equation.

$$\frac{v}{\omega^2 \times r} = \frac{m_0(1-\upsilon \times \rho)}{f} = s \qquad \text{Equation 4.11}$$

This is termed as sedimentation coefficient and it is considered as a physical property of a molecule, a constant like its molecular weight or diffusion coefficient. The unit for s is second. For biological molecules and cell organelles s-value ranges from 1 x 10^{-13} to 10,000 x 10^{-13} sec. Therefore, a new expression, Svedberg unit (S, 1S = 1 x 10^{-13} sec) has been defined for numerical convenience. Note that the Svedberg unit is an uppercase S while the symbol for the sedimentation coefficient is a lowercase s. S is written with no space between the number and symbol (e.g., 70S). Bigger particles exhibit higher S-values implicating faster sedimentation rate. The Svedberg is not additive: a particle formed from two 5S particles does not show a sedimentation value of 10S; since the sedimentation rate is associated with the size of the particle, when two particles bind together there is inevitably a loss of surface area. Thus, when measured separately they have Svedberg values that do not add up to that of the particle formed when they bind together. A prokaryotic ribosome is 70S (70 x 10^{-13} sec) which is composed of a 50S subunit and a 30S subunit.

Equation 4.11 may be further illustrated into its simple form (Equation 4.12) by incorporating molecular weight (M, mass of 1 mole of substance expressed in grams or molar mass) and Avogadro's number (N_A, $N_A = M/m_0$)

$$s = \frac{v}{\omega^2 \times r} = \frac{m_0(1-\upsilon \times \rho)}{f} = \frac{M(1-\upsilon \times \rho)}{N_A \times f} \qquad \text{Equation 4.12}$$

Now, it is possible to incorporate a new parameter, the diffusion coefficient into Equation 4.12, using the relationship $D = \frac{RT}{N_A \times f}$ (where, D is the diffusion coefficient, R is the gas constant, and T is the absolute temperature) to get two new equations (Equation 4.13 and 4.14). While a particle is reaching its terminal velocity, it is also affected by diffusion, usually in a direction opposite to its movement under force. This is a movement of Brownian motion and can be mathematically stated as a constant, the diffusion coefficient (D), which is the speed of the molecule divided by time. It is expressed in units, Ficks, which are 10^{-7} cm^2/sec.

$$s = \frac{M(1-\upsilon \times \rho)}{\frac{RT}{D}} = \frac{M \times D(1-\upsilon \times \rho)}{R \times T} \qquad \text{Equation 4.13}$$

$$M = \frac{s \times R \times T}{D(1 - \upsilon \times \rho)}$$ Equation 4.14

A correction of the experimental s-value to a standard state of water at 20 °C is essential to compare the data obtained from different laboratories under different experimental conditions because sedimentation coefficient is influenced by temperature, solvent density and solvent viscosity. The corrected equation for standard sedimentation coefficient ($s_{20,w}$) is given as Equation 4.15.

$$s_{20,\,w} = \frac{s_{T,Solv}(1 - 0.9982 \times \upsilon)_{20,W}}{(1 - \upsilon \times \rho)_{T,Solv}} \times \frac{\eta_{T,Solv}}{\eta_{20,W}}$$ Equation 4.15

Where, $s_{20,w}$ is the sedimentation coefficient expressed in terms of standard solvent of water at 20 °C, $s_{T,\,Solv.}$ is the experimental sedimentation coefficient value obtained at a particular solvent and temperature, $\eta_{T,\,Solv.}$ is the viscosity of the solvent used at an experimental temperature, and $\eta_{20,\,w}$ is the viscosity of water at 20 °C. Density of water at 20 °C is 0.9982 g/cm^3.

Likewise, diffusion coefficient, D must be corrected as Equation 4.16.

$$D_{20,W} = D_{T,Solv} \times \frac{293°K}{T} \times \frac{\eta_{T,Solv}}{\eta_{20,W}}$$ Equation 4.16

Calculation of s from r (the position of the moving boundary) vs. t (time in sec)

From Equation 4.11 it is possible to get a differential equation (Equation 4.17)

$$s = \frac{v}{\omega^2 \times r} = \frac{1}{\omega^2} \times \frac{1}{r} \times \frac{dr}{dt}$$ Equation 4.17

Separating the variables and integrating, one can get Equation 4.18 or 4.19.

$$\int_{t0}^{t} \omega^2 \times s \times dt = \int_{r_0}^{r_t} \frac{1}{r} \times dr$$ Equation 4.18

or, $\omega^2 \times s(t - t_0) = \text{In}\ \frac{r_t}{r_0}$ Equation 4.19

Integrating between $t = t_0$ ($r = r_0$) and $t = t$ ($r = r_t$), r_0 is the boundary position at the start ($t = t_0$) and r_t is the boundary position at later times. Thus, plotting of In $\frac{r_t}{r_0}$ against ($t - t_0$) gives rise to a straight line with a slope ω^2 s **(Figure 4.1)**

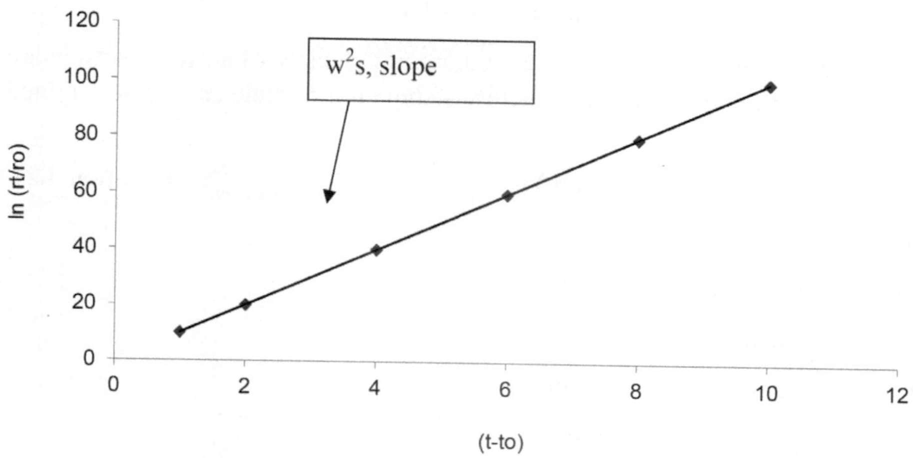

Fig. 4.1: Calculation of s value by plotting the relative position of the moving boundary vs. time

Concentration effect on sedimentation coefficient (s)

Usually 's' decreases with increasing solute concentration because higher concentration of the solute increases the effective viscosity of the solution. This effect is much more pronounced in case of extended molecules like DNA because they bump into one another more often. This phenomenon can be observed by measuring 's' at different solute concentrations and extrapolation at zero-point solute concentration. However, the situation is different, and 's' may also increase with the solute concentration if there exists an equilibrium between different physical states of the solute e.g., monomer and its dimmer state.

$$2\,X \rightleftharpoons XX$$

$$K_{eq} = \frac{[XX]^2}{[X]^1} \qquad \text{Equation 4.20}$$

One can see two major boundaries: slowest moving is for monomer and fastest moving is for dimmer; when the interconversion is very slow. Only one asymmetric boundary is observed during the fast interconversion of the solute. Under the equilibrium condition if the concentration is more, the reaction is directed to the right-hand side forming more dimmer; and if the concentration is less, the reaction is directed to the left-hand side forming more monomers. Therefore, it is possible to study the interactions between the subunits in multi-subunit macromolecules by monitoring the sedimentation behaviour of the same at different concentrations.

Computation of sedimentation time of a particle

If the density and viscosity of the medium and density of a given particle are known; the time needed to completely sediment a particle can be determined by the Equation 4.21.

$$\text{Time required to sediment (min)} = \frac{(D-L)}{(D+L)} \times \frac{\eta}{r_p^2(\rho-\rho_p)\times rpm^2} \quad \text{Equation 4.21}$$

Where,

D = radial distance in cm for r_{max},

L = radial distance to meniscus,

η = viscosity of the fluid medium,

ρ = density of the fluid medium,

ρ_p = density of the particle to sediment,

r_p = diameter of the particle in cms, and

rpm = rotational velocity (revolutions per minute).

Pelleting time

It is the time taken for complete pelleting of a given particle. The lower is the value of *K*, (Equation 4.22) the shorter is the time required to sediment a given particle.

$$T = \frac{K}{S} \quad \text{Equation 4.22}$$

Where,

T = pelleting time (hour),

K = *K*-factor of the rotor, and

S = sedimentation coefficient in Svedberg unit.

K-factor

It is the pelleting efficiency of a rotor. Small *K*-factor leads to a good pelleting efficiency.

$$K = 2.53\times10^{11}\times\frac{[\ln(r_{max})-\ln(r_{min})]}{(rpm)^2_{max}} \quad \text{Equation 4.23}$$

When the rotor is operated at other speed, the *K*-factor is revised as follows:

$$K_{revised} = K\times\frac{(rpm)^2_{max}}{(rpm)^2_{selected}} \quad \text{Equation 4.24}$$

Rotor conversion formula

It is needed to determine the time it requires for the same sample to pellet in a new rotor, when time required for the same sample in an old rotor is known. It is represented as:

$$\frac{T_1}{K_1} = \frac{T_2}{K_2}$$

Or, $T_1 = T_2 \times \frac{K_1}{K_2}$ Equation 4.25

Where,

T_1 = Time to pellet in the new rotor,

T_2 = Time to pellet in the old rotor,

K_1 = K-factor of the new rotor, and

K_2= K-factor of the old rotor.

4.3. Types of centrifuge

Based on the acceleration achieved or speed of the rotor, the centrifuges are classified as (a) Low-speed, (b) Medium-speed, (c) High-speed, and (d) Ultracentrifuges (**Figure 4.2**). Out of these four types, the first three types perform only for preparative purpose i.e., to isolate and sediment relatively heavy materials from biological samples; while ultracentrifuge may be used both for separation and analytical measurements.

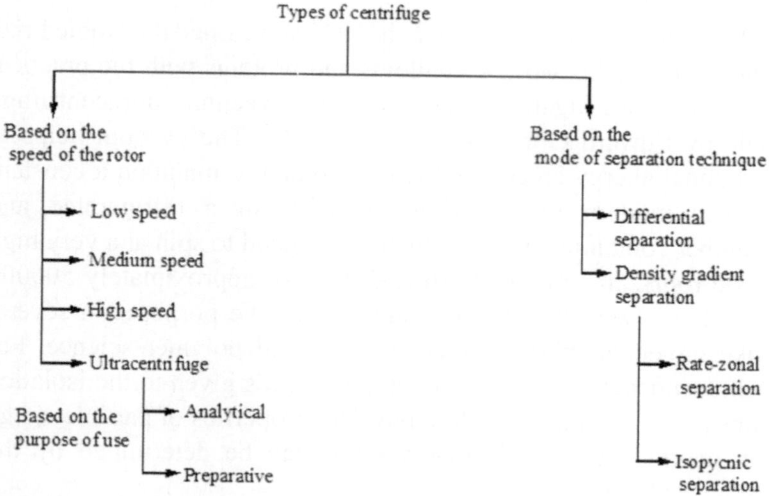

Fig. 4.2: Classification of centrifuge

4.3.1. Low-speed centrifuge

This is operated at room temperature without any provision of temperature controlling accessories with a maximum rotor speed of 5000 rpm and is capable of generating RCF of 6000-7000 x g. Most of the clinical centrifuges are low speed centrifuges used for sedimentation of heavy particles. Commonly low speed centrifuge is accommodated with either fixed-angle or swinging-bucket rotor. It can pellet cells, but not organelles or biomolecules.

4.3.2. Medium-speed centrifuge

Bench top centrifuges fall within this category. 'Microfuges' are the best examples of medium speed centrifuges which are equipped with fixed-angle rotors and operated at 5000-15000 rpm with RCF values up to 10000-12000 x g. They are specially designed for small volumes of sample (< 2 ml) to pellet nucleic acids and denatured proteins.

4.3.3. High-speed centrifuge

High-speed centrifuges are available with fixed-angle, swinging-bucket and vertical rotors. They possess refrigerated rotor chambers to dissipate the high temperature generated due to high speed and frictional resistance of the rotor during the centrifugal operation. The rotor speed ranges from 15000-30000 rpm attaining RCF up to 100000 x g. High-speed centrifuges perform well to pellet the cell debris after homogenization, ammonium sulfate precipitates of proteins, chloroplasts, mitochondria, cell nuclei, most biomolecules, etc.

4.3.4. Ultracentrifuge

Theodor Svedberg invented ultracentrifuge in 1925 and earned the Noble Prize in 1926 in chemistry for his work on colloids and proteins with the practical application of ultracentrifugation. The idea of a vacuum ultracentrifuge was first made by Edward Greydon Pickels in 1946. The vacuum helps to reduce the frictional energy created at high speed and to maintain a constant temperature. The modern ultracentrifuge consists of a refrigerated and evacuated chamber containing a rotor which is designed to spin at a very high speed (1,00,000 rpm), creating a centrifugal force of approximately 500000 to 900000 x g. It is used for analytical and preparative purposes in several disciplines like molecular biology, biochemistry, and polymer science. For preparative ultracentrifugation, the prime importance is given to the isolation of specific substance or a fraction, while physical properties of particles (e.g., sedimentation coefficient or molecular weight) can be determined by the analytical ultracentrifuge.

4.3.4.1. Analytical ultracentrifuge

The purpose of the analytical centrifugation is to characterize key sample properties rather than to prepare or purify a sample for subsequent use. Analytical ultracentrifuge, a versatile and powerful instrument to characterize the biomolecules in their native state under biologically relevant solution conditions, is equipped with an optical detection system that directly measures the sample concentration inside the centrifuge cell during sedimentation and it allows the operator to observe the evolution of the sample concentration versus the axis of rotation profile as a result of the applied centrifugal field. In analytical ultracentrifugation it is possible to visualize the sample in real time during sedimentation as the experiment progresses. The centrifuge tubes are projected with quartz windows that lie parallel to the plane of rotation of the rotor head. As the rotor turns, the images of the substances in the cell are projected by an optical system onto a film or a computer. It is also possible to measure the concentration of the solution at different points in the tube by absorbing light at an appropriate wavelength, which can be achieved by the pen deflection or degree of blackening of the photographic plate. The instrument is configured either with a scanning UV/Vis absorption optic detection system that provides not only the sensitivity for low concentration work but also the selectivity for optimizing detection based on the sample's maximum absorbance, or with an integrated absorbance and Rayleigh Interference Optics (RIO). The addition of RIO increases the opportunity to broaden the research area because it provides the scope to measure the refractive index change resulting from changes in sample concentration. Thus, all biological molecules including biopolymers, such as polysaccharides, may be investigated with increased accuracy. The additional advantage of using RIO is that ligand- or drug-induced changes in protein conformation or association (e.g., ATP or GTP-binding proteins) can be analysed without concern of the fact that the UV absorbance of the ligand/drug will obscure the protein absorption. Further, macromolecular concentration can be increased far beyond the absorbance range, using the RIO system. With the help of analytical ultracentrifuge one can determine the sample purity, subunit stoichiometries, macromolecular conformational changes, and also can measure equilibrium constants and thermodynamic parameters for self- and hetero-associating systems.

4.3.4.2. Preparative ultracentrifuge

The main objective of the preparative ultracentrifuge is to isolate and purify specific particles which can be re-used for further analysis in subsequent investigations. It inherits the same basic design as analytical centrifuge except the optical system. Compared to analytical ultracentrifugation, it is possible to handle a relatively large volume of sample in preparative ultracentrifugation.

On the basis of mode of operation, centrifugation may be broadly classified into two categories: (a) differential centrifugation, and (b) density gradient centrifugation.

4.3.5. Differential centrifugation

Differential centrifugation is a procedure in which a homogenate is subjected to a repeated centrifugation increasing the centrifugal force each time. The theoretical basis of differential centrifugation is to separate particles depending on the differences in the sedimentation rate of the particles due to their difference in size and mass. This is primarily responsible for obtaining partially pure preparation of subcellular organelles and macromolecules in the form of pellets. Initially large particles are sedimented followed by small particles. Likewise, for the particles of the same mass but different density, the highest density particles settle at the bottom of the tube at a faster rate than the low-density particles. This provides the basis for obtaining crude cell organelle fractions by differential centrifugation. The tissues or cells are first disrupted to release the internal contents and then the homogenate is centrifuged repeatedly by stepwise increasing centrifugal force and time to obtain pellets of partially purified organelles. However, it is necessary to wash the pellet, obtained at a given centrifugal force by re-suspending several times in the homogenization medium and re-centrifugation under the same operating condition. By this way it is possible to obtain partially purified cell organelles such as heavy nuclei, mitochondria, lysosomes and microsomes (vesicles of Golgi and endoplasmic reticulum) and/or cytosol (particle free solution of cytoplasm) by subjecting the cell homogenate to sequentially higher centrifugal fields. Because the organelles have much less mass than the cell wall components, the first pellet that forms at low centrifugal force is primarily cellular debris. Usually precipitation of heavier particles requires a low-speed centrifuge. But it is essential to use ultracentrifuge to generate high centrifugal force for the successful separation of lighter cell organelles. Differential centrifugation performs the sedimentation process in a homogeneous medium where the sample is uniformly distributed before centrifugation. Particularly for analytical centrifugation it bears problems in analytical measurements when the sample consists of more than one component. During sedimentation, large components pass through the solvent medium and the small sized components. Therefore, it is difficult to obtain a sharp separation of the components leading to a problem in their analytical measurements. This problem may be overcome by density gradient centrifugation where the sample is centrifuged in a solvent medium that possesses a density gradient. Partially purified cell organelles by differential centrifugation method may serve as the starting material for further purification using density gradient separation.

4.3.6. Density gradient centrifugation

Density centrifuges operate by spinning liquids of different densities within the rotor. The heavier fluid is forced to the inside wall of the rotor with the lighter one to the centre of the rotor. Density gradient centrifugation is a commonly used centrifugation method to purify subcellular organelles. This deals with the centrifugation of a cell fractions in a medium possessing a density gradient of either a continuous or discontinuous (also known as step gradient) mode. The cell fractions are placed on the top of the layer of gradient medium and centrifuged to separate particles based on their individual density or mass/size ratio. Density gradient ultracentrifugation is an accepted and widely used method to fractionate animal, plant, and bacterial cells, viral particles, lysosomes, membranes, etc. Now-a-days this technique (density gradient ultracentrifugation) is significantly utilized to purify immuno-therapy products such as influenza, hepatitis, and rabies vaccines in both batch and continuous-flow zonal modes. Depending on the approach adopted, it may be further classified into two categories: (1) rate-zonal (size) technique, and (2) Isopycnic separation technique.

4.3.6.1. Rate-zonal (size) separation technique

In rate-zonal technique the sample is centrifuged in a pre-formed gradient. This involves the separation of fractions on the basis of their molecular size or sedimentation rate because different particles exhibit different sedimentation velocity, which mainly depends on their size. At first a density gradient of the medium is maintained in such a way that the highest density in the gradient should not be greater than the densest particle to be separated. Density gradient is created by placing layer after layer of gradient medium with the heaviest layer at the bottom and the lightest layer at the top. The substance to be separated is carefully layered over the pre-formed layer of the density gradient with the help of a pipette or hypodermic syringe. Density of the sample solution must be less than that of the lowest density portion of the gradient (top layer). Since the sample mixture is applied on the top of the gradient, during the centrifugation process components pass through the gradient medium according to their size at different rates depending on their *s*-values, facilitating sharp separation of the fractions in the form of bands or zones. Therefore, different components remain separated from others as sharp zones and may be collected separately for their further characterization by careful sucking or puncturing the bottom of the tube. In the puncturing system, faster particles come out first in the early fractions. In the sucking system, the top gradient can be pumped out first introducing a solution relatively more denser than the densest part of the gradient. The gradient is forced out from the top through a special sealing cap and the slowest particles come out first into the earliest fraction. This technique may well be utilized for preparative (isolation of cell components by preparative ultracentrifugation) and analytical (determination of *s*-value by analytical ultracentrifugation) purposes. Rate-zonal centrifugation is commonly used for the separation of DNA, RNA, DNA-RNA hybrids, enzymes, hormones, ribosomal subunits, and other subcellular fractions. Mostly similar types of biological particles are similar in shape and density. For example, antibody classes are of similar density, but they possess different masses. Therefore, separation based on mass separates different classes, whereas separation based on density does not efficiently resolve these antibody classes. Likewise, mitochondria, lysosomes, and peroxisomes possess different densities, but they are of similar size. Thus, they are not likely to be separated by rate-zonal separation technique.

4.3.6.2. Isopycnic separation technique

It is also known as 'equilibrium density centrifugation'. In this technique, separation of particles of different densities may be well performed. This separation process does not depend on the shape and size (the size influences the rate of settling only up to the point of attainment of isopycnic position) of the particles, but it depends mainly on densities. This separation method is time independent. Once the particle reaches its respective isopycnic position in the gradient, the length of centrifugation does not influence the migration of the particle further because the particles keep floating on a cushion of the medium having greater density than its own. Thus, excessive run times have no adverse effect, as it is the case of rate-zonal centrifugation. In isopycnic (isodensity or equal density) separation the sample is layered on the top of the density gradient, which may be of continuous or discontinuous. However, continuous gradient is a preferred one. It should be kept in mind that density of the sample particles must fall within the limits of the gradient densities and the maximum density of the gradient exceeds the density of the densest particle to be separated. In this type of separation, particles of a particular density go on settling during the centrifugation process until they reach a position where the density of the surrounding solution is exactly the same as the density of the particle. At the isopycnic position the sum of the centrifugal and frictional forces becomes zero. Besides layering of sample solution on the top layer of the pre-formed gradient medium, in isopycnic separation it is also possible to use a self-generating gradient which is formed during centrifugation. A good example of such a gradient system is caesium chloride (CsCl) gradient. Dense salt solution of CsCl is a commonly used gradient system with which it is possible to attain an upper density limit of 1.8 g/cm^3. The biological sample solution, after mixing with CsCl solution uniformly, is subjected to ultracentrifugation. Caesium salt redistributes itself in such a way that a density gradient is formed with an increasing density from the top to the bottom and the biological sample molecules move to a region of equal density. Isopycnic separation has wide application in nucleic acid analysis, separation of linear from circular form of DNA, characterization of RNA and DNA based on their *s*-values, plasmid DNA purification and various isotopic studies involving macromolecules. With the help of this method and exploiting the density of ^{14}N-DNA (light isotope) and ^{15}N-DNA (heavy isotope), it is possible to establish that DNA replication is semi-conservative type of replication (M. Meselson and F. Stahl experiment).

Sugars (e.g., Sucrose), polysaccharides (e.g., Ficoll), colloidal silica (e.g., Percoll), and iodinated media (e.g., Nycodenz) are also suitably used as gradient media for isopycnic separations. However, sucrose is not preferred due to its higher viscosity at density greater than 1.1-1.2 g/cm^3 and it exerts high osmotic effect at low concentrations.

Ficoll is a neutral, highly branched, hydrophilic polysaccharide with high mass (radii ranging from 2-7 nm). It dissolves readily in aqueous solutions. It is formed by reacting the polysaccharide with epichlorohydrin and is used mainly in biological laboratories to separate blood to its components (erythrocytes, leukocytes, etc.)

Percoll consists of colloidal silica particles of 15-30 nm diameter (23% w/w in water), coated with polyvinylpyrrolidone (PVP). It is well suited for density gradient experiments for the isolation of cells, organelles, viruses. It shows a low viscosity, low osmolarity, and no toxicity towards cells and their constituents.

Separation of chromosomes on the basis of buoyant density has proved to be difficult due to their high density and sensitivity to high ionic strength. **Nycodenz** is a non-ionic gradient medium which offers significant advantages for separation of chromosomes. Chromosome density labeled with bromodeoxyuridine (BrdU) may be separated from unlabeled chromosomes in Nycodenz gradients. Nycodenz does not appear to alter chromosome morphology or protein complement. The introduction of Nycodenz represents a significant new tool for use in chromosome separation and purification.

4.4. Choice of gradient materials

All gradient materials should qualify some basic criteria for their ideal selection as a gradient medium. First of all, they must permit the desired type of separation and be stable in the solution. They should be inert to the biological or other types of materials to be separated. It is obvious that they should exert negligible osmotic pressure and cause minimum change in pH, ionic strength, and viscosity. Essentially, they should not absorb light of wavelength used for the spectrophotometric determination of the separated molecules and should not interfere in other assaying procedures. However, no material fulfils all the conditions strictly.

4.5. Rotors

A rotor is a large metal object that holds the individual sample tubes and is connected to the spin drive of the centrifuge. The rotors are made of such materials which dissipate heat quickly and they should have sufficient mass to generate momentum so that the rotors require little energy to keep on running, once it starts spinning. It is essential to operate centrifuges under refrigerated condition and vacuum to minimize the heat generated by frictional forces as the rotor spins. Also, rotors should withstand the stress forces through the centrifugal forces created by the rotor spinning. Low-speed rotors, that experience low stress, may be made up of steel, or brass. However, high-speed rotors are essentially made up of aluminium alloy or titanium alloy. Titanium alloy is much preferred one compared to aluminium alloy from the point of its resistance to chemical corrosion and metal fatigue. Now-a-days carbon fibre rotors are available for a variety of ultra-speed, high/super speed, low speed, and tabletop centrifuges. Carbon fibre rotors are extremely durable and designed to give many years of continuous service in all types of laboratory. The carbon fibre rotors are easy to use, since they are of up to 50 – 60% less weight than equivalent metallic rotors. They bear less wear and tear on refrigeration systems and are autoclavable for sterile use. They do not corrode with alkali, caustic aqueous solutions, and most organic solvents. Another advantageous point of these rotors is that as they do not face any stress during centrifugation, tube adaptors can be used in the rotor cavities for processing smaller volumes in the same rotor. The centrifugal force generated is proportional to the rotation rate of rotor (rpm) and the distance between the rotor centre and the centrifuge tube. Hence, one may use a centrifuge with multiple rotor sizes which provide flexibility in choosing centrifugation conditions. Each centrifuge is provided with a special graph, a nomograph, or a table which relates rotation rate (rpm) to centrifugal forces (x *g*) for each size of rotor it accepts. *Centrifuging* rotors are classified into four main categories (**Figure 4.3**): (a) Swing-bucket, (b) fixed-angle, (c) vertical, and (d) zonal rotors; depending upon whether the sample is held at a given angle to the rotation plane or allowed to swing out on a point and into the plane of rotation, or with an aim to feed a large volume of sample solution into the rotor equipped with a special provision of inlet and outlet ports for separation of component(s).

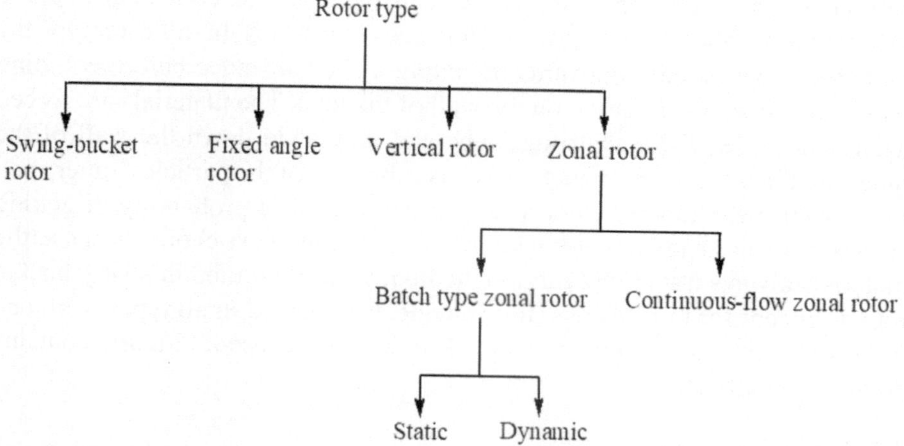

Fig. 4.3: Classification of rotors

4.5.1. Swing-bucket rotor

It is also known as the horizontal rotor. This type of rotor holds the sample tubes into individual buckets hanging vertically at the rest position of the rotor. However, when the rotor starts spinning, the buckets along with the centrifuge tubes swing out to a horizontal position. As a result the solution in the tube is oriented perpendicular to the axis of rotation and parallel to the applied centrifugal field. The line of centrifugal force goes out radially from the centre of rotation and drives molecules into and down the wall of the tube. During centrifugation, the substances to be separated travel the entire length of the tube through the medium and longer path-length allows good separation of a mixture into individual components. The swinging-bucket rotor exerts low relative centrifugal force and for this reason long operation time to separate the mixer is required. Sometimes the convection current generated by some particles striking against the tube wall and sliding down the tube base disrupts the sample bands. This effect may be minimized by slow acceleration and deceleration of the rotor. This rotor is not suitable for pelleting. Since it consists of a number of moving parts, it is very susceptible to wear and tear with extended use.

4.5.2. Fixed-angle rotor

This type holds the sample tube at a fixed angle of the rotor cavity to the vertical axis and commonly used for pelleting purposes and is, therefore, truly preparative. However, it is also useful for isopycnic separation of nucleic acids. This type of rotor comparatively works very fast to precipitate the substances due to its increased relative centrifugal force exerted at an angle to the centrifuge tube wall for a given rotor speed and radius. In this rotor type the particles move radially outwards travelling a short distance before colliding with and hence, precipitating on the wall of the tube. The materials are forced against the side of the centrifuge tube and they slide down the wall of the tube. But this action may sometimes cause abrasion of the particles' outer shell (e.g., outer membrane of chloroplast). When there is a probability of getting such a problem, a swinging-bucket rotor is then the best choice because the materials always travel through the medium within the tube in swing-bucket rotors. Except for metal stress (this may be experienced in all types of rotor), it does not undergo any major mechanical failure because it virtually contains no moving parts on it.

4.5.3. Vertical rotor

It is also known as a fixed zero-angle rotor. It holds the sample tubes in a vertical position during spinning. This rotor type provides quick separation of particles compared to swinging-bucket or fixed-angle rotor because the particles travel across only the diameter of the tube representing a short path-length. This is not suitable for pelleting applications. If it is allowed to form pellets, pelletting occurs along the entire length of the tube wall and there is every possibility that the pellet will be dismantled into the solution after centrifugation. In ultracentrifugation with a gradient medium in this rotor type, the gradient reorients by 90° when it spins. Therefore, acceleration and deceleration must be performed slowly so that gradients can reorient without any mixing. Its applications include plasmid DNA, RNA, and lipoprotein isolations through isopycnic separation techniques.

4.5.4. Zonal rotor

Zonal rotors are designed for large scale separation of particles from a large volume of a sample on a density gradient medium in a single central cavity rather than in tubes. This is used by cell biologists. Some zonal rotors are capable of dynamic loading and unloading of samples while the rotor is spinning at high speed. Gradient medium and samples are introduced into the rotor through a specialized port system. There are two types of zonal rotors, depending on their mode of operation: (a) batch, and (b) continuous-flow type.

4.5.4.1. Batch type zonal rotor

This is a cylindrical hollow container consisting of a centrally situated rotor core and a vane assembly which divides the cavity into four compartments. Through the ducts present in the vane assembly, gradient solution is pumped to the periphery of the rotor from the rotor core. Again, based on the operation mode of the rotor core, batch type zonal rotors are further classified as (a) static, and (b) dynamic type. However, both are equally useful to display a good resolution. The static mode involves the loading and unloading of the gradient and sample solution at the rest position of the rotor, while the dynamic mode deals with the loading and unloading at the spinning position of the rotor. In the static mode after layering the sample solution on the top of the gradient at the rest position, the preliminary spinning of the rotor is employed to accelerate the rotor to nearly 1000 rpm so that the sample and gradient layers approach to reorient vertically. After that, the centrifugal force is gradually increased. The zones are absolutely vertically oriented at the operating high speed and the particles are separated according to the differences in their density. After a sufficient run-time the rotor must be decelerated again to 1000 rpm from its operating speed and then very slowly to its rest position. This performance helps the gradient layers with the sample particles to get back to their original horizontal position without inter-mixing of the layers. In the dynamic mode, at first the lightest density layer of the gradient solution is introduced into the rotor through a fixed or a removable seal of the core, when the rotor is already in a spinning position (nearly 2000 rpm). As a result the lightest layer forms a thin uniform layer orienting vertically at the periphery of the rotor wall under the influence of centrifugal force. This is followed by pumping of denser to densest liquid one by one into the system, which causes the movement of the lightest layer towards the rotor core by centripetal displacement. After the formation of the gradient, a fluid cushion layer made up of a denser solution than the heaviest layer of the gradient is incorporated into the system. Lastly, the sample solution which is lighter than the lightest layer of the gradient is pumped into the rotor cavity by the fluid line located at the centre of the rotor and finally low-density solution is put on the sample solution layer in such a way that similar amount of fluid cushion get displaced from the periphery. The centrifugal force is then increased with increasing the rotation speed of the rotor and after centrifugation for a definite period of time the rotor is decelerated to 2000 rpm and the rotor contents (light density layer followed by heavy density layer sequentially) are collected separately by introducing extra amount of cushion to the rotor.

4.5.4.2. Continuous-flow zonal rotor

When the sample size is large, the use of typical rotors is limiting. To overcome the problem related to the total capacity of rotors, the use of continuous-flow zonal rotors bears distinct and significant advantages over the typical rotors. This is particularly designed to separate small quantities of solid material contained in a huge volume of a liquid phase as suspension. In this case the sample suspension is continuously fed into the rotor accompanied with a continuous removal of particle free liquid encouraging solid particle enrichment and this can be achieved using density gradient or without any gradient. When a density gradient is used, a sample is introduced through the top surface of a rotor core into a sample zone adjacent to a gradient zone in such a fashion that the sample passes substantially 360° and the particles remained in the sample are separated in the said gradient zone due to the differences in their density. The particles are removed after separation through the bottom surface of the rotor core by introducing a fresh liquid gradient through the top surface of the rotor core to the gradient zone to displace the sample gradient. The sample gradient is passed out of the core, while the rotor is spinning, through a passage at the core axis at a reduced pressure. This type of continuous-flow zonal ultracentrifugation is now established as an invaluable technique for rapidly concentrating, and largely purifying milligram amounts of infectious virus from very large volumes of tissue culture fluid. The obtainable quantities of virus are significantly utilized for physical and chemical studies of its protein and nucleic acids or for production of vaccines. When no gradient is used during separation, the sample is allowed to pellet and is recovered after the completion of the whole operation. Limnologists utilize this separation technique to separate plankton from a huge quantity of lake or pond water.

4.6. Care of Rotor

To ensure good centrifuge practices, few necessary safety precautions must be followed carefully. The rotor must be cleaned regularly after use for safety purposes and for maximization of the rotor life. Particular care must be taken in cleaning and decontaminating rotors after they are used with hazardous materials like radioisotopes or infectious organisms. Carbon fibre rotors can be washed with laboratory detergent solutions, even alkaline detergents. Other types of rotors contain metal components which may corrode depending on the exact alloy present. Some special and environment-friendly rotor cleaning solutions are available now. In case of deposits of the material in the rotor it is necessary to use a soft brush to dislodge them from the surface. However, care must be taken so that under any circumstances there should not be any scratches left on the protective surface of the rotor. There are a number of commercially available detergents for removing radioisotopes from centrifuge rotors. When the extent of radioisotopic contamination is not so high, contamination is removed simply by wiping the affected part of the rotor with a tissue paper soaked in the cleaning agent. In the case of high levels of contamination, it may be necessary to soak the rotor overnight to achieve complete removal of the radioisotope. Centrifuge rotors may also become contaminated with one or more of the infectious agents, such as fungi, bacteria, viruses, and prions depending on the application that the rotor is being used for. It is essential that after any work with samples that may be associated with infectious agents, the rotor is immediately disinfected before further use. When the rotor is regularly used with infectious agents, a clear schedule of decontamination should be instituted. Autoclaving (at 121°C, 15 psi for 15 min), use of UV-irradiation and chemicals [Formaldehyde (37%), Glutaraldehyde (2%), 70% ethanol, or phenolic solutions] may be the common means for disinfection of rotors from the biological agents, such as fungi, bacterial cells, bacterial spores, lipid viruses, non-lipid viruses, prions, etc. However, the chemical susceptibility of rotor material(s) restricts the chemical disinfection process and UV light irradiation is not very practical for centrifuge rotors. Therefore, the final choice of disinfection used for the rotor depends primarily on the nature of the infectious agents and also the nature of rotor materials. In fact, autoclaving is the preferred method for disinfecting rotors because it does not involve chemicals which may be corrosive to the rotors.

Proper and correct use of the centrifuge machine including rotor within the recommended limits may avoid many problems. Rotors on high-speed centrifuge and ultracentrifuge units are under strong mechanical stress that may lead to rotor failure. Improper loading and balancing of rotors may cause the rotors to break while spinning. Therefore, it is essential to balance the rotor properly so that the load is distributed evenly around the rotor axis. The centrifuge tubes are balanced with the medium that is identical to that being centrifuged. Samples should be diametrically and evenly distributed around the rotor. A small imbalance may create a catastrophe, particularly in ultracentrifuge which stores a huge kinetic energy while it is operated. It is advisable not to leave the centrifuge until it reaches its operating speed. It is also necessary to 'de-rate' the rotor (limiting the maximum speed at which the rotor is used to some level below the maximum speed listed for the rotor when it was new) based on the duration of its use. For this, it is necessary to keep a detailed history of the rotor use in an ultracentrifuge. If the rotors are used properly following the manufacturer's instructions and the user-guidelines, all rotors are supposed to last a lifetime. They should not require any maintenance other than routine cleaning. However, if the rotor has been overstressed or if the rotor has been dropped onto a solid surface, it should be returned to the factory for specialist inspection.

4.7. Rotor tubes

The sample containers (centrifuge tubes) in a centrifuge are available in different shapes, sizes, and materials. Selection of the correct centrifuge tubes for a specific rotor is very crucial. In case of glass tubes, it is necessary to use adaptors so that the cushioning effect prevents them from their direct contact with the metal rotor and eliminates the probability of their shattering or cracking. Except zonal and continuous rotors, external centrifuge tubes are necessary to hold the sample solution in the rotor. These centrifuge tubes are made of regular glass, Corex glass, nitrocellulose, polypropylene, Teflon [poly(tetrafluoroethylene)] or polyallomer. Usually regular glass tubes are used in clinical centrifuges with low speed. High speed generating high RCF damages it. High speed (18000 rpm) is handled with Corex glass. However, these corex glass tubes also shatter when the speed exceeds 18000 rpm. Polypropylene or polyethylene tubes can withstand a speed up to 20000 rpm. Nitrocellulose tubes are most suited to ultracentrifugation. For ultracentrifugation it is advised to use new tubes for each run because repeated use may lead to tube collapse due to internal molecular stress which may not be always detected from its outer look. Polyallomer tubes are expensive, but chemically inert and re-usable. Due to the slipperiness of the tube material, the molecules slide down the walls of the tube easily, and thus polyallomer tubes are best suited for precipitation centrifugations. Before use, centrifuge tubes must be carefully checked whether they bear any stress fractures, because the tubes with such fractures may hold the liquid medium before centrifugal operation but the cracks will open wide under centrifugal force.

4.8. Model questions

i. What is centrifugation? Write down its principle.

ii. How do you compute the sedimentation time of a particle? How do you enumerate the time required by a sample to be pelleted in a new rotor, when you know the time for its pelleting in an old rotor?

iii. Describe the different types of centrifuges, based on the speed in detail.

iv. Describe the different types of centrifuges, based on the mode of separation techniques in detail.

v. Mention the advantages obtained using Rayleigh Interference Optics (RIO) in an analytical ultracentrifuge. Why is sucrose not preferred as a gradient medium for isopycnic separation?

vi. What is *K*-factor? How is it related to pelleting time and sedimentation coefficient?

vii. Why is carbon fibre rotor preferred over using rotors made up of other materials? What are the different types of rotors available, based on their mode of functioning? Why is more time required for the swing-bucket rotor to separate mixture?

viii. Under which occasion use of swing-bucket rotor is better than fixed-angle rotor? Differentiate the modes of functioning of static and dynamic batch type of zonal rotor.

ix. Under which occasion Continuous-flow zonal rotor is used? What are the different ways to decontaminate the rotors after its use with the biological agents?

x. Illustrate the density gradient centrifugation technique in detail.

4.9. Suggested readings

(i) Bowen, T. J. (1970) An introduction to ultracentrifugation. Wiley International, John Wiley, London.

(ii) Graham, J. (2001) Biological Centrifugation. CRC Press, Taylor and Francis Group, London.

(iii) Laue, T. M.; Stafford, W. F. (1999) Modern applications of analytical ultracentrifugation. Anal. Rev. Biophys. Biomol. Struct. 28 : 75-100.

(iv) Rickwood, D.; Ford,T.C.; Steensgaard, J. (1994) Centrifugation: Essential Data (Essential Data Series). Wiley–Blackwell.

(v) Svedberg, T.; Pedersen, K. O. (1940) The ultracentrifuge. Oxford University Press, London.

(vi) Van Holde, K. E. (1971) Physical Biochemistry. Englewood Cliffs, Prentice-Hall, N.J.

5

Electrophoresis

5.1. Introduction

Electrophoresis is an important separation technique for the charged molecules. Biologically important molecules, such as amino acids, peptides, proteins nucleotides, nucleic acids which carry electrical charge as they possess certain ionizable functional groups may be separated and identified from a mixture with the help of this technique. Some compounds that are non-polar in nature and do not possess any charge may be transformed to charge carrying derivatives. Thus, phosphate derivatives of carbohydrates may be separated and characterized utilizing electrophoretic techniques.

5.2. Principle

Molecules to be separated by electrophoresis are allowed to exist in solution as electrically charged species. The charged molecules migrate to their corresponding electrodes depending on the nature of charge they possess (cations move to cathode and anions move to the anode) when they are exposed to an electrical field. Thus, positively charged molecules are separated from negatively charged molecules. Due to the difference in molecular weight, molecules possessing similar charge (same sign) exhibit different charge/mass ratios. As a result, different molecules show differences in charge magnitude. This difference leads to the differential migration of charged molecules resulting in a separation of individual charged components from a mixture.

The rate of movement for anions and cations depends upon the balance between impelling force of the electrical field (x) on charged (q) ions and opposing force between migrating molecules and surrounding medium by frictional resistance (f) to give a terminal velocity (v).Thus, following equation (Equation 5.1) is obtained.

$$f.v = q.x \qquad \text{Equation 5.1}$$

The event of frictional resistance may be calculated considering Stoke's law that assumes the particle as a sphere. If the radius of the sphere is 'r' and the viscosity of the suspending medium is 'η', then we can write as

$f = 6\pi.\eta.r$ Equation 5.2

Combining Equation 5.1 and 5.2,

$q.x = 6\pi.\eta.r.v$

$v/x = q/6\pi.\eta.r$ Equation 5.3

Therefore, electrophoretic mobility (v/x), migration of molecules per unit field strength is directly proportional to charge of the molecules but inversely proportional to its size and viscosity of the suspending medium.

In general, one may conclude that electrophoretic mobility of charged molecules depends on the nature of ionizable groups present in molecules and magnitude of the charge carried by the ionizing groups. The whole phenomenon is dependent on composition, pH, and ionic strength of the suspending medium.

5.3. Factors influencing mobility of charged molecules

There are several factors that affect migration rate. By manipulating these factors, desired separation of charged molecules may be achieved from a mixture through electrophoresis.

5.3.1. Electrical field

The migration rate of charged molecules in an electrical field is directly proportional to current and voltage, and inversely proportional to resistance which, in turn, varies with ionic strength of the buffer, and type and size of the supporting medium. Normally a direct current (DC) source is used to maintain a constant voltage and current, so that reproducible results may be obtained. Heat is generated during electrophoresis. If the electrical field strength is more; heat, thus generated, may lead to decomposition or denaturation of compounds. Furthermore, solvent evaporation from the supporting medium may also occur due to increased heat. As a result, buffer solution from the buffer tank flows to the supporting medium to replace lost water and causes a displacement of the zones, thus leading to a misleading interpretation. The advantageous side of high voltage application is to get a rapid separation of low molecular weight compounds, such as amino acids and sugars in the form of compact and narrow bands avoiding excessive diffusion. But this high voltage electrophoresis should be done under a cooling device so as to obtain a minimal evaporation loss.

5.3.2. Buffer solution

Ionic concentration, pH, and composition of buffer solution influence the migration rate. The pH directly affects the ionization of compounds. If pH increases, ionization of organic acids increases while that of organic bases increases with decrease in pH. Therefore, pH to be chosen depends upon the chemical nature of the compounds. But for amphoteric compounds like amino acids which possess acidic and basic properties, both increase and decrease in pH of buffer solution cause ionization of compounds. the addition of alkali and acid to the zwitterion form of amino acid may be represented as follows (**Figure 5.1**):

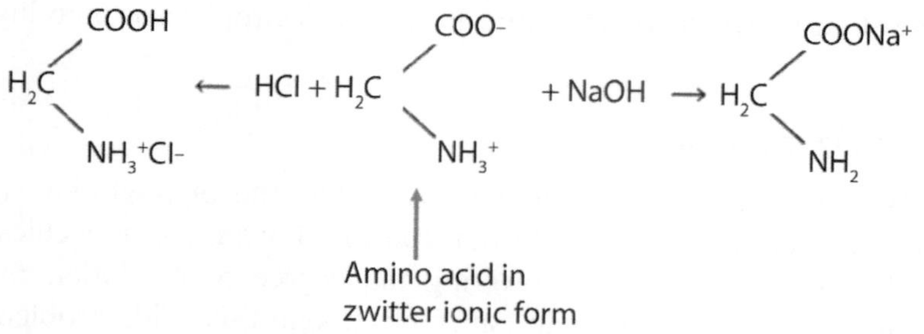

COOH
H_2C
$NH_3^+Cl^-$
← HCl + H_2C
COO-
NH_3^+
+ NaOH → H_2C
COONa+
NH_2
Amino acid in zwitter ionic form

Fig. 5.1: Charge behavior of amphoteric compounds in presence of acid or alkali

The proton on the amino group dissociates to neutralize hydroxyl ion added, while the ionized carboxyl group accepts protons which are added during acidification. Therefore, the direction of migration of the amphoteric compounds is pH dependent. The pH of the environment, in which electrophoresis occurs, must be such that the macromolecules being separated are charged but not denatured. Usually for proteins, the outside limits are 4.5 and 9.0. The ionic strength and concentration of the buffer must be considered. If the electrolyte concentration is too low, the migrating macro-molecules conduct a large portion of current. Therefore, the migrating molecules spread into diffusive zones. They do not migrate as sharp bands and thus, resolution of the method is greatly decreased. High ionic strength of the buffer causes more heat production with decreased rate of migration, whereas the reverse applies for low ionic strength. Heat produced can denature the proteins being separated. As a compromise between the two cases, ionic strength is selected within a range of 0.05 – 0.15 mol dm^{-3}. Tris, acetate, formate, phosphate, and citrate buffers are usually used. In practice, buffers are not expected to interact and cause any chemical change to the molecules under examination. Interaction between buffer and macro-molecules being separated, involving addition or neutralization of a charge on the macromolecule, might change the rate of migration in an electric field. Even, an artifactual appearance of two migrating species in place of one may be obtained due to the interaction. Pure albumin in the borate buffer shows two distinct bands. When one is recovered and isolated band is subjected to a second time electrophoresis, it gives two bands again. This leads to the conclusion that multiple bands are the result of an interaction between some of the albumin molecules and the borate buffer.

5.3.3. Sample under investigation

Electrophoretic mobility of a molecule depends on its size and net charge in an opposite manner. Larger molecules migrate slowly than smaller molecules for more retarding forces by the medium. But migration rate is directly related to the net charge of the molecule. If net charge increases, electrophoretic mobility increases.

5.3.4. Supporting medium

Supporting medium exerts a profound impact on the rate of migration. If the supporting medium possesses adsorption capacity toward the molecules, a 'tailing' effect may be observed lowering the degree of resolution for separation. This may be evidenced in paper electrophoresis. This problem may be overcome by using cellulose acetate in place of cellulose only. In fact, paper electrophoresis is a time-consuming separation technique as adsorption of the separating molecules in simple cellulose paper lowers the overall rate of migration. The supporting medium may also be a 'gel' of starch, agar, and polyacrylamide, in which samples may be separated based not only on their difference in electrophoretic mobility but also on their shape and size because gel as a supporting medium offers a sieve-like structure where pore size may be manipulated with increasing or decreasing the concentration of supporting medium.

5.4. Instrumentation and operational procedure

5.4.1. Instrumentation

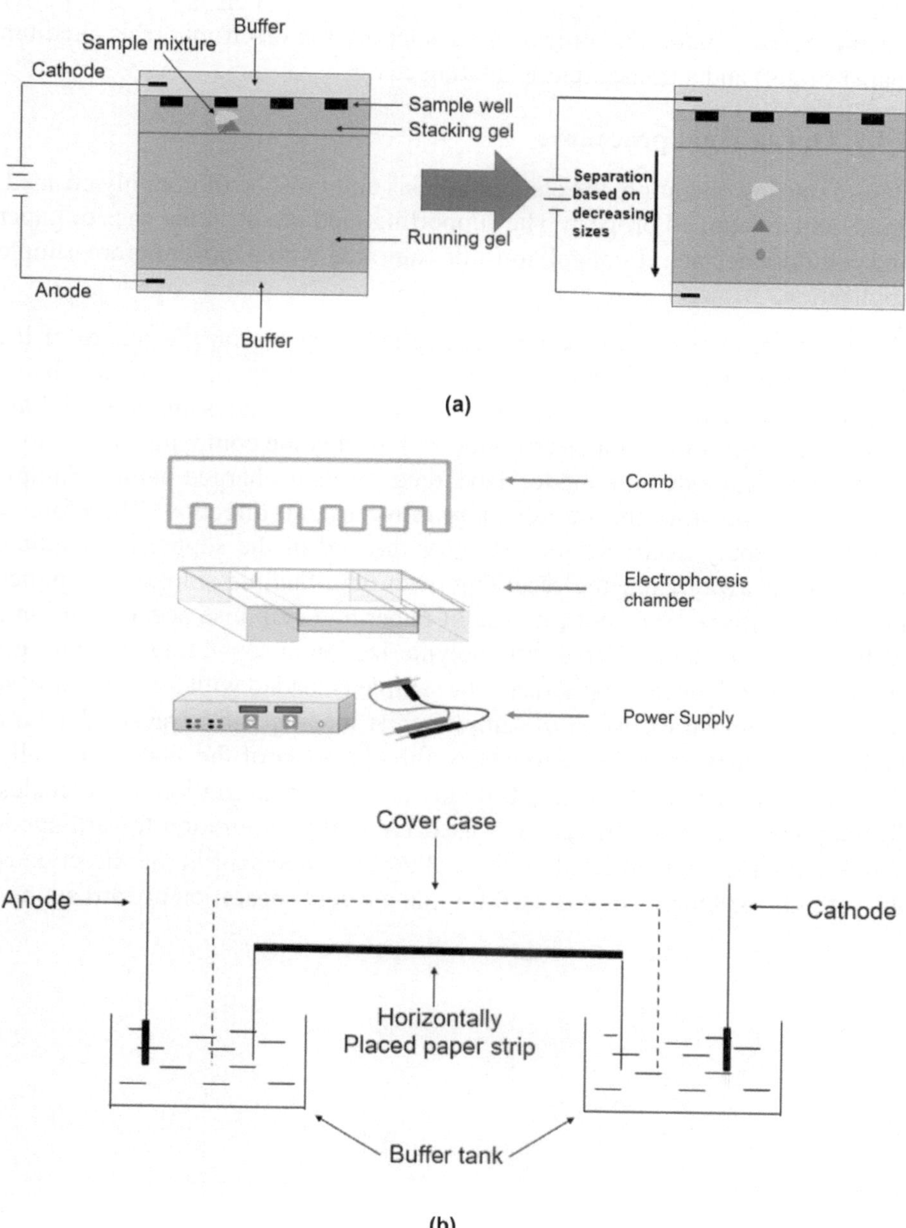

Fig. 5.2: Sketches of vertical gel electrophoresis (a) and horizontal paper electrophoresis (b) unit

Depending on the positioning of support for the electrophoresis medium, there are two types of electrophoretic apparatus: vertical (**Figure 5.2a**) and horizontal (**Figure 5.2b**). In both cases, there are two units: (i) power pack to obtain direct current, and (ii) electrophoresis unit that consists of two buffer reservoirs, two platinum electrodes, a support for electrophoresis medium (paper or gel) and a transparent insulating cover.

5.4.2. Operational procedure

Before starting operation, the electrophoresis unit must be thoroughly cleaned, dried, and assembled properly.The supporting medium as in the case of paper and cellulose acetate, if not gel, must be saturated with a buffer before sample application.

The location, at which the sample is applied, depends on the nature of the sample. The sample solution containing components of opposite charge is applied with a micropipette near the middle of the sampling medium (particularly for horizontal electrophoresis), so that the components can move toward both cathode and anode, depending on their charged nature. Sample molecules containing the same charge move in one direction. Therefore, it is better to apply a sample solution near the end of the supporting medium which is far away from the receiving electrode. Sample solution is applied in a narrow streak or a spot (in case of paper and cellulose acetate) and in a well that is made by a comb during polymerization of separating medium (gel slab). In case of gel electrophoresis, the sample is loaded with 5 – 10% sucrose solution to prevent diffusion of sample while loading and a tracker dye with high electrophoretic mobility to ensure the operation of the system. Usually, the choice of tracking dye depends on the direction of migration of molecules. Bromophenol Blue or Orange G is used for protein migration toward anode and Bromocresol Green or Methylene Blue for movement in the direction of the cathode. Xylene-cyanol is used for nucleic acid separation toward anode.

Table 5.1: Molecular weights of standard protein markers

Sl.No.	Proteins	Molecular weight of polypeptide chains (Da)
1	Cytochrome C	12,000
2	α-Lactalbumin	14,200
3	Hemoglobin subunit	16,000
4	Trypsin inhibitor soybean	20,100
5	Trypsinogen	24,000
6	Carbonic anhydrase	30,000
7	Pepsin	34,500
8	Albumin, egg	43,000
9	Bovine serum albumin	66,000
10	Glyceraldehyde-3-phosphate dehydrogenase (rabbit muscle)	144,000
11	γ- Globulin	150,000
12	Phosphorylase a (muscle)	188,000

It is essential to make, other than unknown sample lanes, a marker lane where a mixture of proteins of well characterized molecular weights may be applied to obtain reliable molecular weight estimates of unknown protein bands. A list of commercially available protein standards is given in **Table 5.1**.

After loading the sample on the separating medium, current is turned on. The whole system is kept under stabilized current and voltage supply for a necessary period of separation. This period, in case of gel electrophoresis, can be estimated by monitoring movement of bromophenol blue. When bromophenol blue reaches the other end of the gel, the power is turned off.

The separating medium is removed carefully from the system, after the run is complete.The separating medium is then subjected to staining. A good number of stains are used to stain specific compounds. Few stains are listed in **Table 5.2**.

Table 5.2: Staining reagents used in electrophoresis for visualization of compounds

Sl.No.	Stainer	Composition/Remarks
Protein		
1.	Coomassie brilliant Blue R250	Coomassie Blue + methanol + water + glacial acetic acid + or Coomassie Blue in water containing 20% sulphosalicylic acid or TCA
2.	Ponceau S	Ponceau S + TCA + water
3.	Nigrosin	Nigrosin + TCA (or acetic acid)
4.	Amido Black	Amido Black + acetic acid + methanol + water
5.	Naphthalene Black	Naphthalene Black + methanol + acetic acid + glycerol + water
6.	Bromophenol Blue	Bromophenol Blue + ethanol + mercuric chloride [Blue band]
Lipoprotein		
1.	Lipid Crimson	Lipid Crimson + ethanol + water
2.	Sudan Black or acetylated Sudan Black	Sudan Black + ethanol
Glycoprotein		
1.	PAS (periodic acid-schiff)	Periodic acid + schiff's reagent + sodium metabisulphite [Brilliant red band]
Haemoglobin and Haemoglobin-Haptoglobin complex		
1.	Dianisidine dye	O-dianisidine in sodium acetate-acetic acid buffer, pH 4.6) [Brown Band]
2.	Benzidine dye	Benzidine + acetic acid + water + H_2O_2 + ammonium chloride [Blue band]
Acid mucopolysaccharide		
1.	Toluidine Blue	in acetic acid
Polysaccharides		
1.	Iodine	Iodine
Nucleoprotein		
1.	Toluidine Blue	in acetic acid
Peptides		
1.	ClO_2 or NaOCl chlorination followed by KI-starch or benzidine-acetic acid	Reacts with all amino compounds
Nucleic acid		
1.	Methylene Blue	-
2.	Acridine Orange	_
3.	Pyronine Y stain	in acetic acid
4.	Ethidium bromide	followed by UV-light

To visualize distinct zones or bands of the separated components, background staining must be minimized. For this, stained separating medium must be destained by repeated immersion in dilute acetic acid. As the individual components of minute quantities are stained faintly, the destaining process should be done very carefully and stopped at the stage when maximum numbers of bands are visualized. The separated zones may also be identified with the help of some specific reagents by enzyme substrate coupling reactions, immunoprecipitation, autoradiography, fluorography, blotting methods, etc.

The separating medium showing distinct bands is then photographed and it is kept for permanent record after drying under vacuum.

The quantity of the separated components may be determined directly by using a scanning densitometer. The density of staining is proportional to the amount of a particular component and at the same time there is an inverse relationship between amount of light reaching the densitometer detector and amount of stained material present in the medium. Cutting the corresponding area of the separating medium and dissolving the component in a suitable solvent, separated component may also be quantitatively determined by a spectrophotometer or by other convenient methods.

5.5. Types of electrophoresis

There are different forms of electrophoresis which may be classified into two main groups: a) Free or moving boundary electrophoresis and b) Zone electrophoresis. The advantages and disadvantages of different types of electrophoresis are discussed separately in **Table 5.3**.

5.5.1. Free or moving boundary electrophoresis

The concept of electrophoresis was first introduced by Swedish biochemist, A. Tiselius in the 1930s. In the classical method of free boundary electrophoresis, rate of migration of the charged molecules is very fast as frictional resistance between the charged molecules and free solution and the electrostatic forces which are exerted by the surrounding medium are minimal. A sample of a complex mixture of proteins mixed in a buffer solution is kept in a simple U-shaped glass column and covered with a pure buffer solution. Care is taken to maintain a constant temperature and to avoid vibration. Two electrodes, anode and cathode, are attached to the two arms of the U-tube separately and an electric field is generated. As a result, charged molecules move toward their corresponding electrodes. Molecules of the same charge but of different charge/mass ratio move separately due to their different electrophoretic mobility and form bands with boundaries. One may get an idea related to the direction and relative rate of migration of proteins in the mixture by measuring the refractive index of the solution along the arm of the U-tube, as the refractive index of the pure buffer is different from the proteins. But with this method a complete separation of proteins is very difficult due to mixing of boundaries by convection.

5.5.2. Zone electrophoresis

In place of pure buffer solution introduction of a solid support infused with buffer solution improves resolving power in separating mixtures. As solid support has mechanical rigidity, the effect of convection and vibrational disturbances may be minimized. In zone electrophoresis substances migrate as distinct zones and a complete separation of a mixture is possible. Relatively inert material is used as the supporting medium. Commonly used supporting media in zone electrophoresis are filter paper, cellulose acetate, starch gel, agar, polyacrylamide, etc.

5.5.2.1. Paper electrophoresis

Paper electrophoresis is the most convenient electrophoretic technique. Simple filter paper, which is chemically cellulose, is used as solid support. But it gives a poor resolution during separation due to its adsorption behavior, when it is compared with other zone electrophoretic methods like cellulose acetate or gel electrophoresis. The adsorption of the sample by paper leads to a 'tailing' effect. As a result, the sample moves in the shape of a comet rather than as a distinct band. It is also a time-consuming process. Commonly paper electrophoresis is used to analyze small molecules. It is difficult to use paper electrophoresis for macromolecules, as for example proteins. This is due to the adsorption and surface tension phenomena that denature the macromolecules, leading to poor resolution.

5.5.2.2. Cellulose acetate membrane (CAM) electrophoresis

In this electrophoresis, cellulose acetate is used as a supporting medium. Compared to the paper, cellulose acetate gives better resolution in separating compounds from samples of small quantities due to minimal adsorption of the molecules on cellulose acetate. It is a quicker process than paper electrophoresis. This is particularly useful for protein analysis because cellulose acetate does not denature protein. Cellulose acetate membrane has large size pores. Therefore, separation by CAM electrophoresis is entirely dependent on charge density.

5.5.2.3. Gel electrophoresis

Electrophoretic resolution is improved when support used is starch or agar or polyacrylamide gel of different porosity that may be controlled by regulating concentration of gels. Compounds may be separated on the basis of both electrical charge and molecular size by gel electrophoresis. Several chemical agents, such as starch, agar, polyacrylamide, etc. are usually used for making gel slabs. Starch consists of two components, amylose and amylopectin.

Starch gel is prepared by heating and cooling in a buffer solution, so that starch grains are partially hydrolyzed and forms gel. Pore size can be adjusted by regulating starch concentration in the solution. Agar is a mixture of two galactose polymers, agarose and agaropectin. Agarose is a polysaccharide obtained from seaweeds. Agarose gels are used to separate and purify very high molecular weight proteins, protein aggregates, DNA and RNA because of their pore sizes. Since proteins are smaller molecules than nucleic acids, agarose gel is preferred for separating nucleic acids and polyacrylamide gel is used for proteins. However, nucleic acids with molecular weight of <500 Kda may be tried with SDS-PAGE (Sodium Dodecyl Sulphate-Polyacrylamide Gel Electrophoresis), possessing 6-8% gel. Polyacrylamide gel is prepared by polymerizing acrylamide with a cross-linking agent (bisacrylamide) in the presence of a reaction initiator (persulfate ion in the form of ammonium persulfate) and a catalyst [TEMED (N, N, N′, N′-tetramethylethylenediamine)] (**Figure 5.3**). Ammonium persulfate is used for chemical polymerization, while riboflavin for photo polymerization. When ammonium persulfate is dissolved in water, it forms free radicals (SO_4^- •) in the presence of TEMED. These free radicals react with acrylamide with their preservation within the acrylamide molecule. In turn, the 'activated' acrylamide reacts with successive acrylamide molecules to synthesize a straight chain of a polymer. Gel solution must be degassed; otherwise, oxygen may trap free radicals and thus hinders polymerization. But polymerization does not help to form gel because long straight chain polymers can slide past one another. Gel formation needs cross-linking of these straight chain polymers to one another by N, N′-methylene bisacrylamide which is composed of two acrylamide molecules coupled head-to-head at their non-reactive ends. As shown in **Figure 5.3**, carrying out polymerization in this manner yields a 3-D network of acrylamide chains forming pores that act as molecular sieve. Acrylamide polymerization may also occur using riboflavin in place of ammonium persulfate. Riboflavin produces free radicals when it photodecomposes in the presence of ultraviolet light. The free radicals, thus produced, act in the same way as it is governed by ammonium persulfate.

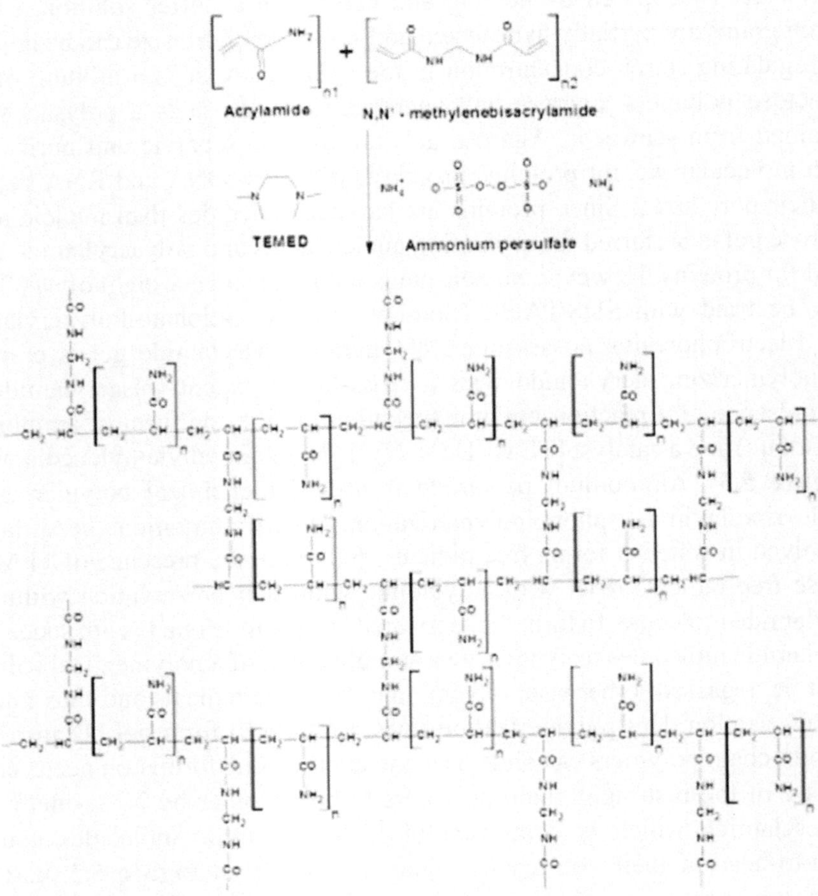

Fig. 5.3: Reaction showing polyacrylamide matrix formation

The pores formed during gel formation are not essentially of equal size, rather of random nature. Some are big and some are small.The porosity and the average pore size of gel is controlled by the relative proportion of acrylamide monomer and cross-linking agent, bisacrylamide. It must be emphasized that if gel pore size is smaller than that of the sample macromolecules, the macromolecules do not move into the gel irrespective of their charge and field strength. Thus, gel pores act as molecular sieve that enhances the resolution of polyacrylamide gel electrophoresis (PAGE). Therefore, Equation 5.3, discussed earlier in this chapter, is not appropriate to define electrophoretic mobility of a charged molecule through polyacrylamide gel as this factor (average gel pore size) has not been considered. It has been seen that relative mobility of proteins is inversely related to the total gel concentration. This also confirms that the average pore size of a gel decreases as gel concentration increases.

Rodbard and Chrambach developed a mathematical model to relate gel concentration and electrophoretic mobility.

$\text{Log } M = \log M_0 - K_R T$ Equation 5.4

Where,

M = electrophoretic mobility,

M_0 = free mobility in a sucrose solution,

T = total gel concentration and

K_R = retardation coefficient (slopes of the line where relative mobility of molecules is inversely related to the total gel concentration)

The retardation coefficient may be expressed as

$K_R = C(R + r)$ Equation 5.5

Where,

C = constant,

R = geometric mean radius of macromolecule,

r = radius of the gel fibers (this is based on the assumption that gel fibers are much longer than macromolecules)

Derivation of these two equations (Equation 5.4 and 5.5) is essentially based on statistical considerations due to the random nature of polyacrylamide gel pore formation.

Gel porosity is regulated by the relative quantity of acrylamide monomer and cross-linker used to form gel. This may be expressed as

Total acrylamide concentration (T, %) $= (X + Y)\, 100 / Z$

Degree of cross-linking (C, %) $= Y \cdot 100 / (X + Y)$

Where,

X = acrylamide monomer content (g),

Y = Bisacrylamide content (g),

Z = solution volume (ml)

Pore size can be progressively increased by reducing %T at a fixed %C. It is observed that regardless of the total amount of acrylamide per unit volume, average pore size reaches a minimum when 4-5% of total acrylamide used is Bisacrylamide. Therefore, it is a general practice to manipulate the pore size by keeping a fixed Bis content (4-5%) and varying total acrylamide content. However, increasing %C at a fixed %T leads to formation of bead-like structure within the gel increasing the pore size. This is particularly suitable for iso-electric focusing (See IEF, Section 5.6.) Two mathematical relationships have been established between pore size (P) and T under different conditions as follows:

$P \propto 1 / \sqrt{T}$ [when amount of crosslinker is low (%C = 1-5)]

$P \propto 1 / \sqrt[3]{T}$ [when amount of crosslinker is high (%C = 15-25)]

Gel electrophoresis is classified as Horizontal and Vertical type depending on the gel casting technique. Vertical gel electrophoresis may utilize both continuous and discontinuous buffer systems, while discontinuous buffers are not applicable in horizontal one. It should be remembered that oxygen inhibits acrylamide polymerization. In a horizontal system, gel is exposed to atmospheric oxygen during its casting in a tray. Thus, acrylamide is not used in horizontal systems and PAGE is always a vertical type of electrophoresis.

5.6. Iso-electric Focusing

Isoelectric focusing (IEF) is an advanced electrophoretic technique to separate amphoteric molecules based on their isoelectric point (PI), exposing them in a stable pH gradient and electrical field. Isoelectric point of an amphoteric molecule is that pH where it does not migrate in an electric field since it has no net electric charge at that pH (oppositely charged sites are balanced). Therefore, when a potential difference is applied across a medium containing a pH gradient, amphoteric substances move toward their corresponding electrodes. As soon as they reach their PI region, they lose their electric charge and are immobilized there. Proteins may be separated from each other with this technique as different proteins possess different PI values. IEF is a highly sensitive technique and possesses high resolving power. It is used for both analytical and preparative purposes to study protein micro-heterogeneity. This is unique in studying isozymes. IEF may be carried out both under native and denaturing conditions (urea). Non-ionic detergents, such as Triton X-100 may also be used for IEF to enhance the solubility of samples. As cyanate ions emerging from urea may carbamylate protein leading to an artifact effect, it is essential to take care in minimizing the potential of cyanate ions. It is also noted that the use of charged detergents such as sodium dodecyl sulfate (SDS) is incompatible with IEF. During IEF processing, salt concentration in the samples leads to band distortion, generates more heat and extends focusing time. This is why it is necessary to remove salt from the sample by either dialysis or gel filtration.

A pH gradient may be maintained for a short time by putting different buffer solutions one upon another. But the gradient formed by this way is unstable due to the movement of buffer ions in an electric field. Commercially available complex mixtures of synthetic polyamino polycarboxylic acids are usually used to generate a stable desired pH gradient.

Isoelectric focusing may be performed in a fluid medium using sucrose solution or in a gel of polysaccharide containing ampholytes. The supporting medium carrying ampholytes may be packed in a tube or may be used as a thin layer. The upper end of the tube or gel slab is connected to a buffer reservoir containing a strongly acidic solution (e.g., phosphoric acid) and the lower end connected to a reservoir containing a strongly alkaline solution (e.g., ethanolamine). Upper reservoir will be connected to the anode and the lower reservoir to the cathode. When current is switched on, ampholytes form a desired pH gradient between anode and cathode. At this stage, a sample is applied at the upper end. The individual charged components start moving and they are focused on their corresponding PI, where they have no net charge and form a sharp stationary band.

To detect protein bands obtained by isoelectric focusing the gel must be free from carrier ampholytes, prior to staining. Otherwise, ampholytes interact with dye and they become stained too. This is not desirable. There is a practice to treat the focused gel with trichloroacetic acid solution with occasional shaking. More likely, 5-6 times changing of trichloroacetic acid solution remove the carrier ampholytes. The focused gel, free from ampholytes, is then stained with dye followed by destaining. Trichloroacetic acid solution, being a denaturing reagent, precipitates proteins into the gel.

In IEF, proteins are supposed to migrate freely according to their charge under an electric field. Theoretically, there should not be any molecular sieving effect. But when polyacrylamide in a low percentage (3-5% T level) is used as a medium for IEF, its porosity limits the free movement of larger particles, exerting a sieving effect. This is the main disadvantage in using polyacrylamide; although it has low endosmotic flow, high optical clarity, and anti-convection properties. It is observed that pure bisacrylamide gel provides a highly cross-linked gel which is composed of spherical units and allows a free passage of high molecular weight particles. But care should be taken in using large amounts of bisacrylamide (it should not be more than 15%) because thus formed gels are opaque and brittle. N, N'-diallyl L-tartardiamide (DATD, **Figure 5.4.**) can also be used as a cross-linker. It has been seen that gels containing 5% T and 15% C as DATD are non-restrictive for molecules up to 0.5 x 10^6 Daltons.

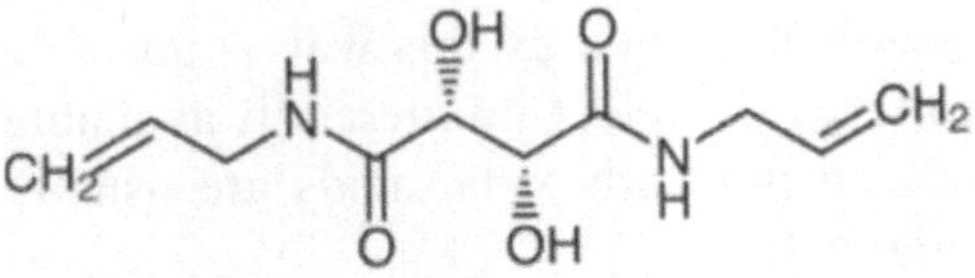

Fig. 5.4: N, N'-diallyl L-tartardiamide

The long-term instability of pH gradient, which is another disadvantage of IEF due to the electroendosmosis properties of IEF gels, may be overcome by introducing immobiline reagents into polyacrylamide gels. This ensures an immobilized pH gradient (IPG). Immobiline reagents are a series of acrylamide derivatives containing either carboxyl (acidic immobilines) or tertiary amino (basic immobilines) groups. As the buffering groups are covalently linked to the polyacrylamide (via vinyl bonds); a pH gradient, which is relatively more stable, is achieved.

5.7. Discontinuous or Disc Electrophoresis

Discontinuous or Disc electrophoresis is a modification of zone electrophoresis. In this method, instead of using continuity in pH of the buffer and porosity of the gel, the protein mixture moves from a more porous to less porous gel and encounters a change in pH. The discontinuity is also due to the ionic strength of the buffer and the nature of ions in gel and in the electrode buffer. The portion of gel containing large pores is called 'spacer' or 'stacking gel' where the sample is stacked in a concentrated and sharp band before it enters the main 'separating gel'. The portion of gel containing small pores is called 'running gel' or 'separating gel' where the mixture is separated into individual components as thin and sharp bands due to molecular sieving and charge effect. Disc electrophoresis is better than continuous one to attain a good resolution. It is so named because of the discontinuity of the buffer system and porosity of gel.

The two gel systems in a single tube or slab are illustrated in **Figure 5.2a**. The running gel (5-10% acrylamide) buffer contains a weak amine base (Tris), adjusted to proper pH (pH 8.9) using HCl. The stacking gel contains a lower amount of acrylamide (2-3%) than the running gel. Stacking gel leads to concentrating all separating molecules in one band, so that they start moving in the resolving gel (running gel) from that point. The molecules are separated in the running gel in accordance with their molecular weight. Buffer system in the stacking gel is also of Tris / HCl and approximately 2 pH units lower than that in the running gel (pH 6.9). Sample buffer should be the same as that of stacking gel. Bromophenol blue, an ionizable tracking dye, and sucrose are also added into the sample buffer. The purpose of using tracking dye is to monitor the electrophoretic run and sucrose is used to settle down the sample into loading well. The upper and lower reservoir contains a Tris / glycine buffer system (pH 8.3).

After loading the sample into the loading well, current is passed through the gel. When the whole system is exposed to an electric field; proteins, bromophenol blue, chloride and glycinate anions start moving toward anode. Glycinate anions are protonated to form zwitterions, as they enter into sample buffer and spacer gel that are of low pH (pH 6.9); thus, creating a deficiency in mobile glycine anion concentration. But at pH 6.9 proteins possess a greater mobility than the glycine ions due to their considerable negative charges and chloride ions because of their full negative charges move faster toward anode than proteins. As a result, relative ion mobility is found as Chloride$^-$> bromophenol blue > proteins$^-$> glycinate$^-$. This enables each protein to accumulate and stack into discrete disc or bands at the end of stacking gel with trailing glycinate ions immediately behind the last protein band. Bromophenol blue and chloride ions which are of relatively greater mobility enter into the running gel from stacking gel. As the disc of protein encounters running gel, its migration is slowed down by small pores of the gel. This permits small glycinate ions to catch up with the proteins. Glycinate ions cross the interface between stacking and running gel. They become fully charged at pH 8.9 and overtake proteins to form a sharp boundary at the chloride ion interface. As a result, the pH of the running gel changes. In the running gel Tris-HCl is converted to Tris-glycine buffer system of pH 9.5. Each protein molecule moves through the running gel at this pH as a function of its unique charge with an extra effect of molecular sieving by polyacrylamide gel. Once bromophenol blue reaches the bottom of the gel, the gel slab is taken out from the electrophoretic system after turning the power off. Gel is then treated with an appropriate stain followed by de-staining. This system is applicable to most protein mixtures because most proteins are anions at pH 8.9.The advantage of disc electrophoresis over zone electrophoresis is that the protein sample enters the separating gel as a narrow zone. The resulting protein bands are, therefore, more compact providing a good resolution.

5.8. Sodium dodecyl sulfate polyacrylamide gel electrophoresis (SDS-PAGE)

When proteins are used in their native state (tertiary or quaternary) during PAGE separation, the separation technique is commonly known as Native-PAGE. It is observed that molecules do not separate from each other during native electrophoresis if they are tightly bound together. Core RNA-polymerase, composed of three non-identical subunits, shows a single band during electrophoresis under native conditions. However, it is possible to separate the subunits if sodium dodecyl sulfate [SDS, $CH_3 - (CH_2)_{10} - CH_2OSO^-_3Na^+$], an anionic detergent, is used. Potassium dodecyl sulfate should be avoided because of its low solubility. Polyacrylamide gel electrophoresis in combination with SDS is now widely accepted method to separate protein mixtures, to facilitate characterization of polypeptide chains and determination of their molecular weight by co-electrophoresis and to analyze 'pure' proteins for their number and size of the molecular subunits. Initially the buffered protein sample is boiled with SDS and β-mercaptoethanol or dithiothreitol in boiling water. As a result, thiol reduces and disrupts all disulfide bonds present in the protein tertiary structure to sulfhydryl groups; while negatively charged SDS molecules bind tightly to the hydrophobic regions of polypeptide chains of proteins, unravel all intra-molecular protein association, and lead to denaturation of protein. One SDS molecule binds for every two amino acid residues in the polypeptide chain. Thus, SDS binding imparts a large net negative charge to the denatured polypeptides, masking the original native charge on the molecule. SDS-protein complex, during electrophoresis in a buffered environment containing thiol, SDS and a high concentration of polyacrylamide gel, moves toward anode and separation is based on the size of protein.

5.8.1. Interpretation of results and estimation of molecular weights

Migration distance of each separated molecule is measured relative to marker dye and relative mobility is defined as:

Relative mobility (R_m) = distance traveled by component molecule/distance traveled by tracking dye

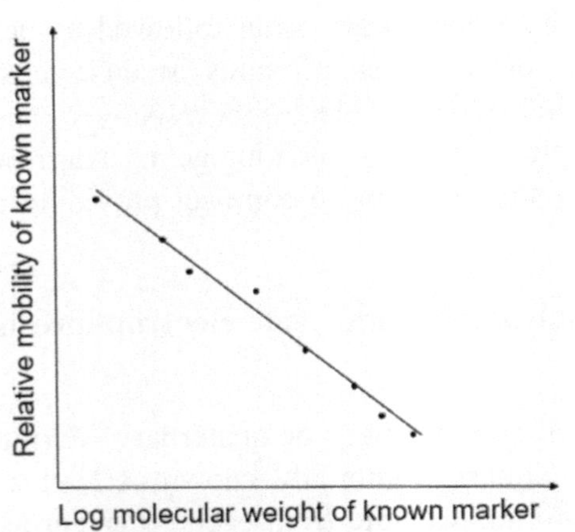

Fig. 5.5: A linear relationship of relative mobility versus log molecular weight of known marker

It is observed that, in SDS-PAGE, mobility of a protein is inversely related to its log molecular weight. Therefore, an unknown molecular weight of a protein may be obtained from its mobility, using a plot of relative mobility of marker proteins against logarithmic value of their known molecular weights (**Figure 5.5**). Thus, a linear regression equation is obtained to estimate the mass of the unknown protein as follows:

$Y = mx + c$

Where,

Y = log molecular weight,

m = slope,

x = relative mobility of the unknown, and

c = intercept

5.9. Native-PAGE

The chief disadvantage of SDS-PAGE is to miss completely the biological and biochemical activity of the separated proteins due to the presence of SDS which denatures protein. Native-PAGE overcomes this problem and separates proteins under native condition in absence of SDS (non-denaturing condition), keeping other operational procedures the same as SDS-PAGE. Since under native conditions proteins exist in their quaternary structure and are bigger molecules, a gel slab of larger pore size (low percentage of acrylamide, 7.5%) is used. After the electrophoretic separation under native conditions, proteins may be characterized easily by the process of enzyme staining, antibody binding, receptor activity, etc. that are not possible with SDS-PAGE. Limitation to the use of native electrophoresis is that this technique is not applicable to those proteins which are insoluble and tend to precipitate or aggregate during electrophoresis.

5.10. Gradient gel electrophoresis

Gradient gel electrophoresis involves separation of molecules on an acrylamide gel, where acrylamide concentration varies uniformly from 5% at the top of gel to 20% acrylamide at the bottom, hence possessing different pore size from top to bottom of the gel. As molecules migrate from the upper part to the lower part of the gel slab, pore size decreases with corresponding increase in acrylamide concentration. Thus, the formed porosity gradient enables the gel to be used for attaining higher resolution in separating a much greater range of relative molecular mass values, which is not possible on a fixed percentage gel.

5.11. Pulsed-field gel electrophoresis (PFGE)

Pulse field gel electrophoresis is a new generation gel electrophoretic technique, introduced by Charles Cantor and Cassandra Smith, usually to separate high molecular weight DNA molecules on an agarose gel slab. With conventional gel electrophoresis using agarose (0.1%), it is difficult to separate DNA molecules beyond 10^5 bp. But the introduction of pulsed-field gel electrophoresis enables a separation of DNA molecules of molecular weight up to the magnitude of 10 million bp. In PFGE, unlike one pair of electrodes as it is found in conventional electrophoresis, two or more pairs of electrodes are placed around the periphery of an agarose slab gel at different angles. The electrode pairs are alternatively activated for different periods (0.1-1000s). Activation period of each electrode pair depends on the size of the molecule. In PFGE, with the first electrical field produced from the pulsing of the first electrode pair, the helical structure of the DNA molecule is first stretched up in the horizontal plane and migrates toward anode. Now, this field is interrupted, and direction of the field abruptly changes with the second electrode pair. As a result, DNA molecules reorient themselves along the new direction of the field and then start to move again in the new direction. Time required for renewed stretching, reorientation and further movement depends on the molecular size. The larger the molecule, the slower it reorients and takes time for further movement. Small sized molecules take less time to do so. Thus, in PFGE pulse length (electrode activation period) and electrode distributions are two important factors affecting migration of the DNA molecule: shorter DNAs migrate faster than larger DNAs, leading to their separation. Large molecules need long pulse time and small molecules need short pulse time for their movement. Orthogonal field alternation gel electrophoresis (OFAGE), field inversion gel electrophoresis (FIGE), transverse alternating field gel electrophoresis (TAFE), contour clamped homogeneous electric field electrophoresis (CHEF) and rotating field electrophoresis (RFE), where the basic principle involved is same, are few examples of PFGE.

5.12. Two-dimensional electrophoresis (2-D electrophoresis)

Sometimes two or more polypeptides commingle due to similar size, charge, and their intermolecular interactions and show a single band on conventional gel electrophoresis. This problem may be overcome using two-dimensional electrophoresis employing independent physicochemical properties in each dimension, which are at right angles to each other. This ensures their complete dissociation, separation and thus characterization of complex protein mixtures. A classical two-dimensional electrophoretic procedure involves utilization of two physicochemical parameters, iso-electric point (PI) and molecular weight, employing IEF in the first dimension and SDS-PAGE in the second dimension, respectively. Basically, IEF (first dimension) is done with polyacrylamide gels in narrow tubes containing urea (8 M) and non-ionic detergent (0.5% W/V, Triton X-100 or Nonidet NP-40). A flat-bed gel may also be used; after electrophoresis it may be cut into strips and loaded onto the support of the second dimensional electrophoresis. The gel rod or strip containing separated proteins is equilibrated in a buffer along with SDS, so as to bind the proteins to SDS and subjected it to the gradient SDS-PAGE in the second dimension. Finally, a pattern of spots is obtained ensuring the separation of more than 1000 proteins at a time. The interpretation of the result is a laborious and complicated process. Therefore, a computerized system is essential to interpret the 2D-electropherogram. 2D electrophoresis is known to impart highest resolving power and purity of proteins, compared to other electrophoretic methods.

Table 5.3: Advantages and disadvantages of different types of electrophoresis

Type	Advantage	Disadvantage
Free or moving boundary electrophoresis	-	1. It is difficult to separate molecules completely due to mixing of boundaries by convection and diffusion. 2. Vibrational disturbances exist.
Paper electrophoresis	1. This method is cheap and easy to handle. 2. Convectional and vibrational disturbances are minimized.	1. It provides poor resolution. 2. It is applicable only to separate small molecules, not for higher molecular weight molecules. 3. A 'trailing' effect is observed due to adsorption of molecules on paper. 4. It is a time-consuming process.
Cellulose acetate electrophoresis	1. It offers a better resolution compared to paper electrophoresis. 2. No 'trailing' effect is observed due to minimal adsorption of molecules on cellulose acetate. 3. It ensures a clear separation of a mixture into discrete zones. 4. It is a quicker process than paper electrophoresis. 5. It has a wide clinical application to separate serum proteins and hemoglobin and is also used for immunoelectrophoresis.	1. It is more expensive than paper electrophoresis. 2. Due to large pore size, it hardly exerts any sieving effect on proteins. 3. Limit of detection is low.

Type	Advantage	Disadvantage
Gel electrophoresis	1. The support medium is transparent, enabling photometric scanning. 2. Molecular sieving is possible. 3. Manipulation of gel pore size is possible by increasing or decreasing medium concentration. 4. It is useful to separate higher molecular weight molecules with a good resolution. 5. It may be carried out with a continuous (continuous zone electrophoresis, CZE) or discontinuous (multiple zone electrophoresis, MZE) buffer system in combination with a linear or gradient gel pores, leading to migration of molecules in the form of a sharp zone. 6. Mixing of zones does not occur by convection.	-
IEF	1. It separates amphoteric molecules based on their pI value. 2. It possesses a high resolving power. 3. It is unique to study the protein's micro heterogeneity.	1. There is no molecular sieving effect. 2. A long-term instability of pH gradient is observed.
SDS-PAGE	1. It offers high resolution, sharp zones and limited diffusion due to the presence of restrictive gel. 2. SDS solubilizes almost all proteins. 3. Pure proteins can be analyzed for their number and size of the molecular subunits. 4. It is an easy method for molecular weight determination. 5. This is applicable to those proteins which are insoluble and tend to precipitate or aggregate during electrophoresis.	1. Biological activity of protein is lost. 2. Glycoprotein migrates slowly in SDS-PAGE because sugar moiety does not bind to SDS.

5.13. General applications of electrophoresis

Electrophoresis is a widely used modern technique for separation, purification, characterization, and identification of many biologically important charged molecules such as amino acids, peptides, proteins, nucleotides, nucleic acids, purines, pyrimidines, phenols, indoles, imidazoles, etc. Electrophoresis is helpful particularly to identify specific nucleic acid sequences (southern and northern blot technique), specific protein to confirm the presence of antibodies (western blot technique), for isoenzyme analysis, etc. With this technique, 'pure' proteins can be analyzed for the number and size of the molecular sub-units. It is also possible to quantify a particular protein, nucleic acid, etc. at microgram level by densitometric scanning of the gel or paper strip. Electrophoretic separation of blood proteins and subsequent assay for isozyme distribution currently are routine operations in hospitals and clinical laboratories.

5.14. Model questions

i. Write down the principle of electrophoresis technique. Explain it with mathematical derivation. What is a 'tailing' effect in electrophoresis?

ii. What are the consequences faced if high voltage electrophoresis is not carried out under a cooling condition? How does buffer solution influence the electrophoretic mobility of charged molecules? How do the size and net charge of the separating molecules affect their mobility in electrophoresis?

iii. Mention the general operational procedure of separation of molecules by electrophoresis. Why does CAM electrophoresis possess an edge over the normal paper electrophoresis?

iv. Illustrate Rodbard and Chrambach mathematical model describing the relationship between gel concentration and electrophoretic mobility. How is pore size regulated by total acrylamide concentration?

v. How do ammonium persulphate, TEMED and N, N′-methylene bisacrylamide work during polyacrylamide gel matrix formation?

vi. What is an isoelectric point? Write down the principle of isoelectric focusing (IEF). Why are salts needed to be removed from the sample during IEF? Why is trichloroacetic acid used in IEF prior to staining? Why is immobiline introduced into polyacrylamide gel in IEF?

vii. Mention the principle of disc electrophoresis. What is the function performed by bromophenol blue and sucrose used in sample buffers? How are sample mixtures separated into individual components? Describe the operation events in detail.

viii. How can you estimate the molecular weight of an individual component separated by electrophoresis? What are the functions performed by SDS and β-mercaptoethanol in SDS-PAGE?

ix. What is PFGE? In which occasion 2-D electrophoresis is employed and how?

x. Mention the merits and demerits of IEF and SDS-PAGE. Assume that a protein contains four subunits of differing molecular weights. If you perform SDS-PAGE, how many distinct bands should you see on the gel and why?

5.15. Suggested readings

(i) Clark, J. M. (Jr.); Switzer, R. L. (1977) Experimental biochemistry (2nd Edition). W. H. Freeman and Company, San Francisco, USA.

(ii) Harris, E. L. V.; Angal, S. (1989) Protein purification methods – a practical approach. In: The Practical Approach Series, Series editor Rickwood, D.; Hames, B. D. Oxford University Press, Oxford.

(iii) Smith, I. (1976) Chromatographic and electrophoretic techniques, Vol. 2 (4th edition), William Heinemann Medical Books Limited.

(iv) Walker J. M. (1994) Electrophoretic technique. In: Wilson, K.; Walker, J. (Eds) Practical Biochemistry – principles and techniques. 4th Edition, Cambridge Univ. Press, pp 425 - 460.

(v) Westermeier, R. (2005) Electrophoresis in practice – a guide to theory and practice. VCH Verlagsgesellschaft mbH, Weinheim, Germany.

(vi) Williams, B. L.; Wilson, K. (1975) A biologist guide to principles and techniques of practical Biochemistry. Edward Arnold (Publishers) Limited, London.

(vii) Work, T. S.; Work, E. (1976) Laboratory techniques in Biochemistry and molecular Biology (Vol. 5). North-Holland Publishing Company, Amsterdam.

6

Spectroscopy

6.1. Introduction

Interaction of radiant energy, exhibiting the properties of both waves and particles, with matter is dealt in spectroscopy. Matter may be defined as material, existing in solid or liquid or gaseous phase, which is composed of molecules, atoms, or ions. Atoms, molecules, and ions, irrespective of their physical state, remain in constant motion in a matter. Molecules may rotate, vibrate, and move from place to place in space. While radiant energy interacts with matter, these molecular motions are influenced to some extent. It is observed that radiations from ultraviolet and visible range of electromagnetic spectrum lead to the movement of bonding electrons from lowest energy state (ground state) to higher energy state (excited state) in the molecules, while absorption of IR radiation causes the molecule to vibrate more. Such changes, referred to as *transition*, are responsible for changing the molecular energy level. Thus, while interacting with electromagnetic radiations, vibrational transitions, rotational transitions, and electronic transitions in molecules may be observed. However, when atoms are concerned, we see that atoms cannot rotate or vibrate but they can move in space and their electronic movement within various energy states is also observed. Electronic transitions require more energy than vibrational or rotational transitions. The relative energy of transitions in a molecule is governed by the order of rotational < vibrational < electronic. During interaction of electromagnetic radiation and matter, radiation may be absorbed, transmitted, reflected, scattered, diffracted, or even re-emitted after its absorption.

It is unique that compounds differ from each other by their light absorbing properties. For this reason, some compounds are colourful and some are colourless. When a compound does not absorb all wavelengths of visible range equally, it exhibits a colour. Water is colourless transparent because it absorbs UV wavelengths that lie outside of the visible light spectrum. Spectroscopic study makes it possible to identify compounds by analyzing and characterizing the pattern of wavelengths or frequencies they absorb because a specific molecule or a specific atom absorbs or emits only certain frequencies of radiation at a specific intensity or amount. Thus, the degree of absorption or emission of a radiation of specific frequency or wavelength is a unique character of a molecule or an atom. This uniqueness becomes the basis for identification of a compound. From the absorption spectrum of a substance, one can gain knowledge on the nature of photons (frequencies / wavelengths / energies) absorbed by the substance and at the same time the absorption spectrum focuses on the amount of photons absorbed at each frequency or wavelength. Similarly, an emission spectrum is constructed during release of radiant energy from an excited atom or molecule while they get back to their lowest energy ground state because excited states are short-lived and energetically unfavourable. This type of light emission after the excitation process is known as luminescence. Depending on the rate of the light emission, luminescence is further categorized as fluorescence (fast emission) and phosphorescence (slow emission). It should be critically noted that luminescence or fluorescence or phosphorescence specially refers to the event that starts after the absorption of photons. These terminologies are not used for the emission of radiant energy which is generated by other excitation sources; in those cases simply 'emission' is used. Actually, apart from absorption of electromagnetic radiation, excitation may occur by many other means such as, transfer of energy due to collisions between atoms and molecules, addition of energy through thermal treatments, from electrical discharges, etc. Several types of emission spectroscopic techniques are available based on the mode of excitation at atomic or molecular level.

Spectroscopic study also enables to quantify a substance in a medium, particularly in solution state, utilizing the proportional relationship between the amount of light absorbed by the substance and the concentration of the substance, Thus, it is a fact that concentration measurement or partial structure elucidation of a compound in a sample may be done using spectroscopic technique where electromagnetic radiation interacts with the matter resulting in its absorption, transmission, reflection, or other behaviour. Obviously, the interaction is a quantum phenomenon and is a reflection of the properties of the electromagnetic radiation and the specific structural features of the samples under analysis. Depending on these behaviours, typical spectra are produced, and molecular features of the compound can be examined qualitatively and quantitatively by interpreting these spectra. Basically, using different electromagnetic radiations to interact with matter, various spectroscopic techniques are employed in analytical study, such as visible spectroscopy, ultraviolet spectroscopy, fluorescence spectroscopy, infrared spectroscopy, atomic absorption spectroscopy, atomic emission spectroscopy, inductively coupled plasma-mass spectrometry, nuclear magnetic resonance spectroscopy and mass spectrometry. One may get a bird's eye view of these techniques from **Table 6.1**. They are different from each other as they are involved in studying different types of molecular or atomic transitions. Also, they differ within themselves based on choosing the region of the electromagnetic spectrum to interact with matter during analysis and using the type of radiation–matter interaction whether it is absorption, emission, or diffraction.

Table 6.1: Various commonly used spectroscopic techniques

Sl.No.	Spectroscopic techniques	Description
1.	Ultraviolet-visible spectrophotometry (UV-VIS)	It deals with the absorption of UV-Visible light, promotes electrons to higher energy state, thus creating electronic transition. This technique is used to know the presence of double bonds, conjugated π systems, and triple bonds in a molecule.
2.	Fluorimetry	It deals with the relaxation from an excited vibrational electronic energy level to its corresponding ground state.
3.	Nephelometry and turbidimetry	Nephelometric and turbidimetric techniques involve in the analysis of suspended particles in liquid or gas by determining the turbidity through utilizing the intensity of scattered light (Nephelometry) and transmitted light (Turbidimetry), when light passes through a suspension of particles.
4.	Infrared spectroscopy (IR)	Exposure of a molecule to an IR frequency leads to molecular vibrations in covalent bonds, as certain bonds vibrate faster. This is largely utilized to identify functional groups.
5.	Atomic absorption spectrophotometry (AAS)	The technique is based upon the principle that free atoms in the ground state absorb light of a certain wavelength which is specific for an element.
6.	Flame photometry	Measurement of the emitted light intensity when a metal is introduced into the flame is the basis of flame photometric determination.
7.	Mass spectrometry (MS)	It provides information about the molecular fragmentation pattern and molecular mass after bombarding a molecule with electrons.
8.	Nuclear magnetic resonance (NMR)	It informs about the detailed molecular structure and atom connectivity, once the nuclear spin transitions are achieved through the excitation of nucleus by radiofrequency irradiation.

6.2. What are electromagnetic radiation and spectrum?

Electromagnetic phenomena arise primarily from photons that cause electromagnetic radiation in the form of a wave. However, although it possesses a mass of zero, it behaves like a particle and transfers its energy to matter during interacting with matter. Electromagnetic radiation refers to the flow of energy occurring through the waves of alternating electrical and magnetic fields which are lying perpendicular to each other, and also to the direction of wave propagation. The oscillating electrical and magnetic fields facilitate the propagation of the wave in vacuum or space. They do not need any medium for their propagation. Radio waves, microwaves, infrared light, visible light, ultraviolet light, X-rays and gamma rays, characterized by their velocity, frequency and wavelength, represent the electromagnetic radiations. They are named as 'electromagnetic radiation' as the energy flows due to the local fluctuating changes in electrical and magnetic fields. When electromagnetic radiations are arranged in such a way that their wavelengths are decreased along with increasing frequencies and energies, the electromagnetic radiations are collectively referred to as electromagnetic spectrum (**Figure 6.1**). Within the total electromagnetic spectrum, the human eye can sense only the visible (VIS) radiations (wavelength 400-800 nm), not others. It should be remembered that visible light of different wavelengths owes to the different colours, but their combination leads to white light. In the electromagnetic radiation spectrum, the visible rays, infrared (IR) rays, microwaves, and radio waves are not harmful to living bodies due to their long wavelength leading to low frequencies and low energies. On the other hand, UV-rays, X-rays, and γ-rays exhibit high frequencies and consequently high energies. Hence, they pose a threat to living organisms causing health risks. For this reason, we take sufficient care and safety measures while handling these rays.

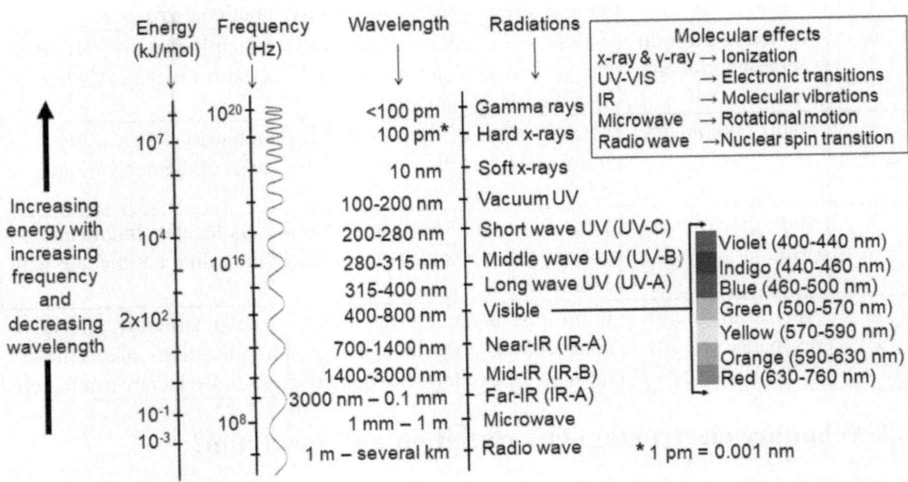

Fig. 6.1: Electromagnetic spectrum with energy, frequency and wavelength

Figure 6.2 represents different properties of a propagating wave. You imagine a circular path of water waves where you see up and down movement of water molecules. When they move up, it forms a crest, and a trough is formed when they move down to the circular path. You can apply the same analogy in case of a propagating light wave. In other ways, we may define trough and crest as the lowest point and the highest point of the wave from its central axis, respectively. Wavelength (λ) of a particular radiation may be defined as the spatial distance between two adjacent crests or troughs of one cycle of a wave (trough – crest – trough or crest - trough - crest). The total number of generated oscillations of a wave in one second at a given point is the frequency (v) of a particular radiation. Frequency is expressed in Hertz (Hz). In the overall electromagnetic radiation spectrum, gamma rays show the highest frequency, and radio waves possess the lowest. The velocity (c) of light in vacuum is 3×10^8 m/s and is determined by the wavelength and frequency. One may get the velocity by multiplying the wavelength and the frequency.

$$c = \lambda v$$ Equation 6.1

Where,

c = the speed,

λ = the wavelength, and

v = the frequency.

Thus, a distinct relationship between the wavelength and the frequency exists in the spectrum. The wavelength decreases with an increase in the frequencies i.e. they are inversely proportional to each other (Equation 6.2).

$$\lambda \propto 1 / v$$ Equation 6.2

It would be convenient to consider light as a stream of particles, photons. Photons are characterized by their energy, *E*. With the help of frequency of light, the energy of a photon may be expressed by the following equation:

$$E = hv$$ Equation 6.3

Where, E is the energy, h is Plank's constant [$6.6607015 \times 10^{-34}$ joule second], and v is the frequency inverse second (Hz).

Using Equation 6.1, Equation 6.3 may be further written as

$$E = h * c / \lambda$$ Equation 6.4

Considering Equation 6.3, it is established that energy increases with increasing frequency and at the same time, considering Equation 6.4, it is also established that energy and wavelength are inversely proportional to each other i.e. when wavelength of radiation increases, its energy decreases.

Since atoms, ions, and molecules remain in certain discrete states with specific energies, a change in their energy state necessitates the absorption or emission of energy. And the energy required for absorption or emission is exactly the same as the difference in energy of two energy levels [$\Delta E = (E_{final} - E_{initial}) = h\,\nu = h.c/\lambda$]. For absorption or emission, ΔE may be positively or negatively signed, respectively. However, an absolute value of ΔE is used during enumeration of frequency or wavelength of the radiation that is responsible for the obtained transition.

Wave nature of electromagnetic radiation is also characterized by three more significantly important properties – *wave number, period,* and *amplitude*. Wave number ($\tilde{\nu}$) emphasizes the number of waves per cm, meaning the reciprocal wavelengths in units of cm^{-1}. Wave numbers are encountered most often in IR spectroscopy. Period (T) refers to the time required in seconds by a wave to complete one wavelength. Therefore, period and frequency are interrelated to each other, where one can say that frequency is reciprocal of period (Equation 6.5).

$$\nu = 1 / T \qquad \text{Equation 6.5}$$

Amplitude refers to the maximum length of the vector: electrical amplitude and magnetic amplitude. Thus, it defines the height of the wave or the distance from the centre line to the peak of the wave. Larger amplitude represents higher energy.

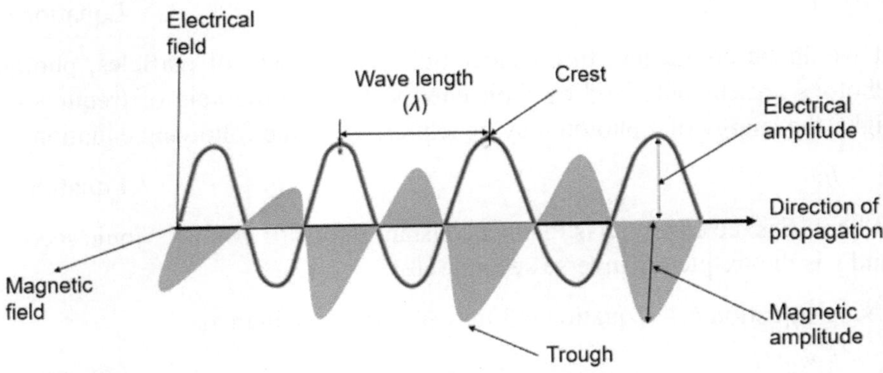

Fig. 6.2: Propagating electromagnetic radiation illustrated with wavelength, crest, trough, and amplitude

Along with wavelike properties, light also possesses particle-like properties. Light is considered as a collection of particles known as photons that carry discrete amounts of energy called quanta. Thus, atoms and molecules gain energy by absorbing photons, while they lose energy by emitting photons that carry energy exactly equal to the loss in energy of the atoms or molecules. This change in energy is directly proportional to the frequency of photons emitted or absorbed (Equation 6.3).

6.3. Why do objects appear coloured?

When electromagnetic radiation, possessing the wavelengths within the range of 400 - 800 nm (white light), falls on the surface of an object and all are absorbed, the object appears black but when all are reflected back equally, it appears as white. On the contrary, if all wavelengths are absorbed except green, the object appears green (**Figure 6.3**). In the similar way, an object shows its red, yellow, or violet colour, when it reflects the colour band of the respective wavelengths, absorbing the other colour bands from the white light.

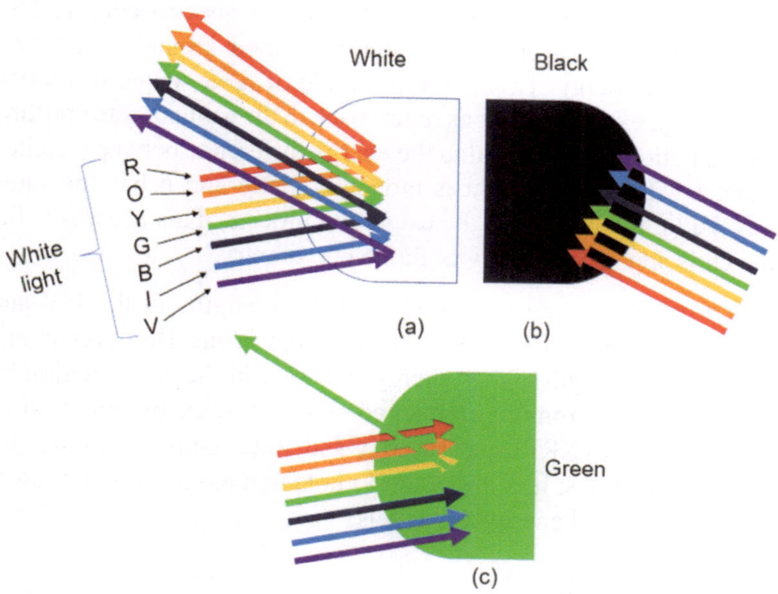

Fig. 6.3: Surface of an object appearing white (a), black (b) and green (c)

6.4. Atomic spectroscopy

Electrons in different orbitals, containing varied energies, surround the nucleus in an atom and for each element the electron number is unique. The electronic energy states of atoms are quantized. The most stable electron configuration of an element is its lowest-energy ground state. Making use of the occurrence of the light absorption, emission or fluorescence process by atoms, atomic spectroscopy is classified as atomic absorption spectroscopy, flame emission spectroscopy (FES) and atomic fluorescence spectroscopy (AFS), respectively. In all of these branches of atomic spectroscopy, gaseous metal atoms at their ground state are produced either by using flame from a burner or other mechanisms. The atoms are then brought into an excited state (promotion of a valence electron to its higher-energy orbital) by absorbing light energy. The absorption magnitude is proportional to the population of ground state atoms. Due to less-stability of the excited state, the electron gets back to its stable ground state by emitting the same quantity of energy as it was absorbed initially during excitation. Due to its unique electronic structure, each element exhibits a unique set of permitted electronic energy levels. Consequently, the wavelengths of light absorbed or emitted by atoms of an element become the unique characteristic of that element and represent the visible and UV regions of the spectrum. The absorption of radiant energy by atoms forms the basis of atomic absorption spectrometry (AAS). Similarly, the absorption of energy followed by the subsequent emission of radiant energy by excited atoms forms the basis of flame photometry and atomic fluorescence spectroscopy (AFS). Since the excited atoms are unstable, they return to their ground state by re-emitting the absorbed energy. This is the principle of AFS. When excitation occurs by heat energy, excited electrons return back to the ground state emitting a light energy (radiation) that is equal to the energy difference between excited and ground states. The emitted light is monochromatic and it has the same wavelength as the light absorbed in the excitation process. This process is the basis of flame emission spectroscopy or flame photometry.

The involvement of absorption or emission of wavelengths in the UV and visible range is entirely due to the valence electron transitions. However, atoms may absorb lower-wavelength higher-energy radiation in the X-ray region by promoting inner-shell electrons to an excited state followed by emission of X-ray. X-ray fluorescence (XRF) spectroscopy and other X-ray techniques are governed by these processes. It is important to note that atoms do not possess any rotational or vibrational energy sub-levels.

6.5. Molecular spectroscopy

In molecular spectroscopy one can utilize rotational, vibrational and electronic transitions and since, similar to atoms, the energy states in molecules are also quantized; various spectroscopic techniques are explored, both quantitatively and qualitatively, using radiations ranging from the radio-wave region to the UV region. Molecules can rotate in space and have rotational energy states. By absorbing appropriate energy they rotate faster as they are elevated to higher-energy rotational states from lower-energy rotational states, generating rotational absorption spectra. However, the energy required for the rotational changes is small and for this reason mostly radiofrequency and microwave regions of the spectrum are engaged in such interactions.

Within a molecule the atoms vibrate in many directions including their bending at various angles. Each vibration represents a characteristic energy. Vibrational energy of a molecule can be increased by absorbing radiant energy in the IR region. The elevation of vibrational energy level in the molecule is associated with an increased molecular rotational energy level because the rotational energy levels are sublevels of the vibrational energy levels. IR absorption in a molecule leads to a change in its rotational and vibrational energy levels. Many vibrational energy states may remain in a molecule of more atoms, giving rise to absorption bands in the IR absorption spectra.

Atomic orbitals in a molecule coalesce to form molecular orbitals, such as sigma (σ) orbitals, pi (π) orbitals, bonding orbitals or antibonding orbitals. Similar to atoms, in molecule also, an outer electron is promoted to a higher-energy excited state from the ground state by absorbing appropriate radiant energy in the UV-VIS range of the electromagnetic radiation spectrum and again goes back to its ground state by releasing UV-VIS light energy as the excited species lacks the stability. As the electronic energy state in molecules possesses rotational and vibrational sublevels, one can expect a change in the vibrational and rotational energies also during the electronical excitement of the molecule. Therefore, total energy change becomes the sum of the electronic, rotational, and vibrational energy changes. Due to the presence of several rotational and vibrational energy states in the molecule, UV-VIS radiation absorption occurs over a wide range of wavelengths which is known as the absorption band. Within molecular spectroscopy, UV-VIS molecular absorption spectroscopy (utilizing the absorption of UV-VIS radiation), and molecular fluorescence spectroscopy and molecular phosphorescence spectroscopy (utilizing the emission of UV-VIS radiation) are commonly used to analyse the chemical species quantitatively as well as qualitatively.

6.6. What are colorimetry, photoelectric colorimetry, and spectrophotometry?

Spectroscopic analysis is usually carried out in solutions but solids and gases may also be studied. A spectrophotometer is involved to measure the absorption or transmittance of light (a specific or a range of wavelengths in a solution / transparent or opaque solid / gas. Often, we use terms like colorimetry, photoelectric colorimetry and spectrophotometry. Although their basic concepts of principle remain the same, they are different to each other based on their mode of operations. In all cases, concentration of a substance in a solution is determined by measuring the relative light absorption with respect to a known concentration of the same substance. When the light used is a white light from a natural or artificial source and the eye is the device to compare the color of an unknown concentration from a set of known concentrations, we refer to *visual colorimetry*. However, one cannot ignore the human error while using the *visual colorimeter* (frequently called as *color comparator* also). But when a photoelectric cell (instead of an eye) is used to measure the color difference, the device is known as a *photoelectric colorimeter*. A narrow band of wavelength from a white light is selected using a *filter* in case of a photoelectric colorimeter. Filter is usually composed of colored glass or gelatin substance that can transmit a specific region of the light spectrum. This is why a photoelectric colorimeter is also known as a *filter photometer*. Colored *glass filters* are wide band pass filters that are obtained by dissolving or dispersing a dye in the glass. *Gelatin filters*, narrow band pass filters, provide better resolution compared to the glass filters. They are manufactured by sandwiching a dyed gelatin film between glass sheets. However, best performance in resolving any peak wavelengths between 200-800 nm and bandwidth between 7-40 nm is obtained from *interference filters*. They contain a coating of a transparent dielectric material (most preferably magnesium fluoride or calcium fluoride due to its hardness) of low refractive index between two semi-transparent silver films which, in turn, again sandwiched between two glass or silica plates. An easy but accurate way to select a filter is to construct an absorbance curve using different filters. The filter that provides maximum absorbance (and thus minimum transmission) should be chosen for further measurements. However, choosing a filter containing complementary colour of the test solution's colour may be an alternative method. A list of transmitted light with its complementary hues is outlined in **Table 6.2**.

Table 6.2: A list of hues and their corresponding complementary hues

Wavelength (nm)	Transmitted hue	Complementary hue
< 400	Ultraviolet	-
380-435	Violet	Yellowish-green
435-480	Blue	Yellow
480-490	Greenish-blue	Orange
490-500	Bluish-green	Red
500-560	Green	Purple
560-580	Yellowish-green	Violet
580-595	Yellow	Blue
595-625	Orange	Greenish-blue
625-780	Red	Bluish-green
> 780	Near Infrared	Red

Source: adapted and modified from Vogel, A.I. (1978) A textbook of quantitative inorganic analysis. Fourth edition, Longman Inc., New York, p.711.

Spectrophotometer is an improved version of photometer, where precision and accuracy of analysis are increased remarkably. In a true sense, a spectrophotometer is a combination of a spectrometer and a photometer. Spectrometer helps to isolate a specific wavelength, possessing a bandwidth of < 1 nm, from a light for its interaction with the substance in solution and for the purpose a prism or grating is used, while transmitted light intensity is measured as a function of substance concentration by photometer. A prism or grating usually splits the incident light beam into different narrow wavelengths, whereas a filter in a colorimeter allows a broad range of wavelengths to pass through and thus exhibits poor wavelength resolution. Typically, filters cannot isolate wavelength ranges smaller than 30-40 nm. This is one of the main reasons for improved precision and accuracy of a spectrophotometer over a filter colorimeter. In a spectrophotometer, when 400 - 800 nm wavelengths are employed, the technique is termed as visible spectrophotometry (VIS-spectrophotometry) and ultraviolet spectrophotometric technique (UV-spectrophotometry) refers to the measurement of a concentration of substrate in a solution using 200 – 400 nm wavelength. UV/VIS spectrophotometry comes under the domain of absorption spectroscopy. Quantitative estimation of metals, nonmetals and organic compounds (<1 - 2 percent) are widely employed in laboratories using spectrophotometric techniques, where the incidence light from a source is passed through substance solution and the intensity of transmitted light and absorbed light (Y-axis) is compared in a graph against a series of the known concentration of the substance (X-axis) in solutions (standard curve or calibration curve). However, qualitative analysis of a compound (identity of the compound) is also possible using a graphical presentation of intensity of transmitted light or absorbed light (Y-axis) against the wavelength (X-axis) where from the absorption spectrum an absorption maximum, which is unique and characteristic feature of the compound, is obtained. In **Figure 6.4**, the absorption maximum of isoprene, 222 nm has been illustrated.

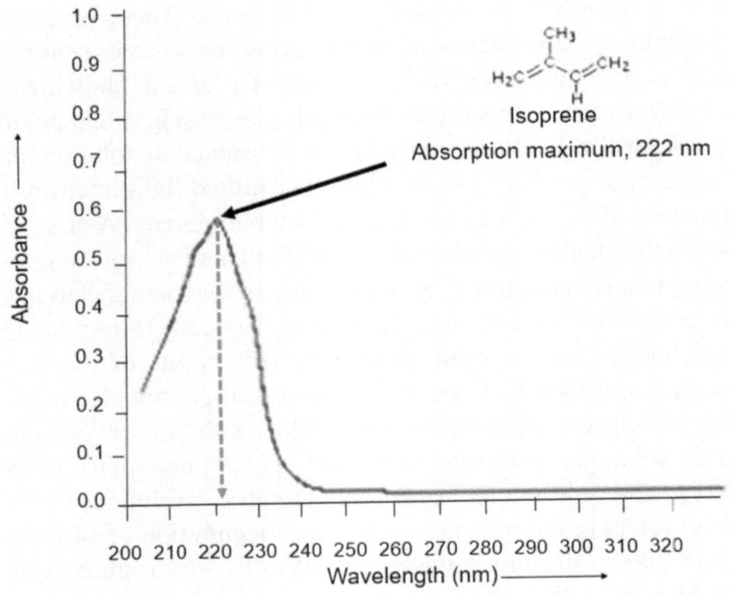

Fig. 6.4: Absorption spectrum of isoprene with its absorption maximum at 222 nm

Albeit its applicability during estimating a compound quantitatively at a concentration of <1 - 2 percent, the sensitivity of UV-VIS-spectrophotometry may be increased using derivative spectrophotometry.

6.7. Basic theory of spectrophotometry

When a light, either monochromatic or heterochromatic, strikes a homogeneous or heterogeneous medium, a portion of it is absorbed within the medium emitting heat energy, a portion is reflected, a portion is transmitted, a portion is scattered, and a portion is diffracted. Thus, the intensity of the incident light (I_o) becomes a summation of the intensity of light in various forms it is distributed. The overall intensity of the incident light is then expressed as the follows (Equation 6.6):

$$I_o = I_a + I_r + I_t + I_s + I_d \qquad \text{Equation 6.6}$$

Where,

I_a = intensity of absorbed portion of light,

I_r = intensity of reflected portion of light,

I_t = intensity of transmitted portion of light,

I_s = intensity of scattered portion of light, and

I_d = intensity of the diffracted portion of light.

In comparison with the intensities of absorbed and transmitted portion, the intensities of reflected, scattered, and diffracted portion of light are negligible and moreover, they are eliminated using a control. Therefore, Equation 6.6 may be re-written as

$I_o = I_a + I_t$ Equation 6.7

6.7.1. Lambert's law

When a monochromatic light travels through a transparent medium, the intensity of the transmitted light (with reference to the intensity of the incident light) decreases exponentially as the thickness of the absorbing medium increases arithmetically. This means that the rate of decrease in light intensity is proportionally related with the increase in absorbing medium thickness (light path). If you have two situations: in one case you are allowing light to pass through 1 cm absorbing medium and in other case, light path length is > 1 cm; then more light absorption is observed in the second case (absorbing medium thickness > 1 cm) than the first case (absorbing medium thickness = 1 cm). This is due to more light interaction with more analyte molecules. As you are allowing light to traverse more paths, more photons are allowed to interact with more molecules leading to more photon absorption.

This is mathematically expressed as

$I_t = I_o \cdot e^{-kl}$ Equation 6.8

$I_t = I_o \cdot 10^{-Kl}$ (changing from natural to common logarithm) Equation 6.9

$I_t / I_o = 10^{-Kl}$ Equation 6.10

Where,

I_t = intensity of transmitted light,

I_o = intensity of incident light,

l = thickness of the absorbing medium,

k = a constant specific for the selected wavelength and absorbing medium,

K = absorption coefficient, and

I_t / I_o = transmitted portion from the total incident light. This is referred to as Transmittance (T).

Another term, Absorbance (A) that is also known as optical density (OD), is most commonly used, and is expressed as

$A = -\log_{10} (I_t / I_o)$, or $A = \log_{10} (I_o / I_t)$ Equation 6.11

$A = -\log T$, or $A = \log (1/T)$ Equation 6.12

6.7.2. Beer's law

Similarly, Beer (1852) also established a relationship between the light intensity and concentration of absorbing analytes. He observed that when a monochromatic light travels through transparent solutions of different analyte concentration, the intensity of the transmitted light decreases exponentially as the concentration of the absorbing medium increases arithmetically. It may be expressed as

$I_t = I_o \cdot e^{-kc}$ Equation 6.13

$I_t = I_o \cdot 10^{-Kc}$ (changing from natural to common logarithm) Equation 6.14

$I_t / I_o = 10^{-Kc}$ Equation 6.15

Where,

c = analyte concentration

6.7.3. Beer- Lambert law

Combining Lambert's law and Beer's law, we get Beer-Lambert law which talks about the relationship among the transmitted light intensity, the light path length, and the analyte concentration and to describe the attenuation of light by a solution. However, we are more interested in deducing the relationship between light intensity and concentration keeping the light path length constant (usually l = 1 cm).

Combining Equation 6.9 and Equation 6.14, we may write as

$I_t = I_o \cdot 10^{-\varepsilon lc}$ that may be further written as

$\log (I_o / I_t) = \varepsilon\, l\, c$, or $A = \varepsilon\, l\, c$ Equation 6.16

Where,

ε = molar absorption coefficient (also known as molar absorptivity or molar extinction coefficient). This is a constant value for a specific analyte at a concentration (c) of 1 mole / dm^3 and a light path length (l) of 1 cm. An analyte may be qualitatively identified by its ε value.

When l is constant (l=1), Beer-Lambert Law may be expressed as

$c \propto \log (I_o / I_t)$

$c \propto \log (1/T)$

$c \propto A$

At a certain wavelength if the absorbance value shows zero, it is imperative to say that the absorbing medium does not absorb light at that wavelength. This can also be explained using Equation 6.11 where the intensity of incidence light and that of the transmitted light remain same. Thus, we may write

$$
\begin{aligned}
A &= \log_{10} (I_o / I_t) \\
&= \log_{10} 1 \text{ (as } I_o = I_t) \\
&= 0
\end{aligned}
$$

When 90% of the light is absorbed, it is obvious that 10% of the incident light is transmitted. Under this situation, the absorbance value shows 1. It can be derived using following relationship:

$$
\begin{aligned}
A &= \log_{10} (I_o / I_t) \\
&= \log_{10} (100/10) \\
&= \log_{10} 10 \\
&= 1
\end{aligned}
$$

Thus, it may be concluded that the absorbance value is not restricted within a range of 0 to 1, it may be of >1.

6.8. Limitations of Beer's Law

Beer's law illustrates the linear relationship between the concentration of the analyte and its light absorption behavior. However, under certain conditions this law does not hold good. One of the major important issues of not obeying the law is the analyte concentration. It has been observed that the law is applicable only to low concentration of analytes (<10mM), not to the higher concentration. At higher concentration the analyte may interact with the solvent (solute-solvent interaction) and the solutes other than the analyte (solute-solute interaction) due to their close proximity that may lead to hydrogen bond formation and may influence the charge distribution of neighboring molecules. Such interactions may cause a shift in the wavelength absorption maxima. Sample's refractive index (η) may also be changed at higher concentration that, in turn, influences the absorbance value and reduces the linearity. In such a case, a correction (Equation 6.17) needs to be followed. However, such correction is not needed at low concentration (<10 mM)

$$A = \varepsilon cl\,(\eta^2 + 2)^2 \qquad \text{Equation 6.17}$$

Molecular size also affects the light absorption phenomenon and exerts its influence to deviate from the theory. Larger molecules at lower concentration may even exhibit the deviation because at lower concentration they show higher absorbance due to their more potential electron transitions. Usually the absorbance value of 0.1 – 1.0 AU (absorbance unit) is considered to minimize the deviation because measurements become noisy and non-reproducible at very low absorbance, while a non-linear relationship is very common at higher absorbance value.

Besides these, the instrument resolution, the signal to noise ratio and the stray light interference also contribute significantly in limiting the linearity of the instrument. Stray light is the light of all wavelengths other than the bandwidth that the monochromator is selecting. This occurs due to the light leak from the outside environment and also due to the diffraction pattern produced by the monochromator.

It has been observed that the law may be deviated due to the chemical effects, such as polymerization, complex formation, dissociation, photochemical reactions, etc. and also due to the temperature effects. It is commonly found a 'red shift' or bathochromic shift in the visible range on dissolution of a gaseous, liquid, or solid phase substance in a solvent containing high dielectric constant, where absorption maximum is shifted more towards longer wavelengths (lower frequency, more energy). A 'blue shift' or hypsochromic shift is common in the ultraviolet region (mostly in the $n \rightarrow \pi^*$ transitions), where absorption maximum shifts more towards shorter wavelengths (more frequency, more energy). Temperature variations (increase or decrease) may change the ionic equilibrium leading to the deviation from Beer's law because the ionic species may differ in their absorption maximum (e.g., $Fe^{+3} \leftrightarrow Fe^{+2}$). Again, this may be the case that the compound during irradiation with photons may tend to be photodecomposed. This causes the deviation from the Beer's law and a source of an error as the absorption maximum of the parent compound and the photodecomposed products are different. The same explanation may be given for dissociation reactions also. As for instance, $Cr_2O_7^{-2}$ (λ_{max}, 375 nm; orange) on dissociation gives rise to CrO_4^{-2} (λ_{max}, 450 nm; yellow) via the production of $HCrO_4$.

$$Cr_2O_7^{-2} \text{ (Orange)} + H_2O \leftrightarrow 2HCrO_4 \leftrightarrow 2H^+ + 2\,CrO_4^{-2} \text{ (Yellow)}$$

6.9. Sources of error

One of the major sources of error is the instrumental one in the form of stray light. We know that an ideal monochromatic light beam, possessing a single wavelength, is the prerequisite criterion for obeying Beer-Lambert law. Obtaining an incident light beam of a single wavelength depends entirely on the quality of the monochromator system being used. In most cases, a monochromator emits a small range of continuous wavelengths, not a single wavelength. Under this situation, if the molecule of interest exhibits the same molar attenuation coefficients (ε) at different wavelengths within the emitted range, the relationship between absorbance and concentration follows Beer-Lambert's linearity. Different ε values of the analyte at different wavelengths within the emitted range forces the relationship to deviate from the Beer-Lambert law. The extent of deviation depends on the extent of difference in ε values. If the difference is more, more is the deviation. Thus, an instrument possessing a high-quality monochromatic system should show a lesser extent of such error.

Temperature alteration over a period of time may be responsible for providing erroneous and erratic results. However, this may be counteracted with updating the baseline by taking new reference measurements.

Changing of solution pH is equally important to play a significant role to give rise erroneous results and could be considered as a source of error during measurement because it may result in structural change of the analyte molecule due to redistribution of electrons in the molecular orbits, that may cause a shift in absorption maximum as for example, litmus gives red color under acidic condition and becomes blue colored under basic condition.

In biochemical analysis, it is a common practice to use various buffer solutions and it is observed that the buffer components may exhibit their own absorbance and may be a source of error. As for example, Triton x-100 [a nonionic surfactant that has a hydrophilic polyethylene oxide chain and an aromatic hydrocarbon lipophilic or hydrophobic group; $C_{14}H_{22}O(C_2H_4O)_n$(n=9-10)] shows a pronounced absorbance in the UV region (276 nm). Under such a situation the error may be reduced by using alternative buffer components that exhibit no or lower absorption, as for instance, hydrogenated Triton x-100 exhibits lower absorption. If there is no other alternative but to use the same buffer component, its concentration should be kept at a minimum.

Optically matched cuvettes for blank, reference and sample measurements, maintaining their proper orientation on the beam path to obtain continuity of the cuvette surfaces and pathlength, should be used to reduce the related errors in measurement. Otherwise, it may be a potential source of error during measurement. It is usually a common practice to measure blank, reference and sample with the same cuvette after thorough cleaning and drying of the cuvette prior to measurements.

6.10. UV-VIS spectrophotometry

6.10.1. Instrumentation

Commercially available normal spectrophotometer works in the range of 200-800 nm wavelengths that are responsible for electronic excitement in many atoms and molecules. For vacuum UV radiation which is also involving valence electron excitation, a normal spectrophotometer cannot be used because it needs air-free light-path. Since oxygen and moisture present in air show their absorption at below 200 nm, it needs special care to make the light-path free form air by purging with helium or creating vacuum.

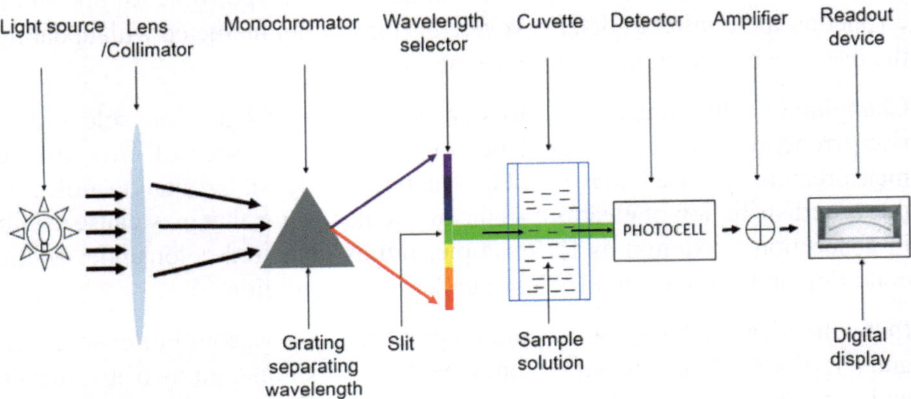

Fig. 6.5: Schematic diagram of a spectrophotometer

A spectrophotometer consists of a radiant energy source, a collimator for straight light beam transmission, a monochromator (prism or diffraction grating resolving the wavelength), a wavelength selector with a slit to isolate the desired spectral band by blocking other dispersed radiation, a sample compartment where sample solution is placed in a glass or quartz cuvette, and a detector (photocell) coupled with a signal-amplifier and a recorder (**Figure 6.5**).

Two types of spectrophotometers are mostly available in the market: (a) single beam, and (b) double beam spectrophotometer.

Single Beam Spectrophotometer

A single beam spectrophotometer (**Figure 6.5**) deals with measuring the relative intensity of the incident light beam before and after the insertion of the test sample. Light directly passes through the sample and reaches straight to the detector in a single beam spectrophotometer. Single beam instrument is simple and not so costly. On the other hand, light is split into two paths in a double beam spectrophotometer (**Figure 6.6**): one going through a reference sample and the other through the test sample. The instrument compares the light intensity between two light paths and allows the subtraction of light absorption of the reference path from that of the sample path.

Double beam spectrophotometer

Modern UV-VIS spectrophotometers, covering a wavelength range from 200 nm to 800 nm, are usually double beam spectrophotometers. This type of instrument is engaged in dividing the monochromatic incident radiation into two identical beams. One beam traverses through the sample solution and the other beam travels through the reference (blank solution).The idea of doing this is to automatically subtract the absorption signal of the reference solution (which is due to the presence of the materials other than the analyte) from the absorption signal of the sample solution, so that a net absorption signal of the sample component is obtained. A Schematic diagram of a double beam spectrophotometer is presented in **Figure 6.6**.

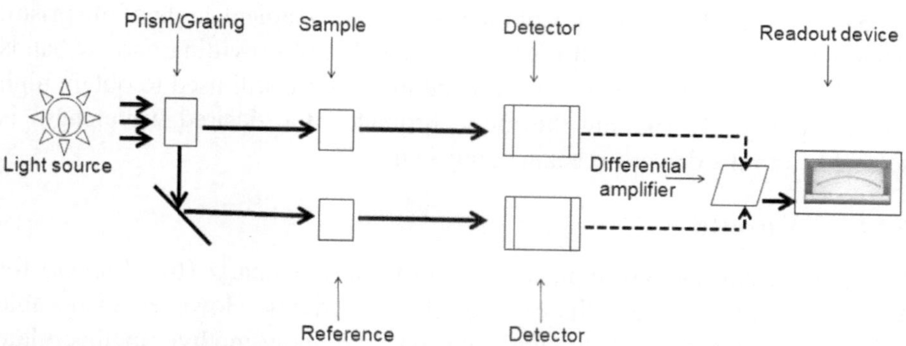

Fig. 6.6: Schematic diagram of a double beam spectrophotometer

6.10.1.1. Radiation source

It is essential to obtain enough radiant energy from the radiation source over the specific region of the electromagnetic spectrum you are selecting, and also it is important to get a constant intensity of light during measurement periods. Deuterium arc lamp as a UV radiation (190-380 nm) source is used to provide powerful and stable light, while tungsten filament lamps are employed for the visible radiation (380-800 nm). However, Tungsten - Halogen lamps, containing iodine as a normal filling gas, are also common to use. In this lamp, the cover is made of quartz to tolerate a higher lamp operating temperature. Xenon flash lamp covers both UV and VIS range, and it is known for its long life span. This is so named as it flashes (up to 80 times a second) instead of glowing constantly. It is also a common practice to use paired deuterium and tungsten lamps in UV/VIS spectrophotometer to cover the full UV and visible spectrum.

6.10.1.2. Monochromator

Prism or diffraction grating is used as a monochromator in a UV-VIS spectrophotometer to resolve the wavelength in a most optically efficient way. However, diffraction grating performs better than the prism. Grating diffracts the various wavelengths of light at different angles when incident radiation falls on closely spaced lines marked on a glass surface (transmission gratings) or from a series of parallel grooves on a metal surface (echelette gratings). Echelette gratings are preferably used to obtain diffraction of UV light. In case of prisms, glass prisms may be employed for working with light in the visible region (400-800 nm). However, glass prisms are not suitable for narrower wavelengths (<400 nm) due to their less transparency. Quartz or fused silica prisms are commonly used in the UV region. The dispersion of light using prism is dependent on the dispersive power (which, in turn, depends on the refractive index of the prism) and the apical angle of the prism. Although mostly all monochromators are capable of providing narrow bands of radiation, a good number of lenses and mirrors are still used to obtain high spectral purity. By rotating the monochromator, the desired wavelength is selected from the dispersed beam using a slit.

6.10.1.3. Cuvette

Rectangular cuvettes are typically made of glass or quartz (fused silica) for VIS or UV-VIS wavelengths of interest, respectively. However, disposable polystyrene cuvettes (VIS applications) and poly-methyl methacrylate (PMMA) cuvettes (UV-VIS applications) are also available. The most commonly used cuvette has a pathlength of 10 mm meaning that the incident light beam covers a distance of 10 mm through the solution. The cuvettes are cleaned meticulously with nitric acid followed by sequentially rinsing with water and acetone. Finally, the cuvette should be dried by wiping with tissue paper.

6.10.1.4. Detector

The function of a detector is to convert light energy to electrical energy in the form of a current output. Higher current output signifies higher light intensity. Several detectors, such as barrier layer photovoltaic cells, photo-emissive tubes, photomultiplier tubes, silicon photodiode transducers, photodiode arrays, photoconductivity transducers, etc. may be used in spectrophotometers. However, photo-emissive tubes, photodiodes and photomultiplier tubes are commonly used.

Photo-emissive tubes

Different types of photo-emissive tubes are available based on their wavelength range, such as blue - sensitive (385-600 nm), red - sensitive (600-950 nm) and broad range sensitive (380-950 nm) photo-emissive tubes. Photo-emissive tube contains a glass bulb and a centrally positioned metal ring. The inner surface of the bulb is thinly layered with cesium or potassium oxide and silver oxide. The oxide layer acts as the cathode emitting electrons on illumination. The metal ring acts as an anode. Photoelectrons are ejected as light strikes the photocathode and photoelectrons are drawn to the anode leading to current generation through an outside circuit.

Photomultiplier tube

It is a combination of a photo-emissive cathode and an internal electron multiplying chain of dynodes, charged at a successively higher potential. A dynode is an intermediate electrode in a vacuum tube that serves as an electron multiplier through secondary emission. Ejected photoelectrons from photo-emissive cathode due to its interaction with the incident radiation are further subjected to an electrostatic field and accelerated towards the first dynode to successive dynodes and finally makes inroads into the anode.

Photodiode

Schematic diagram of a photodiode is depicted in **Figure 6.7**. A photodiode is a PN-junction device made up of photosensitive semiconducting material that converts light energy into electrical energy, working in reverse bias condition. It is used as photodetectors to detect optical signals. It works on the principle of the photoelectric effect. Striking a photon of sufficient energy on the diode creates a couple of electron-hole pairs near the depletion zone of the diode when the energy of the photons is more than the energy gap of the semiconductor material. Due to the electric field at the junction, electrons and holes are separated before they combine. The electrons are forced to migrate towards the n – side and the holes move towards the p-side by managing the direction of the electric field in the diode. Hence, an increase in the electromotive force occurs due to higher number of electrons on the n – side and holes on the p-side. A current is generated when an external load is connected to the system. The current flow is proportional to the created electromotive force; more current is generated with more electromotive force. Again, the extent of electromotive force is directly proportional to the incident radiation intensity. Thus, more current is generated with greater intensity of the radiation.

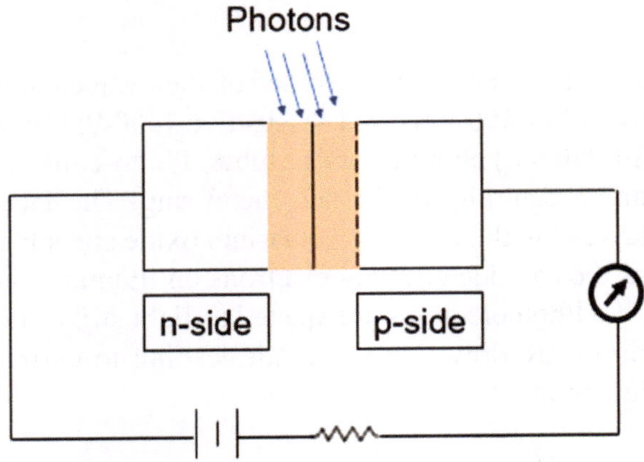

Fig. 6.7: Schematic diagram of a photodiode

6.10.2. Standard curve

When the absorbance value of a solution of known concentration of an analyte is compared with the absorbance value of unknown concentration of the same analyte in solution, it is possible to determine the unknown concentration of the analyte, provided that the absorbance is proportional to concentration according to Beer-Lambert law. For ensuring the proportional relationship between absorbance and concentration, it is imperative to construct a standard curve using a series of dilutions of a standard solution of the analyte of known concentration including a blank solution (without analyte). One should be careful about the selection of solvent for preparing the analyte solution. The solvent should not show any absorbance in the selected wavelength range. The blank solution in the reference cell should be constituted with a pure solvent. Ideally, the reference blank contains all reagents including pure solvent used in sample solution except the test compound because there is every possibility that the reagents and the solvent may also exhibit their absorption maxima that falls within the same absorption range of the test compound. Thus, it is necessary to compare the absorbance of the test solution to a reference blank so that the effects of reagents and solvent on the absorbance value of the test compound are nullified.

A portion of the standard curve in the form of a straight line confirms the proportional relationship and is considered for further computation of unknown concentrations of the analyte. The standard curve is constructed by plotting absorbance (Y-axis) vs concentration (X-axis) for each of the standard solutions and by drawing a best fit line through the points. Once the absorbance of the sample solution of unknown concentration is obtained, its concentration is computed by fitting the absorbance in the standard curve. In **Figure 6.8** the sample solution of unknown concentration exhibits an absorbance of 0.55 that corresponds to the analyte concentration of 4.5 g/L.

However, this is also possible to measure the concentration of a solution using the molar absorptivity from the following expression (Equation 6.16):

$$A = \varepsilon\, l\, c \qquad \text{Equation 6.16}$$

$$\varepsilon = A / l\, c \qquad \text{Equation 6.18}$$

But this is applicable only under 3 conditions: (i) using a cuvette of l = 1 cm, (ii) having access to the value of molar absorptivity, and (iii) assuming that Beer-Lambert Law possesses no limitations over concentration levels. It can work at any concentration range where proportionality relationship between concentration and absorbance is sustained. But this is not practically true. Therefore, construction of a calibration curve is a better idea to measure the concentration instead of using Equation 6.18.

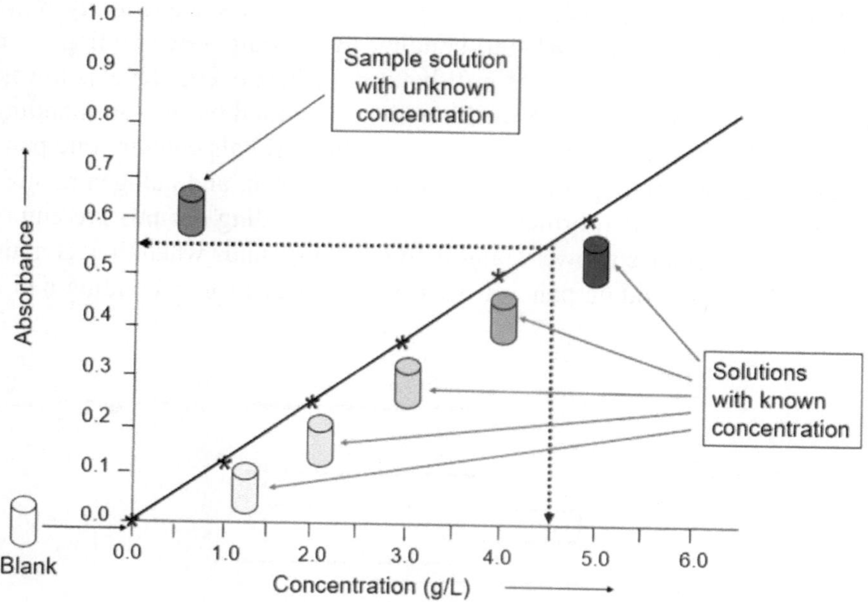

Fig. 6.8: Preparation of a standard curve

6.10.3. What happens when molecules are exposed to the UV-VIS spectrum?

It is to be remembered that the extent of the energy difference between the ground state and first excited states for valence electrons of atoms and bonding electrons of molecules falls within the same range of the energy content of photons from the UV and VIS spectrum. When a compound is exposed to the UV-VIS spectrum, promotion of electrons is observed at their energy levels. A portion of light energy leads to an electron jump from one of the bonding or nonbonding full orbitals into one of the empty antibonding ones. Thus, the electrons are excited through absorption of light energy. A big jump requires greater energy compared to a small one. The possible electron jumps are shown in **Figure 6.9**. Different compounds show different energy gaps between these levels. For this reason, different compounds absorb light at different wavelengths or frequencies depending on the matching with the energy gap. In principle, if the energy gap is more, the electrons need more energy to be excited to their higher energy level. Thus, under the situation of a higher energy gap, light of a higher frequency (consequently containing lower wavelength) is required for an expected electron jump because light with more frequency contains more energy in accordance with Equation 6.3 and Equation 6.4. It is beyond the scope of narrating the orbitals in detail. However, it should be remembered that the electronic energy level increases sequentially if we proceed from **σ**-bonding to **σ*** anti-bonding [**σ** bonding →π bonding → **n** bonding → π* anti-bonding → **σ* anti-**bonding] where energy level is lowest at **σ** bonding and highest at **σ* anti-**bonding. Nonbonding and antibonding orbitals are different from each other. Nonbonding orbitals contain lone pairs of electrons at the bonding level (e.g., oxygen, nitrogen, and halogen atoms). They are stable and filled orbitals while the anti-bonding orbitals are empty orbitals, and they cause a lower stability to the compounds when they contain electrons. Normal bonding pairs of electrons are present in **σ** bonding and π bonding.

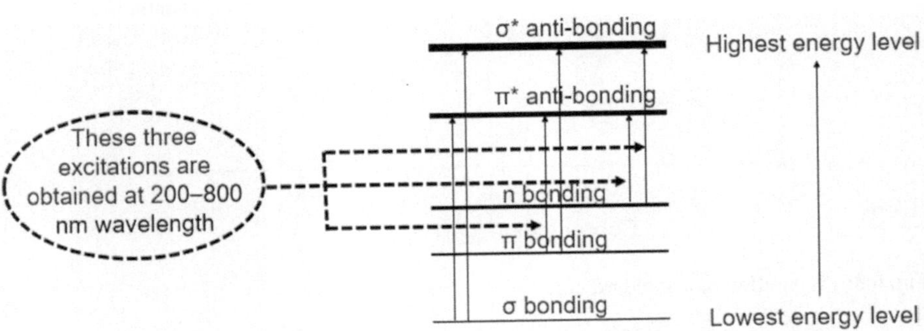

Fig. 6.9: Possible electron jumping from one level to the other

Always remember that a molecule should possess either π bonds or atoms with non-bonding orbitals for light absorption at 200 - 800 nm. The electronic excitations within atoms and molecules due to the presence of unsaturation and/or unbounded electrons in the absorbing molecules are resulted from the absorbance of the UV-VIS radiation. Light absorption at UV-VIS wavelength range is basically associated with electron jump from (1) π bonding orbitals to π* anti-bonding orbitals; (2) non-bonding orbitals to π* anti-bonding orbitals; and (3) non-bonding orbitals to σ* anti-bonding orbitals (**Figure 6.9**). First two jumps involving π → π* and n → π* are considered as very important for UV-VIS absorption spectrometry, while the jump involving n → σ* is less important. Other electronic jumps as shown in **Figure 6.9** are engaged in light absorption at a wavelength range that remains outside the region of the UV-VIS spectrum.

A molecule may contain a single or several chromophoric groups which are basically responsible for light absorption when exposed to visible light. A single chromophore may produce two peaks stating that the group absorbs light at two light wavelengths, as for instance in acetaldehyde carbon-oxygen double bond (C = O) possesses π-electrons and oxygen within it contains a lone pair of electrons. Thus, both π → π* and n → π* are associated with C = O in acetaldehyde giving two peaks at 180 and 290 nm, respectively. Since the non-bonding orbital contains more energy than that of a pi bonding orbital, the jump from an oxygen lone pair into a pi anti-bonding orbital requires less energy, stating that it absorbs light of a higher wavelength (290 nm). The opposite justification is applicable for π → π* jumping. The pi bonding orbital contains less energy than that of a non-bonding orbital. For this reason, the π → π* jumping requires more energy compared to n →π* jumping. Thus, π → π* jumping absorbs light of lower wavelength (180 nm).

When a chromophoric group is attached to a functional group that does not impart any absorption peak above 200 nm, a shift in λ_{max} towards longer wavelength accompanied with increase in intensity occurs. This phenomenon is known as the 'auxochromic effect'. We can define 'auxochrome' as a functional group that alters the capacity of a chromophore to reflect colour but alone fails to exhibit that colour (e.g., -OH, $-NH_2$, -SH, -OR, halogens, etc.). All auxochromes possess one or more non-bonding pairs of electrons.

If you have molecules of simple isolated carbon-carbon double bond and molecules containing conjugated double bonds, you can see a shift of absorbance maximum to its longer wavelengths due to less requirement of energy for absorption as the amount of delocalisation increases. The conjugated double bonds in a molecule refer to the presence of more than one double bond in an alternating fashion with a single bond. Molecules, containing simple isolated carbon-carbon double bonds, exhibit their absorbance maximum at shorter wavelengths whereas molecules containing conjugated double bonds show their absorption maximum at longer wavelengths. This happens due to the fact that more conjugated bonds mean more amount of delocalization of the pi bonding orbitals in the molecules, referring to the associated decrease in the electronic energy spacing. This illustrates the smaller energy difference between ground and excited state. Always remember that with the increase in the amount of delocalisation, the energy gap between the bonding and anti-bonding orbitals becomes less. This is why light with higher wavelength (lower frequency) containing lower energy level is required to be absorbed by molecules containing greater extent of conjugated bonds. If you consider a molecule of massive delocalisation i.e. more numbers of conjugated carbon-carbon double bonds, you will find that the compound absorbs light of higher wavelength in the visible region of the spectrum, and the compound is coloured to look at, as for example red colour in tomato is due to lycopene that contains several conjugated carbon-carbon double bonds (**Figure 6.10**). It is observed that lycopene in hexane displays three absorbance maximums at 443, 471 and 502 nm in the visible range.

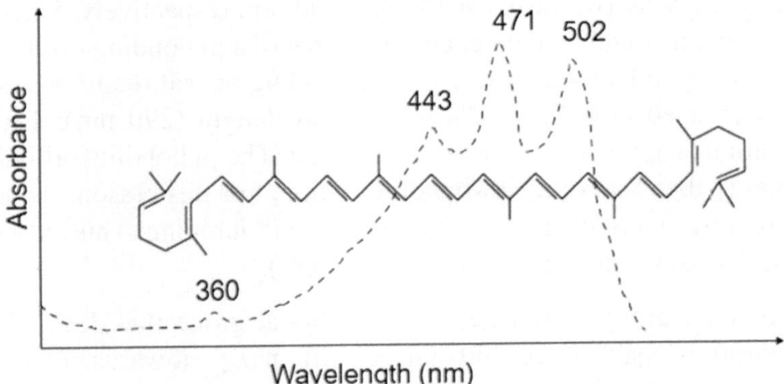

Fig. 6.10: Chemical structure and absorbance peak of lycopene

6.10.4. Applications of UV-VIS spectroscopy

UV/VIS spectroscopy is commonly practised to analyse organic molecules, biomolecules, metal ions or complexes both qualitatively as well as quantitatively by obtaining the absorption spectra of substances at ultraviolet (UV) and visible (VIS) region using a spectrophotometer. Absorption spectra of substances are all unique in nature because of the specificity of light-substance interaction. Since specific functional groups in a molecule absorb specific wavelengths, they exhibit specific peaks in the absorption spectrum. Thus, the peak positions in the spectrum throw lights on the molecular structure of the sample.

However, it is to be remembered that the UV/VIS spectrum does not characterize the compound molecule as a whole because the ionic or molecular spectrum in the UV/VIS electromagnetic wave region only involves dealing with transitional energy levels of certain functional groups of the molecule. The study may be performed both in liquid (solution) and solid (films) states. However, analysis in the solution phase is very common. It is possible to estimate the analyte concentration in solution easily by measuring the absorbance at a specific wavelength. Later, concentration may be enumerated using the absorbance value of the sample. This has been described earlier in *Section 6.10.2*. UV-VIS Spectrophotometer is a single instrument that can cover both UV and VIS range. The spectrum also reveals the specific physical properties of a substance as for example, extinction coefficient of a compound may be calculated from its UV/VIS spectrum. The rate of reaction may be determined by monitoring the absorption spectra as a function of time. Absorbance value at a single wavelength is monitored at periodic intervals, followed by its plotting against time. With this, the reaction progress may be assessed. By checking the peak position and the peak profile of a compound, one may conclude whether or not the substance contains any impurity. Broadening of peaks or shifting of peaks along with additional peaks confirms the presence of impurity in a compound.

6.11. Fluorimetry

6.11.1. Introduction

Fluorimetry is also referred to as spectrofluorimetry or fluorescence spectroscopy. This is utilized to measure and analyze the fluorescence emitted from a sample. The emission of the electromagnetic radiation (fluorescence) is due to the result of relaxation from an excited vibrational electronic energy level, which is earlier achieved by the activation process via UV-VIS radiation absorption, to its corresponding ground state. Molecules cannot fluoresce if they do not exhibit any absorbance in the UV/VIS region (e.g., saturated hydrocarbons). This technique involves a simultaneous occurrence of activation and deactivation processes. Before studying fluorimetry it is imperative to gain a basic knowledge on luminescence properties of a substance. Luminescence refers to any emission of light (electromagnetic waves) from a substance that does not arise from heating and thus, it is distinct from incandescence. Different types of luminescence may occur depending on the energy sources, such as photoluminescence (excitation by absorption of photons), chemiluminescence (by chemical reactions), electroluminescence (by electrical field), radioluminescence (by ionizing radiations), mechanoluminescence (by mechanical stress), thermoluminescence (by heating after prior storage of energy), sonoluminescence (by sound waves) and cathodoluminescence [by cathode rays (electron beams)], that have different applications in various fields of science and industry. Fluorimetry is one such application where the phenomenon of photoluminescence is engaged. Photoluminescence is light emission through photo-excitation from any form of matter after the absorption of photons. Fluorescence is one type of photoluminescence occurring in gas, liquid or solid chemical systems through absorption of photons in the singlet ground state promoted to a singlet (The excited state of the molecule with paired electron spins is known as a singlet state) excited vibrational state in a higher electronic state followed by its relaxation to its singlet ground state. Fluorescence occurs on vibrational relaxation of an atom or molecules to its ground state. Fluorescence is mostly found due to $\pi^* \rightarrow \pi$ and $\pi^* \rightarrow n$ transitions (less energetic) rather than $\sigma^* \rightarrow \sigma$ transition (more energetic), signifying the occurrence of fluorescence from the lowest vibrational level of the first excited electronic state to one of the vibrational levels of its ground state. However, the probability of occurrence is more common in $\pi^* \rightarrow \pi$ transitions than $\pi^* \rightarrow n$ transitions. Unlike phosphorescence (another type of photoluminescence), in fluorescence the spin of the electron is still paired with the ground state electron, which means that the electronic spin in the excited level is still opposite to that of the electron in the valence level. When the excited species comes to its ground state, again it emits photons. The emitted photon is usually of longer wavelength accompanied with lower photon energy compared to the absorbed radiation. This phenomenon of shifting wavelength is known as 'Stokes shift'. Therefore, it may be the case that a fluorescent compound after absorbing invisible electromagnetic radiation in the UV-region emits light in the visible range. However, 'resonance fluorescence' produces the emission radiation of the same wavelength as it is with excitation radiation (e.g., sodium vapour). Fluorescent substances stop glowing when it is removed from the incident radiation source. This is opposite to the phosphorescence character of a substance where a phosphorescent substance continues to glow for a period even after the removal of incident radiation and may extend from fraction of a second to several weeks. It is observed that in fluorescence the absorption and re-emission occur within a characteristically short span of time of 10^{-12} - 10^{-9} second as the excited states are short lived, while in phosphorescence re-emission takes place with a time delay of > 10^{-8} second. Phosphorescence is significantly a slower process than fluorescence as it deals with a change in the electronic spin. In phosphorescence, an excited singlet state undergoes a radiationless intersystem crossing (ISC) transition to the excited triplet state (parallel electron spins: the electronic spin in the excited level is spinning in the same direction to that of the electron in the valence level) by reversing the spin of the excited electron. The molecule then relaxes to the ground state from the triplet state with a consequent photon emission. Therefore, phosphorescence involves in excitation of the molecule by absorption of light to an excited singlet state at first, followed by attaining its triplet state through ISC in the next step, and finally emitting a photon due to a triplet–singlet transition to reach the ground state. Compared to the fluorescence, the photon obtained from the phosphorescence process is of longer wavelength, possessing lower energy.

It is also possible that excited molecules may attain its ground state without emitting any radiation through collision with other molecules, more particularly with solvent molecules. However, this quenching effect may be minimized by decreasing the temperature and increasing the viscosity, so that the frequencies or numbers of collisions are reduced considerably. The process is known as collisional deactivation, one form of quenching effect which is more prominent in the phosphorescence process. As phosphorescence involves the long stay of molecules in their excited triplet state, there remains more probability of such collision.

6.11.2. Basic theory

Since the fluorescence intensity or power is linearly proportional to the concentration of the fluorescent molecule which in turn is proportional to the radiant power absorbed by the sample, we can use the fluorescence spectroscopy to determine the concentration of a compound applying the Beer-Lambert law. However, the proportional relationship between the fluorescence intensity and the concentration of the fluorescent substance is applicable when the εlc-value approaches 0.01 or less than that. Since the fluorescence is emitted from the substance due to the subjection of another electromagnetic radiation, the law is applied indirectly as follows:

$$F = K(I_o - I_t) \qquad \text{Equation 6.19}$$

Where,

F = power or intensity of fluorescent radiant energy which is assumed to be proportional to the intensity of absorbed radiant energy $(I_o - I_t)$,

K = proportionality constant, referring to the fluorescence efficiency which is defined as the ratio of number of photons emitted as fluorescence to the number of photons absorbed. It is also known as quantum efficiency which is specific for any given system.

I_o = intensity of incident electromagnetic radiation, and

I_t = intensity of transmitted electromagnetic radiation.

It is assumed that F is proportional to the intensity of absorbed radiant energy ($I_o - I_t$).

Beer-Lambert law states that $I_t/I_0 = e^{-\varepsilon lc}$. From this relationship we may get the following equation:

$1 - I_t/I_0 = 1 - e^{-\varepsilon lc}$ (subtracting both side by 1) Equation 6.20

$(I_o - I_t) = I_o (1 - e^{-\varepsilon lc})$ (multiplying both side by I_o) Equation 6.21

Since ($I_o - I_t$) refers to the amount of light absorbed, we may express the fluorescence intensity as follows:

$F = I_o (1 - e^{-\varepsilon lc})$

Since ($I_o - I_t$) refers to the amount of light absorbed, we may express the fluorescence intensity as follows:

$F = K I_o (1-10^{-\varepsilon lc})$ Equation 6.22

$F = K I_o\ 2.303\ \varepsilon lc$ Equation 6.23

$F = k I_o c$ (at low concentrations, εlc becomes <0.05) Equation 6.24

Thus, the fluorescent intensity is directly proportional to the fluorescent substance concentration and also to the intensity of the incident radiation from the source, if other parameters are kept constant. However, the relationship between the fluorescent intensity and incident radiation intensity is applied to a certain limit for those compounds that are photodegradable. At higher concentration, the relationship between fluorescent intensity and fluorescent substance concentration does not follow the linearity because of the self-quenching effect. At higher concentration, fluorescence emitted by the molecules from the inner core of the sample are absorbed by the molecules remaining in the outer periphery of the sample. Thus, the fluorescence intensity is decreased to some extent. However, the problem may be overcome by diluting the solution.

Fluorescence property of a substance is mainly dependent on its two parameters: molecular structure and the chemical environment. It is evidenced that the intensity of the fluorescence is influenced by the pH of the medium, presence of quenchers that mask the fluorescence effect, type of solvent used, temperature, etc. There are several compounds that fluoresce or de-fluoresce on changing solution pH. This behaviour is utilized in acid-base titrations where compounds, such as eosin, acridine, fluorescein, etc. are commonly used as fluorescent indicators. Thus, the types and concentrations of reagent used during fluorimetric determination of any compound should be carefully chosen. Higher analyte concentration results in producing a self-quenching effect because under this situation there is a possibility of formation of a complex, known as 'excimer', due to an interaction between the excited molecule and a molecule remaining in the ground state, leading to a bathochromic shift. On the other hand, dilution reduces this effect.

6.11.3. Instrumentation

A schematic diagram of a fluorimeter is presented in **Figure 6.11**. The instrumentation design of a fluorimeter is nearly the same as it is seen in UV-VIS spectrophotometer except the optical device. But indeed the optical materials remain the same: quartz for the UV and glass for the visible light. Two filters or monochromators are needed in fluorimeter (filter) or spectrofluorimeter (monochromator): one for selecting the proper wavelength from the excitation beam and another for selecting the proper wavelength from the emission beam. A fluorimeter sequentially consists of a radiant energy source, a collimator (for changing the diverging light or other radiation from a point source into a parallel beam), an excitation monochromator (primary filter) that allows the right wavelength to pass for the needed excitation, a sample compartment where sample solution is placed in a glass or quartz cuvette, an emission monochromator (secondary filter) that allows the fluorescent radiation to pass but block the primary radiation by absorbing, and a detector (photomultiplier tube) coupled with a signal-amplifier and a recorder. It is a general convention to align the detector at 90 degree angle with the path of emitted radiation relative to the axis of the excitation beam so that the interference of transmitted and scattered radiation is minimized. Fixing the excitation monochromator, the emission monochromator is rotated to obtain an emission spectrum; while for getting an excitation spectrum, the excitation monochromator is rotated and the emission monochromator is kept fixed.

Same radiation sources as they are used in UV/VIS absorption spectrometry can be adopted for fluorescence. However, as there exists a direct proportional relation between fluorescence intensity and the incident light intensity, it is a common practice to use an intense radiation source, such as laser light sources. Besides, mercury or xenon arc lamps are also used.

It is important to note that a molecule has to attain its excited state of higher energy prior to emitting its fluorescence. The excitation and emission spectrum are nearly mirror images of each other due to approximately the same vibrational energy difference at the ground electronic state and the excited electronic state.

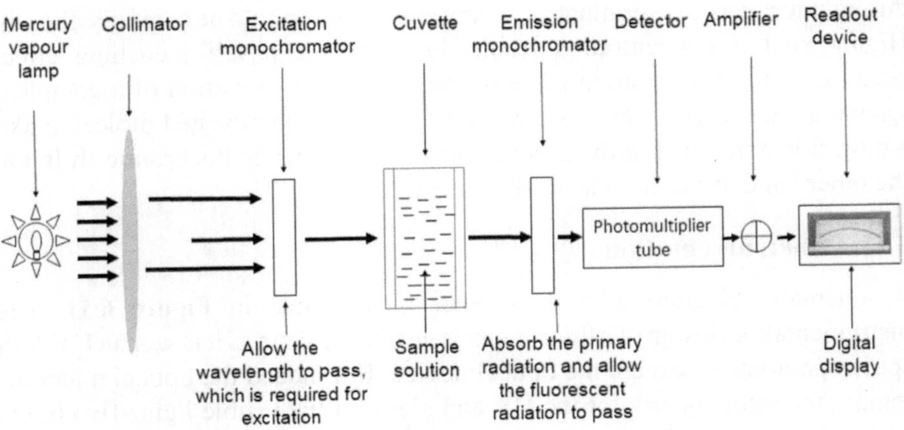

Fig. 6.11. Schematic diagram of a fluorimeter

6.11.4. Intrinsic and extrinsic fluorescence

Quantitative fluorescence spectroscopy is several folds more sensitive than UV-VIS spectrometry. However, a most important limitation of this technique lies in the fact that very few molecules exhibit their fluorescence property. It is customary to tag a non-fluorescent compound with a fluorophore by treating the compound being studied with a fluorescent compound so that molecular events in the compound can be studied. The fluorescence, thus obtained in a non-fluorescent compound, is known as *extrinsic fluorescence*. Dansyl chloride [5-(Dimethylamino)naphthalene-1-sulfonyl chloride], Fura-2 AM [Fura-2-acetoxymethyl ester], SYBR Green [N',N'-dimethyl-N-[4-[(E)-(3-methyl-1,3-benzothiazol-2-ylidene) methyl]-1-phenyl quinolin-1-ium-2-yl]-N-propyl propane-1,3-diamine], fluorescein [3′,6′-dihydroxy spiro{isobenzofuran-1(3H),9′-[9H]xanthen}-3-one] etc., are few examples of commonly used fluorophores. On the other hand, when a compound fluoresces itself and does not need any labelling from other fluorescent compounds, it is known as *intrinsic fluorescence*. In intrinsic fluorescence, the property is native in nature. Even compound with weak intrinsic property may also be bound with certain extrinsic fluorescent probe to enhance the native fluorescence of the weak intrinsic fluorescent compound. As for instance, DNA with weak intrinsic fluorescence signal may be treated with ethidium bromide or SYBR Green to enhance its fluorescence emission.

6.11.5. Application

This technique is utilized to assess the concentration of several metals, organic compounds and biomolecules, such as vitamins, calcium, aluminium, zinc, cadmium, etc. Several metal ions are reported to form organic complexes that fluoresce. Calcium complexes with Calcein (fluorescein iminodiacetic acid) under alkaline conditions and gives rise to fluorescence. Similarly, aluminium is quantitatively estimated by forming its complex with Pontachrome Blue Black R. Fluorescence property may also be utilized by using fluorescent molecules as tags in molecular biology to monitor protein activity, to detect and locate a specific DNA sequence on a chromosome and for many other purposes. The use of FISH (Fluorescence *in situ* Hybridization) is a fast-growing technique in tumor biology, gene mapping, gene amplification, genomics, cytogenetics, prenatal research, and basic biomedical research. Fluorescence spectroscopy is known for its high sensitivity and specificity.

6.12. Nephelometry and turbidimetry

6.12.1. Introduction

It is earlier stated that a portion of electromagnetic radiation may be absorbed or transmitted or scattered or reflected when radiation passes through a solution or a suspension of substances. Nephelometric and turbidimetric techniques involve in the analysis of suspended particles in liquid or gas by determining the turbidity utilizing the intensity of scattered light (Nephelometry) and transmitted light (Turbidimetry), when light passes through a suspension of particles, inorganic or organic sediments, bacteria, virus, plankton, pollen, etc. Both techniques are applied to measure the turbidity of a suspension, however, in different modes. Since the intensity of transmitted light is measured in turbidimetry, turbidimetric analysis is engaged in a direct estimation of substances. If the solution is cloudier, intensity of transmitted light is decreased proportionally. More intensity of transmitted light means lesser concentration of substances. Turbidimetry bears a resemblance to colorimetry because, in both cases, the transmitted light intensity is measured. Conversely, nephelometry is an indirect way of measuring substances because it considers the intensity of bounced off scattered light instead of using the intensity of transmitted light or attenuation due to cloudiness, and it is more sensitive than the turbidimetry. More intensity of scattered light implies more concentration of the substance in suspension. As a whole, high turbidity causes reduction in transmitted light intensity due to more absorption and more scattering; consequently, light attenuation coefficient becomes more. However, in both nephelometric and turbidimetric analysis, particle size is an important parameter to determine the efficiency of the method because the large size of complexes or particles tend to be precipitated and consequently, lead to erroneous results due to decreased light intensity, either it is scattered or transmitted one. Apart from size, other parameters, such as ratio of concentration of solutions mixed, ionic concentrations participating in the precipitation reaction and their rate of mixing, temperature, dispersion stability, time allowed for precipitation and many others also affect the optical properties of suspension. Thus, the results obtained with their reproducibility depend on how the suspensions are prepared and obviously on reproducibility of their optical properties. It is necessary to regulate the experimental conditions properly. Two samples of same analyte concentration may differ in their scattering behavior if they differ in producing same number of precipitating particles or if they differ with respect of precipitating particles size distribution, that are possible only when they are subjected to different experimental conditions, such as temperature, ratio of sample analyte and precipitant, time allowed for precipitation and many other conditions.

A nephelometer and a turbidimeter also differ within themselves from the point of geometric positioning of the incident light source and the detector. It is a usual practice to measure the intensity of scattered light at right angle to the incident light beam in nephelometric technique. But the turbidimetric estimation is done at an angle of 180^0 to the incident light beam. Usually the turbidity of a suspension is expressed as nephelometric turbidity units (NTU). Few NTU levels in environmental water matrices are provided in **Table 6.3.**

Table 6.3: NTU levels in various environmental water matrices

DI or RO	0
Tap water	0.3
Clear surface water	< 5
Drinking water	< 5 (ideally <1) [The WHO (World Health Organization), establishes that the turbidity of drinking water should not be more than 5 NTU, and should ideally be below 1 NTU, while the European standards for turbidity state that it must be no more than 4 NTU]
Turbid surface water	10-50
Water generated from soil erosion	Up to 1000
Suspended water creating problem to aquatic organisms	50-100 and more

6.12.2 Theory

In turbidimetric analysis, the intensity of the transmitted light is considered as a function of the concentration of the suspended particles and in nephelometric analysis, the intensity of the scattered light is considered as a function of the concentration of the suspended particles in a dispersed system. Obviously, the concentration, shape, refractive index and size distribution of the suspended particles influence the quantity or intensity of scattered light. Therefore, it is imperative to prepare the suspension very carefully under controlled conditions. Similar to spectrophotometry, in these measurements also Beer-Lambert law is applicable up to a certain concentration of the analyte in suspension. This necessitates to construct a calibration curve and attempts should be made to keep identical experimental conditions for all measurements to obtain reproducibility of the results.

Both reflection and scattering phenomena play an important role in turbidimetric and nephelometric analysis. It is important to note that the dominance of these two phenomena depends on the relation between the size of the suspended particles and the wavelength of the incident light. Reflection dominates over the scattering when the size of the suspended particles is more than the wavelength of the incident light. Scattering occurs predominantly when the suspended particles are of the same dimension or smaller than the wavelength of the incident light. Thus, for nephelometric measurement the size of the suspended particles should be necessarily smaller than the incident light wavelength so that scattering is more than reflection. However, scattering also depends on the refractive index of the particle, particle geometry, molecular absorption, sample concentration and size distribution. On the other hand, in turbidimetry, particle size may be of bigger or smaller size than the incident wavelength. It does not become a major issue because, in turbidimetry, intensity of transmitted light is measured. Therefore, we are more concerned with the total attenuation of incident light, either it may be by absorption or reflection or by scattering. However, if the suspended particle size is too large, then there is a tendency of precipitation leading to more transmission and low attenuation of the incident light. This gives rise to false signals.

In general, turbidimetry is preferred for concentrated suspension. Conversely, dilute suspension is analyzed by nephelometry.

In nephelometry, the relationship among the intensity of incident light, the intensity of scattered light and the concentration of analyte in suspension is expressed as:

$I_s = K_s \, x \, I_o \, x \, C$ Equation 6.25

Where,

I_s = intensity of scattered light,

K_s = empirical constant that depends on the suspension medium and suspended particles,

I_o = intensity of incident light, and

C = concentration of analyte in suspension

Equation 6.25 states that intensity of scattered light is directly proportional to the intensity of incident light and concentration of analyte in suspension.

6.12.3. Instrumentation

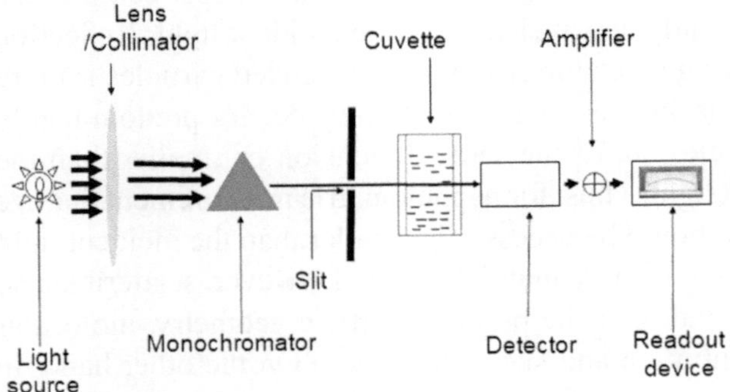

Fig. 6.12: Schematic diagram of a turbidimeter

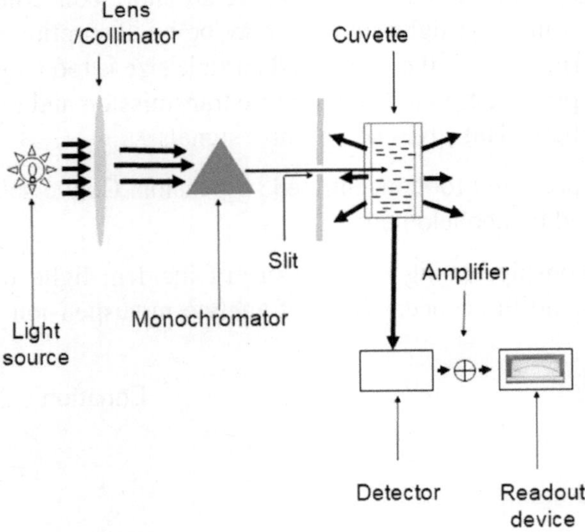

Fig. 6.13: Schematic diagram of a nephelometer

First of all, it is most important to remember that a turbidimeter measures the intensity of transmitted light and a nephelometer measures the intensity of scattered light. Basic components used in turbidimeters and nephelometers are quite similar to a filter colorimeter or spectrophotometer, consisting of radiation source, collimator, monochromator, sample cell, detector, amplifier and readout device. Even a spectrophotometer may be used as a turbidimeter. Regarding instrumentation, a turbidimeter (**Figure 6.12**) and a nephelometer (**Figure 6.13**) differ to each other depending on the geometric positioning of a detector. In a turbidimeter, the detector is positioned in the same direction as the propagation of the primary radiation beam; while in a nephelometer, the detector is positioned at a right angle to the direction of the propagation of the primary radiation beam. In both cases, radiation from a mercury arc lamp or a tungsten filament lamp or a laser light is used. For conversion of polychromatic light to monochromatic light, absorption or interference filters are employed. A sample cell, made up of glass or plastic, with a rectangular (turbidimeter) or semi-octagonal faces (nephelometer) is generally used. Photovoltaic cells or phototubes are preferred for turbidimetric analysis. But sensitive photomultiplier tubes are essentially used in nephelometers because the scattered light is generally of weak radiation intensity. Photomultiplier intensifies the scattered signal by multiplying the primary photoelectrons through secondary emission. In nephelometers, the detector is generally mounted on a circular disc, so that it can be rotated at different angles during measurement. However, using a fixed detector at right angle to the primary beam is most common. Finally, the reading is obtained in the readout device after the processor deciphers the data passed through the detector.

6.12.4. Application

Nephelometry and turbidimetry may be efficiently used to monitor the air and water pollution. Water clarity is one of most important parameters to study the water quality because water may lose its transparency due to the presence of various suspended particles evolved from various sources such as waste discharge, sediments from erosion, phytoplankton, algal growth, etc. Through turbidity measurement one can easily measure the total suspended solids (TSS) in natural water and municipal wastewater. In air, levels of dust and smoke are monitored using these techniques. Polymers evolved from factories and settled down on the earth are also studied for their molecular weight determination by these techniques. In clinical laboratories, it is a common practice to apply these techniques to measure serum proteins' and antigens' concentration, rheumatoid factors, immune complexes, etc. In inorganic domain, these techniques may be employed for quantitative estimation of sulphate, chloride, fluoride, cyanide, calcium, carbonate and zinc, forming precipitates of $BaSO_4$, AgCl, CaF_2, AgCN, oxalate, $BaCO_3$, ferrocyanide, respectively. Nephelometric estimation of phosphorus in trace quantity, in biological and industrial sciences, is commonly practiced using reagents consisting of nitric acid solution of strychnine and molybdic acid. Besides, the techniques are also used in determining the clarity of beverages, liquid pharmaceuticals, and drinking water very efficiently. Nephelometers are specially used in global warming studies, for measurement of visibility in air, and in the detection of smoke. In biochemical analysis, turbidity may be used for quantitative estimation of amino acids, vitamins, antibiotics, etc. It is also used for continuous monitoring of the bacterial growth in a nutrient medium.

6.13. Infrared spectroscopy

6.13.1. Introduction

Within the electromagnetic spectrum, infrared (IR) region is found in between visible and microwave regions and it possesses three distinct sections: Near IR (energy, 150-50 kJ/mol; wavenumber, 12800-4000 cm^{-1}; wavelength, 0.78-2.5 μm), Mid IR (energy, 50-2.5 kJ/mol; wavenumber, 4000-200 cm^{-1}; wavelength, 2.5-50 μm), and Far IR (energy, 2.5-0.1 kJ/mol; wavenumber, 200-10 cm^{-1}; wavelength, 50-1000 μm). Infrared spectroscopy deals with the analysis of a substance by studying the interaction of a molecule with IR radiation. It is colloquially called IR spectroscopy. It is the branch of spectroscopic science where absorption of electromagnetic radiation by matter involves in the infrared region which specifies the radiation exhibiting lower energy than visible light due to their comparatively lower frequency and longer wavelength. If a matter is exposed to infrared radiation, it absorbs the radiation only when the vibrational frequency of bonds in the molecule is the same as the frequency of the incident radiation. The prerequisite of IR absorption by molecules is that the frequency of the radiation must satisfy $\Delta E = h\nu$ (where ΔE is the energy difference between the vibrational states involved). IR irradiation causes molecular vibrations in covalent bonds. However, all covalent bonds do not show bands in the IR spectrum. The covalent bonds possessing polarity can only exhibit IR bands and that too depending on the magnitude of the dipole moment associated with them (**Table 6.4**). A polar bond interacts with IR radiation, and presence of separate areas of the molecule's partial positivity and negativity facilitates the electric field component of the IR wave to excite the molecule's vibrational energy. As symmetrical non-polar bonds (N≡N, O=O, etc.) do not interact with an electric field, they do not exhibit IR absorption. These are applicable to the Raman spectrum. Thus, polarity of the bond influences the intensity of the absorption. The IR spectroscopy can be performed by measuring reflection, emission, and absorption. Primarily, in organic and inorganic chemistry, the functional groups of molecules are studied by IR spectroscopy, measuring absorbance or transmittance. IR emission is usually used to study heated samples. On heating samples, many molecules reach excited vibrational states and return to their ground state by emitting radiations that are characteristics of the vibrational levels of the molecules and thus, samples can be characterized by their IR emission spectrum.

Table 6.4: Magnitude of polarity in covalent bonds and IR band intensity

Sl.No.	Magnitude of polarity	Band intensity	Example
1.	Weakly polar and symmetric bonds	Weak band or non observable band	HC≡C-CH_3 (Weak), ↑ H_3C-C≡C-CH_3 (Not observable), ↑
2.	Moderately polar and asymmetric bonds	Medium bands	-C≡N ↑
3.	Strongly polar	Strong band	C=O ↑

For convenience, an IR spectrum is split into two regions: functional group region (4000 cm^{-1} to 1300 cm^{-1}) and fingerprint region (1300 cm^{-1} to 400 cm^{-1}). The bands, present in the functional group region, identify the functional group of an unknown compound. In the fingerprint region several bands are observed leading to complexity and difficulty in interpreting the data reliably. This region is meant for comparing the spectra of one compound to another.

6.13.2. Modes of vibration

There are two main vibration modes: stretching and bending (also referred as *deformations*) vibrations. Compared to bending, stretching vibration occurs at higher frequencies. Stretching is responsible for a change in interatomic distance because it leads to a change in bond length. On the other hand, bending causes a change in the bond angle. Again, within stretching mode there are two types of stretching modes: symmetrical stretching and asymmetrical stretching; when two same types of atom are present in a group of three or more atoms. Within these two stretches, asymmetric stretches more likely leads to a change in dipole moment and thus, absorption of IR radiation. Likewise, there are four types of bending vibrations: scissoring, rocking, wagging, and twisting. It will be clear if we analyze these modes considering a CH_2 group as an example (**Figure 6.14**). In symmetrical stretching, both H atoms move away from the C atom, while in asymmetrical stretching, one H atom moves away from the C atom and the other moves toward the C atom. In the case of bending, scissoring and rocking are in-plain bending modes, whereas wagging and twisting (*oop* bending) are out-of-plane bending modes. In case of scissoring and rocking types, the H atoms remain in the same plane as the C atom does (plane of the page). Wagging and twisting involve in bending above (marked with + sign) and below (marked with - sign) the plane of C atom. In-plane and out-of-plane (oop) bending show only one band in spite of the fact that they both are IR active because both occur at the same frequency (666 cm^{-1}). They are referred to as *degenerate*.

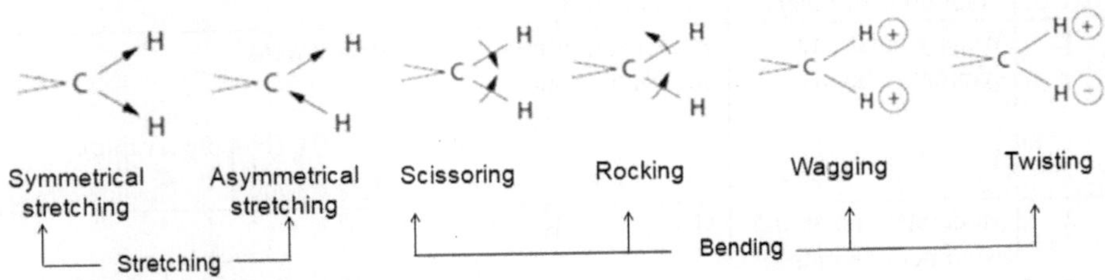

Fig. 6.14: Different types of vibrational modes

Sometimes vibrational modes are also referred to as vibrational degrees of freedom. There is a general rule that a linear molecule with N number of atoms exhibits vibrational modes of $(3N - 5)$, whereas for a non-linear molecule it is $(3N - 6)$. H_2O (a non-linear molecule) has 3 (3 x 3 – 6 = 3) modes of vibrations. CO_2 (a linear molecule) exhibits 4 (3 x 3 – 5 = 4) modes of vibrations.

Occasionally, fewer peaks than calculated are found that may be attributed to IR-inactive vibrations, degenerate vibrations, and very weak vibrations. However, apart from fundamental absorption bands (most likely and common transition due to excitation from the ground state v_0 to the first excited state v_1; the frequency of radiation is same as the initial frequency of vibration of the bond), additional peaks are also usually found in the spectrum, which might be due to overtone (excitation from the ground state to higher energy states v_2, v_3, etc.; occurring at approximately integral multiples of the frequency of the fundamental absorption; however, the event is less favored and consequently, the intensities of these absorptions are much lower than that of the fundamentals), combination (due to interaction of the vibrational modes generating a new frequency $v_3 = v_1 + v_2$), and difference bands (if the coupling of two frequencies is made in such a way that v_3 becomes $v_1 - v_2$).

6.13.3. Principle

The basic principle of IR spectroscopy is based on the fact that when the frequency of the vibration of bonds in the molecule matches the specific frequencies of IR radiation; the molecule absorbs it, leading to a faster rotation or a more pronounced vibration. According to this behavior of the molecule, upon its exposure to IR radiation, some frequencies are absorbed while some others are transmitted without any absorption. Under this situation, an IR spectrum may be obtained by plotting the percent absorbance or transmittance (intensity of the absorption or transmission, Y-axis) versus frequencies (X-axis) in units of inverse centimeters (wavenumbers). An IR spectrum may be presented both in transmission and absorption mode. However, transmission mode is a more commonly used representation than the absorption mode. Here, it is to be noted that molecules absorb specific frequencies of light that are characteristic of the corresponding structure of the molecules. Different bonds, such as C-C, C-O, O-H, N-H, C=O, C≡C, etc. possess different specific vibrational frequencies. For this reason, each band exhibited by the spectrum denotes a specific characteristic functional groups and bonds of the substance. Thus, the IR spectrum of a substance can be utilized as a fingerprint for its identification by detecting the presence of the bonds within it. Since the bonds within a molecule possess characteristic stretching and bending frequencies, and the stretching needs more energy than the bending (as it applies in the spring also), one finds the stretching absorption of a bond at higher frequency and the bending absorption of the same bond at lower frequency. For example, the absorption band near 3000 cm^{-1} specifies $C\text{-}H_{(stretch)}$ [or we call $C\text{-}H_{str}$] absorption and the absorption band nearly at 1400 cm^{-1} represents $C\text{-}H_{(bend)}$ absorption.

The energy level for any molecular vibration can be expressed by the following equation (Equation 6.26):

$$E = (\upsilon + 1/2)(h/2\pi)\sqrt{k / \frac{m_1 m_2}{m_1 + m_2}} \qquad \text{Equation 6.26}$$

Where,

v = vibrational quantum number,

h = Planck's constant,

k = force constant of the bond, and

m_1 & m_2 = masses of the individual atoms.

Thus, from Equation 6.26, it is clear that vibrational energy or frequency is directly proportional to the bond strength but inversely proportional to the mass of the molecular system. This explains the reasons for vibrations at different frequencies by different chemical functional groups.

In the IR spectrum, the bands are classified as strong (*s*, covering nearly whole portion of the *Y*-axis), medium (*m*, covering nearly half portion of the *Y*-axis), or weak (*w*, covering nearly one third portion or less of the *Y*-axis), based on their relative intensities.

6.13.4. Sample preparation

In IR spectroscopy, samples may be examined in any of its states: solid, liquid, or gaseous state. There are many ways to prepare solid samples. Salts, as for instance, Sodium chloride (NaCl) / Potassium bromide (KBr) / Calcium fluoride (CaF_2) / Cesium iodide (CsI) / Cesium bromide (CsBr) / Zinc sulfide (ZnS) / Zinc selenide (ZnSe) / Silver chloride (AgCl) / Silver bromide (AgBr) / Cadmium telluride (CdTe) are preferred as carrier materials since they do not exhibit any IR absorption (100% transparency). In 'mull technique' samples are crushed into fine powder, mixed with mulling agent (Nujol or mineral oil) homogeneously in a pestle-mortar to make a thick paste. A thin film of this paste is then applied onto the salt plates. In 'solid run in solution', samples are dissolved in a non-aqueous solvent. For this purpose, a solvent (chloroform, carbon tetrachloride, cyclohexane, etc.) should be chosen carefully, so that it must not interact chemically with the sample (especially showing a tendency to form H-bond) and obviously it must not absorb any IR radiation. Solute solution is then dropped on an alkali metal disc and is left for its dryness with complete evaporation of the solvent, allowing a left-over residue of the sample solute on the surface of the alkali metal disc for further study. However, this may also be possible to keep the solute solution directly in a liquid sample cell and then mounted on the IR beam. But this method shows limitations for its use because no single solvent which exhibits 100% transparency towards the IR beam is available. Samples of resins or other amorphous solid compounds are typically prepared to form solid films on the salt surface using solvent followed by its evaporation. Hot pressing of polymers on the salt plate is also a very common practice. Another widely used technique for sample preparation is 'pressed pellet technique' where a solid solute sample is first made into a fine powder which is then mixed with KBr (1:100) homogeneously. KBr must be free from moisture, otherwise a broad absorption is found at 3500 cm^{-1}. A pellet of 1 cm diameter and of 2 mm thickness is formed from the mixture, using hydraulic pressure. Pressure must be suitably adjusted; otherwise, very high pressure may result in polymorphic changes in crystallinity. Pressed pellet technique is considered to be superior to others.

If the sample is in liquid state, it can be kept in a rectangular cell of KCl or KBr. Mostly, sample cell thickness is 0.1-0.5 mm to allow transmittance of 15-20%. The normal practice is to dispense a drop of liquid sample on the polished surface of a salt plate, followed by covering the liquid content by another salt plate to allow the uniform distribution of the liquid substance in the form of a thin film. Both plates are then clamped together. Alternatively, the sample is dissolved in CCl_4, CS_2 or $CHCl_3$ and is kept in a solution cell. Pure solvent is kept in the reference beam to negate the solvent absorption. Consequently, the spectrum obtained in the recorder is only due to the solute molecules.

Cells possessing a long photon path length (10 cm) are usually used for gaseous samples due to their low concentration and small sized particles.

6.13.5. Instrumentation

In IR spectroscopy, particularly for mid-IR, dispersive instruments and Fourier transform (FT) instruments are commonly used. In a dispersive instrument, a monochromator disperses the frequencies of the incident radiation and the sample's sequential-absorption of each frequency can be measured by a suitable detector. In a FT-IR device, an interferometer is preferred over a monochromator. Instead of a dispersed radiation as it is done by a monochromator, in a FT instrument, a simultaneous arrival of all wavelengths at the detector occurs followed by a mathematical treatment to convert the signal obtained into a typical IR spectrum.

6.13.5.1 Dispersive type

A schematic diagram of an IR spectrometer is presented in **Figure 6.15**. Initially, IR radiation, coming from a source, is split into two beams, a sample beam (obtained after passing through the sample) and a reference beam (obtained after passing through the blank). Both beams separately and alternately go to the monochromator (prism/grating) through a chopper, a rotating segmented mirror. The chopper rotates @ approximately 10 times per second so as to reflect two beams alternately to the monochromator. Alternate frequencies are then received by the detector. The detector then detects the transmitted frequencies, where IR energy is converted into electrical energy. Two types of detectors: thermal detectors (*thermocouples*, *bolometers*, *thermistors*, and *pyroelectric devices*) and photon-sensitive semiconducting detectors are used in IR spectroscopy. The signal then reaches to the amplifier that amplifies only the signal of alternating current. Thus, the system measures the difference in intensities between the reference and sample beam at each wavelength. Alternatively, we may say that the absorption is measured as a function of wavelength with a spectrometer and the obtained IR spectrum can be used as a 'molecular fingerprint' to identify organic and inorganic samples. Finally, the data are processed in a readout device, and the values of the absorbed and transmitted frequencies are reported.

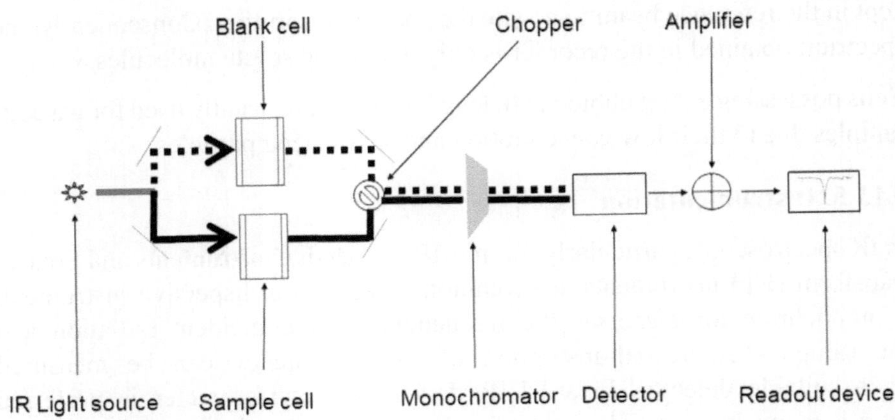

Fig. 6.15: Schematic diagram of an IR spectrometer

6.13.5.2. FT instruments

Now-a-days most of the modern IR-spectrometers are of the FT type, containing an interferometer in it. The interferometer splits the IR- beam into two equal halves using a beam-splitter followed by recombining through reflecting back the split beams with mirrors. The reference beam goes to the detector via a mirror, while the other light beam traverses through the sample, hits a second mirror, back through the beam-splitter and goes to the detector. In dispersive technique, the sample is subjected to different single wavelength irradiation. But in FT-IR the spectral data of all wavelengths are collected in one pass by generating an IR light over a wide range of IR wavelengths (near-IR to far-IR) from a continuous source and then passing through an interferometer. As a result, an initially obtained interferogram is later Fourier-transformed and translated into an IR spectrum to define the light intensity as a function of wave number because the interferogram defines the light intensity as a function of the mirror position inside the interferometer. The Fourier transform, a mathematical function that takes apart waves and returns the frequency of the wave based on time, converts the interferogram into the infrared spectrum graph that is used to identify or quantify the material. Compared to a dispersive system, the acquisition of FT-IR spectra is a faster process and the spectra exhibit higher signal-to-noise ratio.

6.13.6. Application

1. Qualitative identification of organic and inorganic compounds is a primary use of IR spectroscopy. IR spectrum is utilized efficiently to identify the various functional groups present in a molecule and also to obtain an idea on the

working range of frequencies at which the IR radiation-substance interaction occurs. Multifunctional compounds exhibit separate absorption bands for their various individual functional groups, if they do not interact with each other (**Table 6.5**).

Table 6.5: Few prominent bands and corresponding functional groups in organic molecules from IR spectrum

Functional groups	Nature of bond vibration	Wavenumber (cm^{-1})	Remarks
Fingerprint region	-	600-1400	Several bands are observed. It is difficult to interpret.
Alkanes	C-H_{str}	Around 3000	No functional groups, displaying C-C & C-H bond vibration, most prominent C-H bond
Alkenes	C=C_{str} =C-H_{str}	Around 1600-1700 Around 3080	Sharp & medium band. Might be obscured by a broader band at 3000 cm^{-1}.
Alkynes	C≡C_{str}	Around 2100	Sharp but weak band.
Terminal alkynes (triple bond at the end of the carbon chain)	≡C-H_{str} [associated with *sp* carbon (carbon forming part of the triple bond)]	Around 3300	Sharp but weak band. This is not displayed by the internal alkynes (triple bond in the middle of the chain) because they do not contain C-H bond to sp carbon.
Alcohol	O-H_{str}	Around 3000-3700	Broad and strong band.
Aldehyde	C=O_{str} C-H_{str} (associated with the sp^2 carbon of C=O bond)	Around 1710-1720 Around 2700-2800	Sharp and strong band. A pair of medium strength band. One can easily differentiate the spectrum of a ketone from an aldehyde because these bands are not displayed by a ketone due to the unavailability of such C-H bond.
Ketone	C=O_{str}	Around 1710-1720	Sharp and strong band.
Amines	N-H_{str}	Around 3200-3600	Medium to weak but broad band. However, it is not so broad as it is due to O-H_{str} found in alcohol. Primary amines possessing two N-H bonds show two spikes, secondary amines possessing one N-H bond show one spike and tertiary amines possessing no N-H bond show no spike.

Functional groups	Nature of bond vibration	Wavenumber (cm^{-1})	Remarks
Amides	N-H_{str}	Around 3100-3500	Strong and moderately broad band.
	C=O_{str}	Around 1710	Strong and pointed bands. Two, one and no spikes are found in primary, secondary and tertiary amides, respectively.
Carboxylic acids	O-H_{str}	Around 2800-3500	Strong and broad band.
	C=O_{str}	Around1710	Sharp and strong band.
Nitrile	C≡N_{str}	Around 2250	Sharp and strong band.

2. Quantitative estimation of components in a mixture is possible by infrared spectra, however, it is cumbersome and it is not possible to estimate with accuracy of the same magnitude as it is done with UV spectra. The absorbance of a sample is a logarithmic ratio of incident light intensity and transmitted light intensity [A = log (I_0/I_t)] and transmittance is logarithmically and reciprocally related with absorbance [$A = \log(1/T)$]. Usually, IR spectrometers record the spectra where the intensity of the bands exhibits linear function of %T. Thus, band intensity is in logarithmic function of absorbance. This creates complexity to define the relationship between the band intensity and the concentration because concentration is linearly related with A but inversely and logarithmically related with %T. Besides, small error in measuring %T may lead to large error in measuring A. This becomes mostly a fact when %T tends to be 100%. Empirically, measuring A with acceptable accuracy is only possible when band intensity is found within 30-60 %T.

3. When two unknown samples show the same IR spectrum, the samples may be primarily confirmed that they contain the same functional groups. Thus, IR provides a molecular fingerprint to compare samples. However, it cannot produce comprehensive and detailed information on molecular structure and formula. For that, a holistic approach is needed taking care of other techniques.

6.14. Atomic absorption spectroscopy (AAS)

6.14.1. Introduction

Atomic absorption spectroscopy (AAS) is an analytical technique that facilitates quantitative elemental analysis at percentage to parts per million (ppm) levels (Flame-atomic absorption spectrometer) in a sample, even sometimes in ppb levels as evidenced in graphite furnace-atomic absorption spectrometer. Atomic absorption spectrometry was first discovered in 1802 by William Hyde

Wollaston. Later, several scientists [Joseph von Fraunhofer (1817), Gustav Kirchhoff and Robert Bunsen (1860)] contributed a lot in the same domain. Finally, the atomic absorption technique was developed for chemical analysis during the 1950s by Sir Alan Walsh and his team at the Commonwealth Scientific and Industrial Research Organization (CSIRO), Australia and was started for its commercial use from 1960. Since then, the technique is adopted for rapid elemental analysis in most of the laboratories because it is an easier and less time consuming process compared to other analytical methods. However, measurements are made separately for each element of interest. Comparatively less expensive AAS is treated as an alternative to the more expensive Inductively Coupled Plasma Optical Emission Spectrometer (ICP-OES) or Inductively Coupled Plasma Mass Spectrometry (ICP-MS). However, ICP-OES and ICP-MS involve multi-element analysis with a lower detection limit (< µg/l). The atomic absorption spectrometer determines the reduction of the intensity of electromagnetic radiation (ultraviolet & visible light) subsequent to its passage through gaseous atoms. AAS is a widely used technique for metals' and metalloids' detection in environmental samples and the technique enables it to measure more than sixty elements from any matrix. AAS is comparatively more sensitive than atomic emission spectrometers. Its detection limit ranges between 100-0.1 ppb. However, pretreatments of samples for concentration-enrichment by solvent extraction, liquid-liquid extraction or ion-exchanging processes increase the detection limit more.

6.14.2. Principle

AAS is based on the principle of ultraviolet or visible light absorption at a specific and unique wavelength by the free atoms, present in the gas phase, which are generated in an atomizer. The atomizer converts the metallic constituents to be reduced to their gaseous atoms. However, a portion of these atoms may also be ionized depending on the temperature. When a collimated monochromatic light beam passes through the vaporized sample, atoms in the gas phase absorbs light at a specific wavelength following the Beer-Lambert law which states that amount of light absorption is a function of the length of the path traversed and of the concentration of the element of interest. Analogues to the Beer-Lambert law, the more light intensity is absorbed, the higher is the concentration of the analyte in the sample. Absorption of light energy facilitates the electrons to be excited from their ground state to higher energy levels. The absorbed light intensity is proportional to the concentration of the element. Thus, the concentration of a specific element in a sample can be estimated reliably, using a calibration curve where intensity of absorbed light is plotted against known concentration of the element to obtain their proportionality relationship following the Beer-Lambert law. It is a common practice to introduce the sample solution in a flame where it is ionized in the gas phase. Light at a wavelength specific for the concerned atoms to be determined is passed through the flame which makes the atoms to freely dissociated ground state atoms. Subsequently, through absorption of light energy, the electrons in the atoms are elevated to an excited state, and for this event, the energy of light of specific wavelength must correspond to the electronic transition. Based on the Beer-Lambert law, the amount of light absorption and the number of atoms excited from their ground state in the flame are proportional to each other.

6.14.3. Instrumentation

An atomic absorption spectrometer contains a hollow cathode lamp as a light source, a nebulizer to form an aerosol from sample solution by a jet of compressed gas, an inlet each for fuel gas and air, a spray chamber with mixing paddle and other accessories, an atomizer [burner for a Flame-AAS, a graphite furnace (electrothermal atomizer) for a graphite furnace-AAS, a hydride generator for a hydride generator-AAS or a cold vapor system for a cold-vapor-AAS] to convert sample aerosols into an atomic vapor, a monochromator to isolate a narrow wavelength band, a detector and a computer-driven data readout system. Usually, hollow cathode lamps used in AAS may be of element specific (single element) lamps. This type of lamp contains a cathode made of the element being analyzed, and a tungsten anode electrode, that are sealed in a hollow tube filled with a noble gas (neon or argon). However, multi-element lamps are also available now but they are less sensitive compared to element specific lamps. The required wavelengths for few elements of interest are provided in **Table 6.6**.

Table 6.6: Few elements with wavelength settings for their measurement in AAS

Elements	Wavelengths (nm)
As (Arsenic)	193.7
Au (Gold)	242.8
Ca(Calcium)	422.7
Cd (Cadmium)	228.8
Co (Cobalt)	240.7
Cr (Chromium)	357.9
Cu (Copper)	324.8
Fe (Iron)	248.3
Hg (Mercury)	253.7
K (Potassium)	769.9
Li (Lithium)	670.8
Mg (Magnesium)	285.2

Elements	Wavelengths (nm)
Mn (Manganese)	279.5
Mo (Molybdenum)	313.3
Na (Sodium)	589.6
Ni (Nickel)	232.0
Pb (Lead)	217.0
Sn (tin)	286.3
Zn (Zinc)	213.9
V (Vanadium)	318.4

The monochromator contains an entrance slit, collimating lens, a grating, a focusing lens and an exit slit through which the selected wavelength band traverses and reaches to the detector. A monochromator selects and isolates a single atomic resonance line from the light spectrum emitted from the interaction of the atoms of interest with the incident light beam, and directs it to the detector, excluding other wavelengths. A photomultiplier tube is generally used as a detector in most of the AASs, which converts the light energy to electrical energy. A photomultiplier tube contains a series of dynodes to amplify the signal. The functioning of the photomultiplier tube has been described in section 6.10.1.4. The output from the detector is then suitably amplified and displayed on a readout device like a meter, digital display or a printer. The system requires supporting software for instrument control, acquiring the data and their interpretation. The overall working outline of an AAS is depicted in **Figure 6.16**.

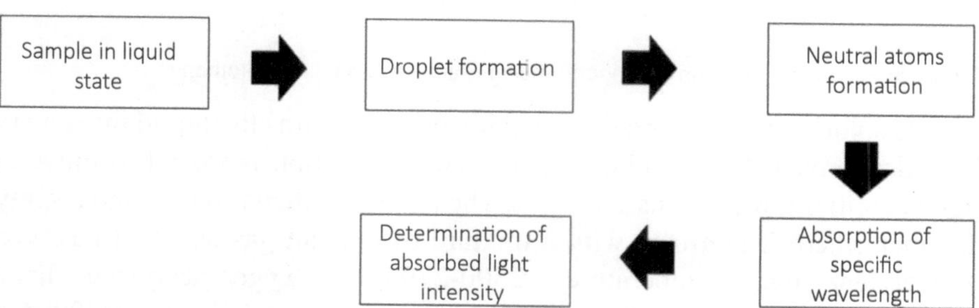

Fig. 6.16: Working outline of AAS

There are two types of AA spectrophotometer: (1) single- (**Figure 6.17**) and double-beam (**Figure 6.18**) spectrophotometer. Single-beam Flame-AA spectrophotometer uses a single optical path through the flame, while double-beam Flame-AA spectrophotometer has two optical paths: one passes through the flame and other is the reference beam. The reference beam directly goes to the chopper, without traversing the flame. The double-beam Flame-AA spectrophotometer involves the determination of changes in light intensity due to the dual optical path. In double-beam Flame-AA spectrophotometer, the sample and reference blank are measured at the same time but in single-beam Flame-AA spectrophotometer they are analyzed separately.

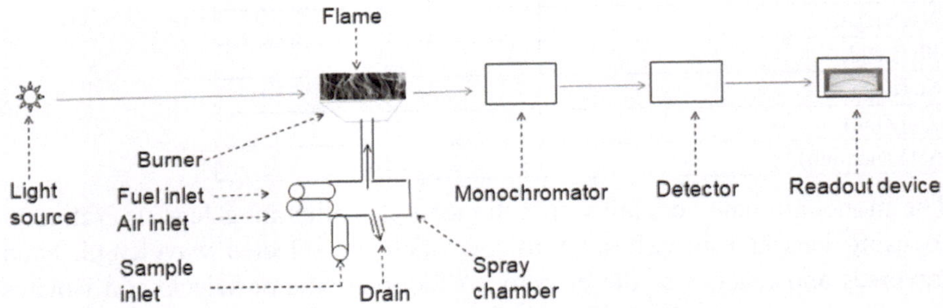

Fig. 6.17: Schematic diagram of a single-beam Flame-AA spectrophotometer

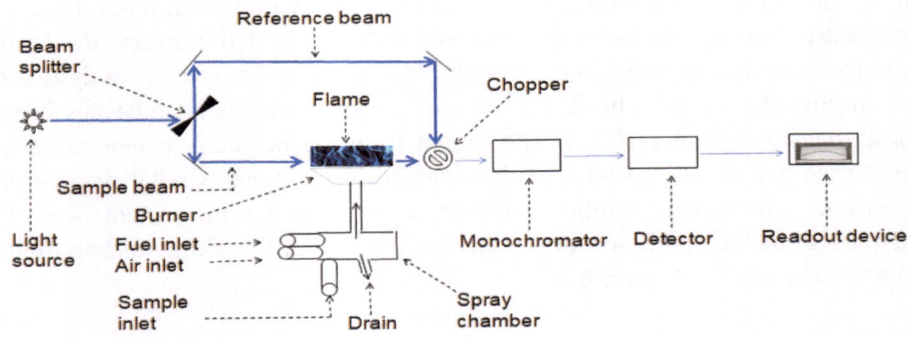

Fig. 6.18: Schematic diagram of a double-beam Flame-AA spectrophotometer

Sample solution is transported to a nebulizer which turns the liquid into a very fine mist/aerosol. In the nebulizer, the sample solution is forced to move at high velocities through a narrow tube. The aerosol is then brought into a spray chamber where it is mixed with a mixture of oxidant gas and fuel gas, viz. air-acetylene (flame temperature = 2300-2700 °C), oxygen-acetylene (flame temperature = 3300-3400 °C), air-hydrogen (flame temperature = 2400 °C), oxygen-hydrogen (2650 °C), air-natural gas (1700 °C), air-propane (2300 °C) or nitrous oxide-acetylene (flame temperature = 2700-3000 °C). Air-acetylene flame is commonly used for the majority of elements. However, nitrous oxide-acetylene, generating a flame with temperature of about 2700-3000 °C, may be used for oxide forming elements because nitrous oxide-acetylene flame maintains a more reducing environment. The spray chamber is designed and operated in such a manner that very fine aerosol droplets (< 10 μm) enter the flame. Next to the spray chamber, the most important component in an AAS is the atomizer. It may be a burner, graphite furnace or a hydride generator. The function of an atomizer is to transform particles into individual molecules which in turn, give rise to atoms. In a Flame-AAS, on the top of the burner head, a flame of about 5-10 cm laterally long is maintained. The aerosolized analyte (< 10 μm) and oxidant-fuel gas mixture enter the burner where few important events occur, viz. desolvation (the analyte is dried and converted to nanoparticles by evaporating the solvent), vaporization (bringing the nanoparticles to the gaseous phase), and finally, atomization (obtaining freely dissociated atoms without causing appreciable ionization). However, conversion of free atoms to gaseous ions also occur to some extent. The gas flow for yielding a flame is regulated to give rise to the maximum concentration of free atoms. It is also necessary to adjust the burner height to pass the light from the hollow cathode lamp through the zone of highest atom cloud density in the flame. These measures obviously increase the instrumental sensitivity.

Flame-AAS is known for its ease of handling, continuous operation, low operational costs and working speed. These advantageous features make it possible to use as an interface with chromatographic systems for speciation. However, flame-AAS has some limitations also. It requires a good quantity of sample solution (0.5-1.0 ml) to be injected for reliable data because sample introduction efficiency is poor. Hardly 5-10% nebulised samples reach the flame. The dispersion of atoms in the flame atomizer across a relatively large volume leads to lower sensitivity of flame-AAS. Apart from these phenomena, erratic atomization may be obtained for viscous samples in a flame atomizer, thus dilution of the sample is much needed to obtain steady data. Albeit the popularity of the flame burner as an atomization source, graphite furnace and hydride generator are also extensively used for atomization.

The atomic absorption spectrometer equipped with a *graphite furnace* as an atomizer is generally referred to as graphite furnace-AAS. In graphite furnace-AAS, samples are deposited in a graphite tube which is then kept in the light path and electrically heated in a furnace filled with inert gas, using high current power supply to atomize the analyte via its vaporization. Electrothermal heating provides a high temperature for complete atomization (100%). In a graphite furnace, the atoms are concentrated in a relatively small volume compared to flame atomizer and thus, it is more sensitive to detect elements, even sometimes at ppb levels. Additionally, handling of typically small volume (about 20 μl of sample solution per measurement) is an added advantage of using graphite furnace-AAS. Graphite furnace-AAS enables to analyze the solid samples directly with a reduced pre-treatment, enabling to reduce total cost of analysis, analyte loss and unintentional contamination during sample handling. It offers an improved selectivity compared to flame-AAS with the possibility of estimating non-metals (e.g., Cl, F, P or S). However, the use of graphite furnace-ASS is to some extent restricted due to sample matrix interferences.

In case of *hydride generating atomizer*, an acidified solution of hydride-forming elements (As, Se, Sb, Bi, Ge, Sn, Te and Pb) is mixed with sodium borohydride solution to yield corresponding hydride gas which is electrothermally or flame heated in a fused silica glass tube. Subsequently, neutral metal atoms are generated from hydrides on decomposition at high temperature. This atomizer also exhibits a low detection limit (<µg/L) due to higher sampling efficiency. The added advantage of using this atomizer is that matrix-interferences can be avoided due to the formation of analyte specific hydride gas.

The AAS may be equipped with another atomizer, *Cold vapor system,* which is typically applicable to analyze mercury because mercury exhibits significant vapor pressure at room temperature. Since mercury can exist in the free ground state at room temperature, AAS may be employed for its quantification using no heat. The sensitivity of the cold vapor technique is many-folds more than the flame-AAS due to its 100% sampling efficiency. The detection limit for mercury in cold vapor technique is about 0.02 µg/L. Mercury containing samples which are organic in nature are first treated with an oxidizing mixture, H_2SO_4 acid and $KMnO_4$, to remove organic matter and to form mercuric sulphate. Stannous chloride or sodium borohydride in a closed reaction system reduces mercuric sulphate generating mercury vapor. Mercury vapour then enters a quartz cell (15 cm length and 0.75 cm diameter) through which collimated monochromatic light passes. In exceptional cases but not always, the cell is heated to avoid water condensation.

Compared to the conventional flame vaporization, flameless vaporization process increases the sensitivity due to the longer residence time within the radiation beam. Thus, more analytes are supposed to be transformed into free atoms. Contrastingly, strong thermal current reduces the residence time in flame vaporization. Instrument automation features, such as using auto-sampler, automatic online dilution of over-range samples, etc. remove the human error and make the technique more comfortable and user-friendly.

6.14.4. Background correction

The background reading which may be due to the light scattering effect by the fine particles present in the solution as well as due to the light absorption by the undissociated molecules is a major concern in this technique because erroneous results are obtained in a situation if there is no effort for background correction. One of the most common ways to avoid background reading in a single beam AAS is using a blank solution that contains no element of interest. The subtraction of blank signal from the sample signal provides a background-corrected signal. Deuterium background correction technique is common in a double-beam AAS where the signals from deuterium and hollow cathode lamps are alternated using a chopper mirror, and a background-corrected signal is obtained by subtracting the signals from deuterium lamp from that of the hollow cathode lamp. Instead of specific spectral lines as with a hollow cathode lamp, the deuterium lamp produces broadband radiation. Light absorption by analyte atoms and the background is encountered with the specific light from the hollow cathode lamp, whereas the background absorption is estimated with the light from the deuterium lamp. However, applying a magnetic field around the hollow cathode lamp to alternately polarize light is a widely used technique (Zeeman technique), particularly in graphite furnace-AAS, by which spectral and non-spectral interference can be easily minimized. The analytical measurement is made by putting the magnetic field off, while activating the magnetic field facilitates the background measurement. The subtraction of the two signals provides the correct results.

6.14.5. Sample preparation

Sample preparation is considered as the most important and vital step in atomic absorption spectroscopic analysis. The type and nature of the samples, which may be of organic or inorganic solids and liquids, aqueous solution, and gases, direct them to be pre-treated differently before the final analysis. Weighed samples must be dissolved in appropriate solvent or digested using various techniques followed by dilution of the solution. Aqueous solutions (e.g., biological fluid, sewage, water, river water, etc.) and organic liquids (petroleum products etc.) need little preparation and may be directly sucked in the mixing chamber of the AA spectrophotometer. Elements in urine and blood are often measured simply after a dilution. In some cases, samples are dissolved in organic solvents and then analyzed; in that case, care must be taken about proper ventilation and burner control. Inorganic solids (e.g., fertilizers, minerals, rocks, ores, etc.) are usually made into aqueous solution using hot water acidic medium. They may also be acid digested (perchloric acid, hydrochloric acid, nitric acid, sulphuric acid, etc.) for long duration to obtain an aqueous solution. Organic solids, such as plant and animal tissues, food materials, etc. are commonly wet digested. However, their dry-ashing in a muffle furnace followed by dissolution in acid is also a common practice. Microwave dissolution of samples in acid at high temperature and pressure is also recommended. If the sample contains analytes at low concentration or contains objects which may interfere with the light absorption process and may lead to inaccurate results by increasing or decreasing signal, the samples are pre-treated by extraction and concentration process, such as solvent extraction, liquid-liquid extraction, ion-exchange method, etc. For viscous samples, dilution of the sample with suitable solvent that does not exhibit any light absorption at the specific wavelength of interest, is necessary to minimize and adjust the sample viscosity.

6.14.6. Interferences

In atomic absorption spectroscopy, interferences refer to the inaccurate signal due to its big or low size obtained during the analysis. This is caused by the presence of some species in the sample other than the analyte. They may be of solvent droplets, compounds of matrix or other atoms or molecules apart from the elements of interest. There may be of spectral and chemical interferences. Spectral interferences may occur when the absorption wavelength of the matrix component coincides or become very close to the analytical line. This can be avoided either by selecting another analytical wavelength or by cleaning up the sample solution from the matrix components. Sometimes, presence of some components in the sample matrix leads to the formation of new compounds (possessing low volatility) of the analyte through chemical reactions (*chemical interference*). This causes lowering the signal. Attenuation of calcium absorbance is observed with the presence of phosphate ions in the sample solution because of the formation of calcium pyrophosphate that remains undissociated in the flame leading to decreasing light absorption by calcium. It has been evidenced that if phosphate anionic concentrations are increased in the sample, light absorption by calcium goes on decreasing. However, this type of chemical interference effect may be minimized by increasing flame temperature or by adding releasing agents. Calcium is released by addition of lanthanum (La) because lanthanum reacts with phosphate more easily, releasing calcium. For this reason, addition of lanthanum nitrate solution in sample solution containing calcium as the element of interest is a common practice to obtain calcium absorbance independent of the amount of phosphate.

In Flame AAS, flame temperature plays a significant role in the determination process because it is the flame temperature that causes the ground state atom production. It has been evidenced that using cooler temperatures leads to more interference problems due to its lower thermal energy which is not capable of atomizing the several sample analytes completely. For this reason, it is a common practice now to use pre-mix flames generated from air-acetylene and nitrous oxide-acetylene (hotter than air-acetylene) for several refractory-forming elements. However, this is also true that using hot flames leads to further ionization of the ground state atoms in many cases because the additional energy of the hot flames removes an electron from the atom. Therefore, it is very crucial to select elements of interest for their determination by nitrous oxide-acetylene; otherwise, ionization interference, another type of chemical interference, may occur. But this is experimentally proved that air-acetylene flame may also result in ionization interferences in easily ionized elements, the alkali metals and alkaline earths. However, this problem may be overcome by incorporating in the system more amounts of ionization suppressants (potassium, rubidium, and cesium salts) which are easily ionized elements that produce electron abundance to prevent the analyte to be ionized. The ionization suppressants are also referred to as ionization buffers. Number of free atoms in the gas phase decreases during both the formation of stable compounds and the event of ionization; consequently, absorbed light intensity is decreased.

Use of chemical modifiers in modern electrothermal atomic absorption spectrometry (graphite furnace-AAS) is also very common to change the thermochemical behavior of both the analyte and the sample matrix. There are mainly two options to select chemical modifiers in electrothermal atomization:

1. Modifier reacts with an interfering substance in the sample to produce a compound which is comparatively more volatile than the analyte compound. This enables it to remove the interfering matrix component at low temperature stage during atomization, keeping behind the analyte component. Working on this principle, NH_4NO_3 acts as a classical example of a chemical modifier to remove chloride salts from a sample matrix by volatilization ($NaCl + NH_4NO_3 \rightarrow NH_4Cl \uparrow + NaNO_3$). The decomposition temperature for NaCl, $NaNO_3$, NH_4Cl, and NH_4NO_3 are about 1400 °C, 380 °C, 330 °C and 210 °C, respectively. Generated volatile NH_4Cl is removed from the system during the ashing step, while no interference to the analyte signal from $NaNO_3$ is observed during the atomization. Now-a-days mixture of palladium and magnesium nitrate is used as a mixed modifier, but not universally, with superior results for many graphite furnace determinations.
2. Another method is to increase the volatility of the analyte during atomization; so that analyte atomizes before the matrix at a lower atomization temperature and consequently, background signal is reduced during atomization.

However, the chemical type and concentration of the modifier must be optimized. Otherwise, selection of a wrong chemical modifier or its use in a large volume may exhibit negative impact providing high background signal, corrosion of the graphite tube, sensitivity reduction, increasing cost of analysis and so on.

6.14.7. Application

AAS is efficiently applied to monitor and quantify trace elements, heavy metals and metalloids in soil, plant, water, industrial effluent, municipal sewage, ocean, river, and in many other environmental samples. This is utilized to analyze additives and purity in steels and other metal alloys. It is considered as the simple and reliable technique for quantification of lead in paint, to determine metals in various petroleum products, to check various elemental contaminants in foods, soft drinks and beer, to obtain an idea about the presence of toxic impurities and catalyst used for drug manufacturing in pharmaceutical industry. The presence as well as the concentration of different elements in metallurgical and geochemical samples (e.g., earth, rocks and minerals) is routinely tested using AAS. In the mining industry, the amount of metals (e.g., gold, etc.) in the ore is checked by AAS to confirm their presence whether it is worth mining. AAS is also applied in medical science for the determination of metals in urine, whole blood, blood serum, hair, biological tissues, or saliva fluids.

6.15. Flame emission spectroscopy (FES or Flame photometry)

6.15.1. Introduction

You might have seen an additional yellow flame in the oven for a very short spell during cooking if a pinch of salt (NaCl) is thrown on it. This is due to sodium metal atoms at their free ground state, which originate from the sodium chloride molecules after thermal dissociation. The free atoms then attain their excited higher energy levels deriving thermal energy from the flame. Since the higher energy state is unstable, they come back to their ground state again emitting a light energy (it is of yellow color in case of sodium). A specific metallic element exhibits a specific light emission. Such behavior of metal atoms in a flame is employed to develop *Flame emission spectroscopy* (FES). It is also known as *flame photometry* or *flame atomic emission spectrometry* (FAES according to the recommendation of International Union of Pure and Applied Chemistry (IUPAC) Committee on Spectroscopic Nomenclature). Flame photometry is a simple, selective and sensitive analytical technique for qualitative and quantitative determination of alkali (Group I) and alkaline earth metals (Group II) at a very trace quantity (ppm or even sometimes ppb levels). Flame photometry is also successfully employed to estimate certain transition elements, such as copper, iron and manganese. Being a highly empirical method of analysis, it relies on frequent calibration of the method for obtaining good and reproducible results.

The instrument which is engaged to study the metals utilizing the technique is called as a *flame photometer*. The flame photometer was developed by B. Barnes, D. Richardson, J. Berry and R. Hood during the 1980s for estimation of sodium and potassium at low concentrations. The wavelength of the emitted light points out the nature of the element (qualitative) and the intensity of the light emitted describes the amount of the element of interest in a given sample (quantitative). It is important to note that flame photometry is not suitably applied for estimation of metalloids and nonmetals because they do not yield isolated neutral atoms in a flame, but mostly as polyatomic radicals and ions.

6.15.2. Principle

When the solution of the alkali and alkaline earth metals are subjected to a flame in the form of a mist or aerosol, the compounds through sequential evaporation and vaporization dissociate into free ground state atoms which are then excited to higher energy levels by deriving thermal energy from the flame, leading to the excitation of the outer orbital electrons. Due to their instability at higher energy levels, they again attain their ground state after a short spell (typically within 10^{-6} to 10^{-9} s) by releasing energy in the form of emission of electromagnetic radiation in the ultraviolet-visible range. Under controlled and constant conditions, the intensity of light emission is directly proportional to the metal concentration, following the Beer-Lambert law for a certain concentration range. On heating in a flame, each of the alkali and alkaline earth metals emits a characteristic specific wavelength (**Table 6.7**).

Table 6.7: Characteristic wavelengths of light emitted by different alkali and alkaline earth metals in flame photometry

Elements	Emitted wavelengths (nm)	Colour
Sodium	589	Yellow
Potassium	766	Violet
Barium	554	Lime green
*Calcium	622	Orange
Lithium	670	Red
Magnessium	285	-

**Calcium is measured by using the calcium hydroxide band emission at 622 nm. Main atomic emission of calcium occurs at 423 nm.*

6.15.3. Instrumentation

Primarily, a flame photometer (**Figure 6.19**) closely resembles a flame AA spectrophotometer (**Figure 6.17**) except the light source. A flame photometer does not require any external light sources as it is seen with Flame-AA spectrophotometer. But it contains other components, such as a source of flame (usually air-propane or air-natural gas fuel mixture), a pneumatic nebulizer to transform a sample from a liquid state into a mist or aerosol of finely divided droplets, an optical system, a monochromator to separate the wavelength of interest from other irrelevant emissions, a photodetector (a phototube or photomultiplier tube) for converting light energy to electrical energy and a readout device. The system requires supporting software for instrument control, acquiring the data and their interpretation. The overall working outline of a flame photometer is depicted in **Figure 6.20**.

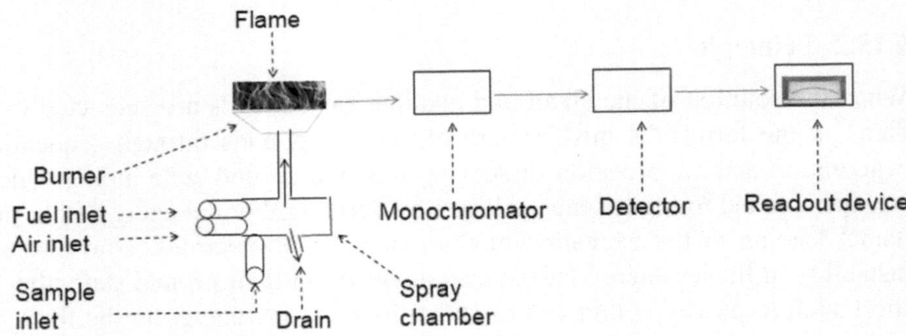

Fig. 6.19: Schematic diagram of a flame photometer

Similar to the AAS system, a sample is introduced in a flame photometer in its liquid state. Thus, aqueous sample solution can be directly aspirated into the nebulizer, while inorganic solid samples are brought into their liquid state using deionized water and for organic solids, organic residues in the sample are eliminated by ashing and the sample remnants, in the form of oxides, are dissolved in acids. In all cases, the final solution must be filtered or centrifuged with a view to eliminating the particulate matter; otherwise, the aspirating tube or needle tends to be blocked.

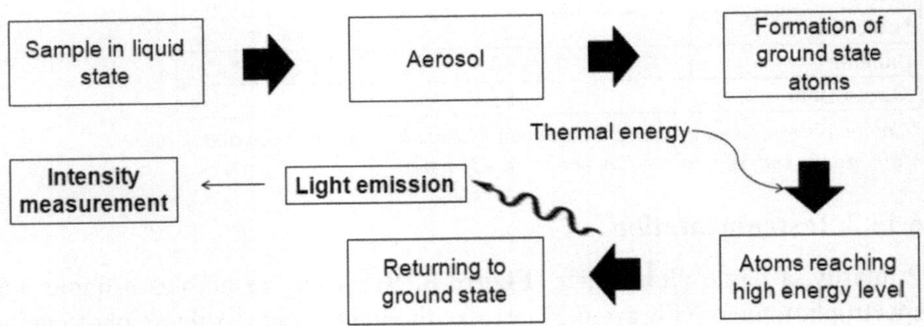

Fig. 6.20: Working outline of Flame photometry

Since Na, K, Li, Ba and Ca are atomized and excited to higher energy levels easily at low temperatures within 1500-2000 °C, flame photometry is employed for their determinations. Air-natural gas, air-butane, and air-propane fuel mixtures are usually used to generate the required temperature in a flame photometer.

6.15.4. Application

In agriculture, presence of potassium in their plant-available form in soil is critical for plant growth despite the fact that soil contains a considerable amount of potassium. Most of the potassium in soil remains in their plant unavailable form. Flame photometric determination provides an easy and a quick assay of soil to check the status of potassium availability so that soils, poor in available potassium, can be supplemented with potassium containing fertilizers. Likewise, in the clinical domain, Na^+ and K^+ ions play a pivotal role in the human and animal body system for various metabolic functions. Therefore, in clinical laboratories, it is a common practice to dilute and aspirate blood serum samples into the flame for understanding the Na/K concentrations quickly. The technique is applied for routine testing of soft drinks, fruit juices, alcoholic beverages and food stuffs to monitor the concentrations of the alkali and alkaline earth metals in the food and drink industry. Flame photometry is used to analyze water for determination of its suitability for drinking, washing, irrigation and industrial purposes. Hardness of water is monitored by estimating calcium concentration in water. Municipal sewage or industrial effluent water is routinely monitored for the metallic presence using flame photometric technique. Even for qualitative purposes also, samples are analyzed by flame photometric technique to know the presence of any specific element in the given sample because a specific element emits a specific characteristic wavelength of an electromagnetic radiation, which is used to identify a specific element.

6.16. Atomic fluorescence spectroscopy (AFS)

The excited atoms, produced by the absorption of electromagnetic radiation and being unstable in nature, immediately return to their ground state by re-emitting the absorbed energy. Subsequently, the emitted fluorescence is measured in AFS, based on the principle that the fluorescence intensity is directly proportional to the concentration of the element of interest. However, it is applicable only at the low concentration of the element. Higher concentration, resulting in higher fluorescence emission, may suffer from self absorption where a portion of the resulting fluorescence is absorbed by the atoms in ground state. Thus, the proportionality is lost. Besides, another drawback of higher concentration is the quenching effect that may be generated due to the inter-atomic collisions.

When the atoms in the excited state collide with each other, their energy is transferred to the vibrational levels of these atoms. This event may occur with more probability when the atomic population in a sample is considerably high.

The main advantage of fluorescence detection over the absorption measurements (as it is the case in AAS) is its greater sensitivity due to the fact that it makes use of distinctive fluorescent spectra of each specific metal and it carries zero or very negligible background radiation noise in fluorescence signals. AFS is commonly used for elemental analysis of antimony, arsenic, bismuth, cadmium, germanium, lead, mercury, selenium, tellurium, tin, zinc, etc. at very low concentrations (ppb levels) in various environmental samples of diverse fields, such as clinical and pharmaceutical, petrochemical, agricultural, geological and metallurgical, etc. It is successfully applied in cancer detection, forensic sciences and toxicological studies.

6.17. Model Questions

i. State the relationship between the light absorbance and the sample concentration.

ii. Calculate concentration of 1, 10 – phenanthroline in a solution from an absorbance value of 0.007 under a light path length of 3 cm. Assume the molar absorptivity of 1, 10 – phenanthroline is 12000 litre mole^{-1} cm^{-1}.

iii. Draw a schematic diagram of a single-beam spectrophotometer and describe each component in detail.

iv. Explain why does ethylene show absorption maximum at lower wavelength (171 nm) and hexa-1,3,5-triene at higher wavelength (258 nm)?

v. What are the blue shift and the red shift in spectra?

vi. Illustrate the operating principle of the photomultiplier tube and photodiode. What are they used for?

vii. Explain the basic principle of fluorimetry. How do you mathematically derive that fluorescence intensity is directly proportional with analyte concentration?

viii. In IR spectroscopy, what are the fundamental, overtone, combination and difference vibrational bands?

ix. How can primary, secondary, and tertiary amines be distinguished by their IR absorption spectra?

x. What are the prerequisites for IR absorption by a molecule?

xi. Discuss briefly the principle underlying the quantitative as well as qualitative analysis by flame photometry.

xii. Enlist the merits and demerits of flame photometry.

xiii. On which principle atomic fluorescence spectrometry does work? Why is AFS considered as a superior technique over AAS? Why does it not hold good at higher analyte concentration?

xiv. Which is the correct sequence during atomization in AAS?

(a) Desolvation → Nebulization → Dissociation → Volatilization → Ionization ion

(b) Nebulization → Desolvation → Volatilization → Dissociation → Ionization ion

(c) Desolvation → Nebulization → Volatilization → Dissociation → Ionization ion

(d) Nebulization → Volatilization → Desolvation → Dissociation → Ionization

xv. Write down the principle of AAS. Describe various atomizers used in AAS. What are the merits and demerits of using flame atomizer?

xv. Why is cold vapour-AAS specifically used for mercury determination? How is background noise minimized in AAS?

xvi. What are interferences in AAS? What are the different ways to minimize the interference effects in AAS?

xvii. How do AAS, AFS and FES differ from each other?

xviii. How do FT-IR and dispersive type IR spectrometry differ from each other? Which one is mostly used and why?

xix. What are the differences between Nephelometry and turbidimetry? Write down their applications briefly.

xx. What is electromagnetic Radiation? How does electromagnetic radiation interact with matter?

6.18. Suggested readings

(i) Hofmann, A. (2010) Spectroscopic techniques: I Spectrophotometric techniques II Structure and interactions. In: Wilson, K.; Walker, J. (Eds.) Principles and Techniques of Biochemistry and Molecular Biology. Seventh edition, Cambridge University Press, Cambridge, UK, pp. 477-551.

(ii) Beer (1852) Bestimmung der Absorption des rothen Lichts in farbigen Flüssigkeiten (Determination of the absorption of red light in colored liquids), Annalen der Physik und Chemie, vol. 86, pp. 78–88.

(iii) https://chem.libretexts.org/Bookshelves/Physical_and_Theoretical_Chemistry_Textbook_Maps/Supplemental_Modules_(Physical_and_Theoretical_Chemistry)/Spectroscopy/Electronic_Spectroscopy/Radiative_Decay/Fluorescence

(iv) Lambert, J. H. (1760) Photometria sive de mensura et gradibus luminis, colorum et umbrae [Photometry, or, On the measure and gradations of light, colors, and shade] (Augsburg ("Augusta Vindelicorum"), Germany: Eberhardt Klett).

(v) Robinson, J. W.; Frame, E. M. S.; Frame II, G. M. (2014) Undergraduate Instrumental Analysis. Seventh Edition, CRC Press, New York.

(vi) Penner, M. H. (2010) Basic Principles of Spectroscopy. In: S. Suzanne Nielsen (Ed.) Food Analysis. Fourth Edition, Springer, New York, pp. 375-385.

(vii) Penner, M. H. (2010) Ultraviolet, Visible, and Fluorescence Spectroscopy. In: S. Suzanne Nielsen (Ed.) Food Analysis. Fourth Edition, Springer, New York, pp. 387-405.

(viii) Wehling, R. L. (2010) Infrared Spectroscopy. In: S. Suzanne Nielsen (Ed.) Food Analysis. Fourth Edition, Springer, New York, pp. 407-420.

7

Mass Spectrometry

7.1. Introduction

Mass spectrometry (MS) is a powerful and versatile analytical technique through which mass of an atom, or a molecule is assessed. It helps for identification and structure elucidation of various complex compounds. The instrument used is known as **Mass Spectrometer**. It serves the purpose of qualitative identification, confirmation as well as the quantification of an atom or a molecule. MS identification is conclusive because the mass spectrum of each substance is unique. In general, MS instruments are composed of a sample inlet, an ionization source, a molecule accelerator, a detector, an amplifier, and a recorder (**Figure 7.1**). Prerequisite of MS is that analysis essentially requires samples in gaseous state. If samples are of solid or liquid state, it is necessary to convert them to gaseous state. For this reason, the sample inlet is put under high temperature (approximately up to 400 °C). The gaseous molecule then enters into the ionization chamber where it collides with a beam of high voltage electrons. The event gives rise to loss of valence electrons followed by the formation of positively charged parent ion along with various well defined positively charged fragment ions which are characteristic to the molecule. The positive ions are then taken from the ionization chamber to the acceleration chamber by applying a small positive charge of several volts to the repeller. In the acceleration chamber the charged particles, the positive ions, are attracted towards negatively charged plates and accelerated at high speed under the influence of an accelerating voltage following the principle of faster movement for lighter ions and slower movement for heavier ions. Thus, the ions are separated according to their mass to charge ratio. To receive all fragment ions in the detector, a varied accelerating voltage is needed because different masses accelerate at different voltage. The whole operation is performed in a high vacuum system which allows unhindered movement of ions.

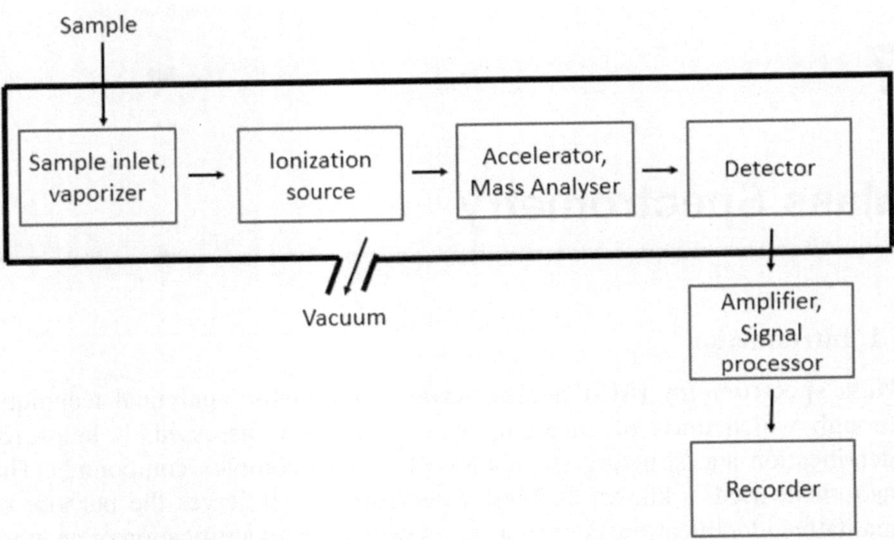

Fig. 7.1: Schematic representation of a mass spectrometer

7.2. MS applications and uses

MS basically measures the molecular mass of a compound. This spectroscopic analysis makes it possible to identify as well as quantify a compound at a very low concentration (10^{-12} g) with high accuracy. Number of isotopes with their relative abundance and exact mass may be determined. It is very convenient to analyse proteins, peptides, oligonucleotides and other macromolecules in the modern biochemical and biotechnological domain by MS. It is used efficiently in pharmaceutical and clinical areas for drug testing, drug metabolism studies, pharmacokinetics, haemoglobin analysis, etc. Moreover, MS has made a major breakthrough in environmental science in the analysis of PAHs, PCBs, various other soil, air, and water contaminants like pesticides, providing chemical and structural information about their degradation metabolites.

7.3. MS Components

There are three major components in a mass spectrometer:

(1) Ionization unit: It enables production of ions from samples by various methods that are harnessed with their own merits and demerits. Compared to neutral molecules, the ions are easier to manipulate. This is why molecules are converted to their respective ions before MS analysis. The selection of ionization methods for a sample depends on the nature of the sample and the type of mass analyser.

The sample is introduced into the ionization unit in two ways: (a) One is involved with direct insertion of sample into the ionization unit and (b) MS is coupled with high performance liquid chromatography / gas chromatography / capillary electrophoresis, so that the sample is firstly separated into its different components which, in turn, directly introduced in the ionization unit one by one for individual MS analysis.

(2) Mass analyser: This unit separates the ions in accordance with their m/z ratio. There are different types of mass analyser available now, namely, Magnetic-sector analyser, Quadrupole analyser, Time-of-flight analyser, FT-Ion cyclotron analyser, Ion-trap analyser, etc. These analysers differ among themselves based on mass accuracy, resolution, the m/z range covered and their compatibility with various ionization techniques.

(3) Detector: The separated ions are detected in this unit. The signal is sent to a data system where the m/z ratios are stored together with their relative abundance for presentation in the format of an m/z spectrum. A plotting of the m/z values vs their intensities is obtained which says about the number of components in the sample along with the molecular mass and relative abundance of each component. Different types of detectors used in MS include Photomultiplier, Microchannel plate electron multiplier, etc.

7.4. Ionization

Ionization is defined as the formation of ions from an atom or a molecule by gaining or losing electrons, acquiring a negative or positive charge, respectively. Ionization can result from the loss of an electron after collisions with subatomic particles, collisions with other atoms, molecules and ions, or through the interaction with light (**Figure 7.2**).

Ionization is considered as the most important step in mass spectrometry, where gas phase ions are produced through introduction of gaseous or liquid samples, which are then resolved in mass analyser. During ionization lone pair electrons are more easily displaced than bonding electrons and electrons in pi-bonds are more easily displaced than those in single bonds. There are two types of ionization, based on the quantity of energy imparted in the molecule: hard and soft ionization. Hard ionization converts the molecule in its ionic forms with large degrees of fragmentation; the opposite is true for soft ionization (**Figure 7.3**).

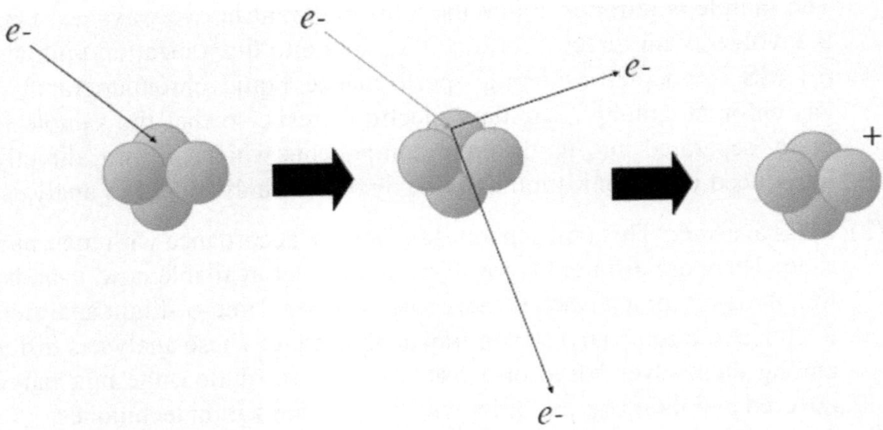

Fig. 7.2: Ionization scheme

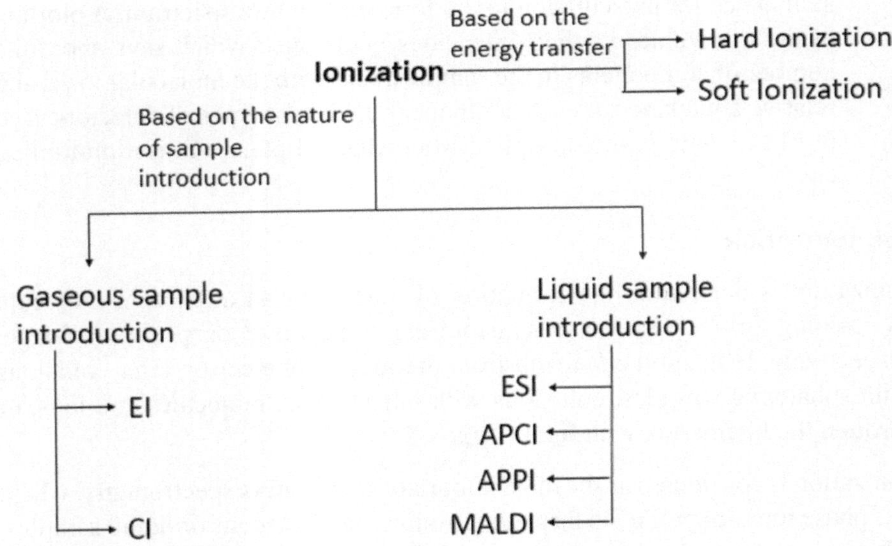

Fig. 7.3: Types of ionization

7.5. Sources of Ionization

7.5.1. Electron Ionization (Electron Impact, EI)

The ionization of the molecule in a gaseous state is made by an electron impact from a high energy electron beam. Electrons are generated from a heated wire filament through an applied voltage. A beam of high energy electrons is produced by accelerating these electrons under the influence of the electric field. Molecules, on colliding with electrons (generally with 70 eV energy), lose electrons. It becomes positively charged radical ion ($M^{+\cdot}$), and/or daughter ions. Therefore, this process provides information on molecular weight as well as fragmentation pattern of the analyte ion, which helps to elucidate the molecular structure. Here, molecular weight obtained is M^+. If the electron energy is reduced, numbers of generated ions as well as the fragmentation are reduced. Reproducible mass spectra are obtained by this method of ionization and as a result mass spectra of a compound in a question may be compared with mass spectral fingerprint in a mass-spectra-library. However, an EI mass spectrum shows weak or no signal for the molecular ion peak in many compounds. Thermal stability and volatility of compounds are necessarily their prerequisite properties for EI mass spectroscopy.

Why does fragmentation occur in EI?

In 70 eV EI, one can expect nearly 1400 kJ/mole energy transfer from ionizing electron to the analyte molecule during the ionization process. Ionization of most organic compounds requires 960 kJ/mole energy, and it is evident that typical chemical bond energy is of 290 kJ/mole energy. This corroborates the intense fragmentation in EI mass spectra. Thus, the amount of fragmentation and the distribution of fragment ions may be varied with changing the electron voltages.

7.5.2. Chemical Ionization (CI)

In this case, an analyte molecule is allowed to react with reactive ionized species to form a protonated molecule species. It is a softer ionization process. For instance, interaction between M and Brønsted acid gives rise to the formation of protonated molecule (MH^+):

Initially, CH_4^+ is generated through EI of methane (carrier gas), which then reacts with CH_4.

$$CH_4^{+\cdot} + CH_4 \rightarrow CH_5^+ + CH_3^\cdot$$

$$M + CH_5^+ \rightarrow MH^+ + CH_4$$

Common reagent gases used in CI mass spectroscopy are methane, isobutane and ammonia, out of which methane is more commonly used as it is the strongest proton donor (**Table 7.1**).

Table 7.1: Proton affinity values of reagent gases used in CI Mass Spectra

Reagent gas	Proton affinity (PA) in eV
Methane	5.7
Isobutane	8.5
Ammonia	9.0

This process informs about the molecular weight of the analyte. Unlike EI, structure elucidation of a compound is not possible by this process. Fragmentation of the molecule is not encouraged because it is a soft ionization technique which only produces ions with little excess energy. Thus, CI Mass Spectra help to obtain molecular weight (MH^+) and it may be used as complementary to the EI spectra that inform about the nature of fragment ions.

7.5.3. Field ionization (FI) / Field Desorption (FD)

A multi-tipped emitter, made up of tungsten or rhenium filament with carbon or silicon microneedles or whiskers grown on its surface which is similar to the FI method, is used in the FD method. FI is employed to analyse volatile liquids/solids introduced by evaporation from a sample vial that remains in close proximity to the ionizing tip or electrode in the presence of a high electric field. In FI, gas molecules are ionized with no or little fragmentation under the influence of high electric field strength in excess of 10^8 Vm^{-1}. However, the FD method bypasses the evaporation of the analyte prior to ionization. In FD, the sample is deposited onto the emitter either by dipping the emitter into an analyte solution or by using a micro-syringe and in FI, samples are introduced in the gas-phase. The ionization process for different analytes is employed through four different mechanisms: (a) **Field ionization** - the sample is ionized by quantum mechanical tunnelling mechanism, on subjecting a high voltage to the emitter. The ions are accumulated on the tip of the whiskers and are later desorbed, resulting formation of positive ions which are radical ions $M^{+\circ}$; (b) **Cation attachment** – cations (H^+/Na^+) are attached to the analyte molecules, followed by desorption of cation attached species $(M + Na)^+$ through the action of heating and high field strength; (c) **Thermal ionization** – the emitter heats the sample considerably for its ionization followed by desorption of the ions from the heated emitter surface; and (d) **Proton abstraction** – this is different from other three mechanisms as this generates negative ions $(M - H)^-$.

FI/FD results in ions of extremely low internal energy (soft ionization) and as a result it exhibits only abundant molecular species and almost no fragmentation or very limited fragmentation. This is why no structural information could be gathered from FI/FD-MS. FI/FD is useful for thermally stable organic molecules, low-molecular-mass compounds, many organometallics, and petrochemical fractions, which do not respond to EI-MS. It is very much sensitive to alkali metal contamination. Solubility of a sample towards a solvent is a major characteristic feature for being chosen for FD-MS analysis. This technique is now mostly substituted by other alternative desorption techniques.

7.5.4. Fast Atom Bombardment (FAB)

It is a soft ionization technique, applicable to those high molecular weight compounds (peptides, proteins, carbohydrates, fatty acids, organometallics, etc. MW not more than 10000 u.) that are not normally accommodated with EI / CI. Thermally and energetically labile polar compounds may be analysed with this ionization source. The process involves dissolution of the compound in a matrix followed by bombardment with fast moving neutral atoms of noble gas, Argon / Xenon (8-15 KeV).

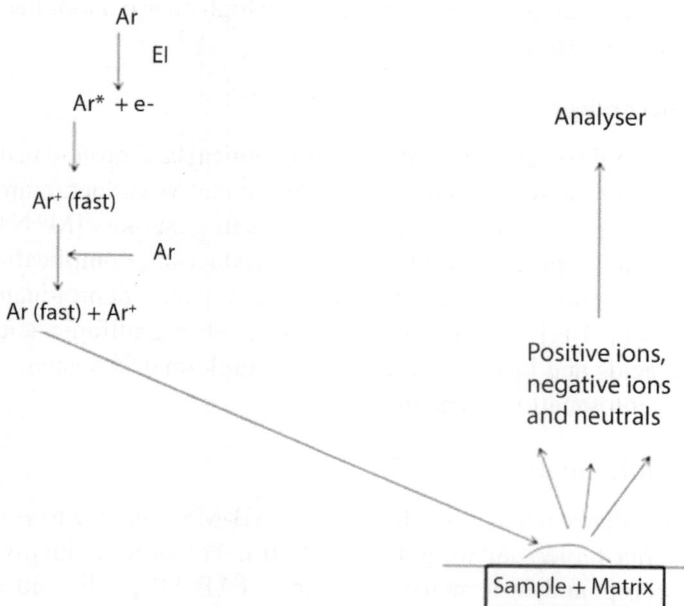

Fig. 7.4: FAB ionization

Ar or Xe ions are generated simply by EI. The accelerated ions are then interacted in the collision cell with neutral atoms, where charge - transfer takes place and the accelerated ions become neutral atoms. In the next step, fast moving atom beams strike on the sample which remains dissolved in a matrix and produces positively and negatively charged analyte ions along with neutrals (**Figure 7.4**). However, neutrals will be ionized afterwards in the plasma just above the sample surface. Thermally labile compounds that are susceptible to decomposition at high temperature are studied satisfactorily with FAB, as the process involves no heat generation.

7.5.4.1. Matrix

Glycerol, thio-glycerol, m-nitrobenzyl alcohol, diethanolamine, triethanolamine, magic bullet (dithiothreitol : dithioerythritol :: 3:1), sulfolane, etc. are commonly used as matrix in FAB. Its low volatility under vacuum, its sample solubility and its non-interference in any type of reaction with the analyte are the main criteria of FAB matrix for its selection.

Matrix reduces the chances of analyte decomposition from high energy atom bombardment because it itself absorbs maximum amount of incident energy, besides encouraging desorption and ionization processes. Moreover, the presence of matrix replenishes and renews the surface continuously with new analyte, where the interaction between sample and the high energy atom beam leads to desorption and ionization.

7.5.4.2. FAB Disadvantage

Additional attention should be given to sort out high chemical background in the FAB mass spectra due to the serious interference of the matrix adduct, matrix cluster ions and matrix ions. Moreover, presence of salt generates $[M+Na]^+$ / $[M+K]^+$ ions that may suppress the $[M+H]^+$ ion production, complicating the mass spectra. Sometimes, pseudo molecular ion peak, deprotonated negative molecular ion $[M-H]^-$, is found dominant (e. g., strong sulfonic acid). Sometimes organic acids and bases are used in the sample-matrix system, so that protonation or deprotonation is encouraged.

7.5.5. FIB (Fast Ion Bombardment)-MS

FIB-MS came into existence when it was found that FAB-MS is unable to work on molecules of higher molecular weight (>10000 u.). The process involves dissolution of the compound in a matrix (similar to FAB-MS) followed by bombardment with fast ions of higher energy (Cs^+ up to 35 KeV). Sensitivity of FIB-MS is comparatively more than that of FAB-MS due to higher energy of Cs^+. This is also referred as Liquid Secondary Ion Mass Spectrometry (LSIMS)

7.5.6. Secondary Ion Mass Spectrometry (SIMS)

The main difference between FAB and SIMS lies in the nature of the primary ionizing beam used to produce secondary sample ions. In FAB, a neutral atom beam is used as the primary ionizing beam, while in SIMS primary ionizing beam is an ion beam from Cesium or Argon. In SIMS, generated monoenergetic noble gas ions are accelerated to 300 – 3000 eV and the ion beam then bombards on the surface of the sample (also known as 'sputtering'), resulting in the production of secondary sample ions by charge transfer mechanism between sample molecules and primary gas ions. Neutral atoms and neutral molecules are also the resultant product of this sputtering. However, the ions are then accelerated along a potential gradient to higher kinetic energy, focused, and transferred into a mass spectrometer for analysis. The SIMS ion source is capable of producing ions from solid samples without prior vaporization. SIMS is of two types: (a) Static SIMS (No damage on the sample surface, a very low-current primary ion beam) and (b) Dynamic SIMS (causing erosion on the sample surface). SIMS is designed to study solid samples across the surface and depth profile in a single location. No matrix is used, and the ionizing beam is focused directly on the sample. Thus, sampling and quantitation are quite difficult in SIMS, even though SIMS is regarded as a powerful tool for studying surface chemistry of a sample, generating high resolution chemical maps.

7.5.7. Matrix Assisted Laser Desorption/Ionization (MALDI)

Sample mixed with organic matrix [e.g.: 3,5-dimethoxy-4-hydroxycinnamic acid(sinapinic acid); 2,5-dihydroxybenzoic acid (gentisic acid); α-cyano-4-hydroxycinnamic acid; 3,5-dihydroxycinnamic acid (caffeic acid); 3-hydroxypicolinic acid; nicotinic acid in 50% aqueous acetonitrile containing 0.1% trifluoroacetic acid) is allowed to co-crystallize in vacuum (mixing of analyte and matrix powder in a mortar is also used sometimes, when analyte and matrix are not soluble in solvent). The crystals are bombarded with a laser beam. Matrix absorbs light energy from high-intensity lasers due to the presence of chromophores in them. The energy obtained in the matrix, in turn, is transferred to the sample analyte triggering generation of analyte ions (more abundantly $[M+H]^+$, and less abundantly $[M+2H]^{+2}$, $[2M+H]^+$), which are then focussed to mass analyser through acceleration in the flight-tube under the influence of high electrical field. Now-a-days MALDI-MS is acting as a key player for identification of bacteria and microbial fingerprinting. It is a soft ionization technique and provides rapid and convenient molecular weight determination. Large organic polymers and biomolecules such as sugars, proteins, DNA, etc. may be satisfactorily analysed by this technique. Time of flight (TOF) mass detector is usually used for MALDI-MS. MALDI is not coupled with quadrupole MS analysers.

Some common lasers used for MALDI are: Nitrogen (λ=337 nm), Nd:YAG μ3 (Neodymium-doped yttrium aluminium garnet laser, λ=355 nm), Er:YAG (Erbium-doped yttrium aluminium garnet laser, λ=2.94 μm), CO_2 (λ=10.6 μm). However, N_2 lasers are considered as the standard.

7.5.8. Atmospheric Pressure Ionization (API)

Atmospheric Pressure Ionization (API) has been designed for mass spectrometric study of thermo-labile, polar, and high molecular weight compounds. Unlike conventional ionization sources viz., EI, CI, FD, FAB, etc. at high vacuum; this source of ionization works at atmospheric pressure employing various techniques such as Electrospray ionization (ESI), Atmospheric Pressure Chemical Ionization (APCI), Atmospheric Pressure Photoionization (APPI) and Atmospheric Pressure MALDI (AP-MALDI). However, a high vacuum is still maintained in the rest of the instrument (mass analyser and detector). Thus, to stitch into the two different conditions, the region in between the ion source and mass analyser is comprised of a series of compartments with differential pressure (from high to low), where skimmers, focusing lenses and flow restrictors are employed to transfer of ions to high vacuum region efficiently and to drive out air, solvent vapours, and other neutral volatile species (**Figure 7.5**). A pressure gradient from source of ionization (high pressure) to mass analyser (vacuum) is maintained. Direct coupling of a high-pressure liquid chromatograph to a MS equipped with ESI or APCI is very common as these ionization sources as well as the HPLC are used efficiently and effectively for polar thermo-labile analytes. Sample ionization at atmospheric pressure exhibits approximately 10^3 to 10^4 times more ionization efficiency as compared to that at a reduced pressure (e.g., CI) due to fast thermal stabilization at atmospheric conditions.

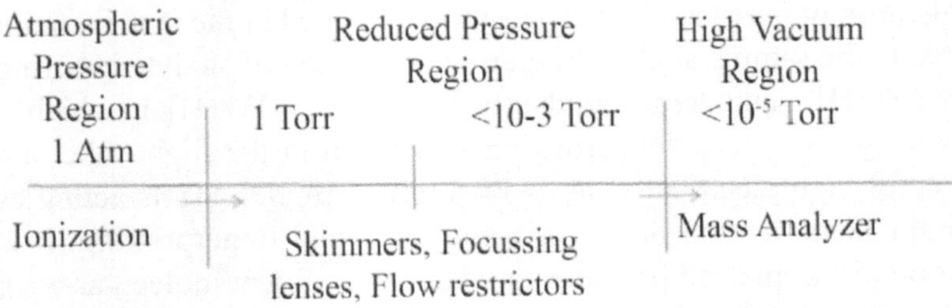

Fig. 7.5: Schematic representation of API

7.5.8.1. Electrospray Ionization (ESI)

Electrospray Ionization (ESI) is a soft-ionization technique, imparting no or little fragmentation. It provides an added advantage that one can always find the pseudo-molecular ion. However, fragmentation may be promoted for molecular identification by adjusting conditions and using a triple quad MS. ESI is efficiently interfaced with HPLC and Capillary Electrophoresis (CE). Critically, ESI is not an ion formation technique. It is to transfer ions from a solution phase to gas phase that are eventually analysed by mass spectrometry. The principle of ESI is based on the formation of highly charged liquid droplets (1–10 μm) that progressively lose solvent to form ionized molecules, by applying a strong electric field to a liquid under atmospheric pressure and by transferring the analyte ions from liquid solution to gas-phase. A sample solution is introduced in the ion source compartment through a high voltage needle (potential difference of 3–6 kV between the needle and the counter-electrode), where ionization takes place. If the wall of the needle is positively charged, positively charged ions stay in the solution and negatively charged ions drift towards the wall. The repelling behaviour of the positively charged ions within themselves in the solution and also with the positively charged needle wall helps to pile the ions into a drift of a cone-shaped flow at the tip of the needle. Finally, droplets containing several positively charged ions lying on the outer surface come out of the needle tip and are subjected to a curtain of heated nitrogen gas (80 °C) for desolvation.

The charge density on the droplet is increased by progressive decreasing of droplet's size on evaporation, which facilitates coulombic splitting of the droplet after attainment of Rayleigh Limit (Rayleigh limit: the maximum amount of charge a liquid droplet could carry). Thus, gas-phase ions are formed [Dole, M.; Mack, L. L.; Hines, R. L.; Mobley, R. C.; Ferguson, L. D.; Alice, M. B. *J. Chem. Phys.* 1968, *49 (5)*, 2240-9.] and are then introduced into the mass spectrometer (**Figure 7.6**). The gas-phase ion formation from the droplet may also be explained by the fact that the droplet reaches a stage when the repulsive force among the ions, being high, overcomes the surface tension (Iribarne, J. V.; Thomson, B. A. *J. Chem. Phys.* 1976, *64 (6)*, 2287-94.). In the same way, negative ions may be created by subjecting the needle wall to negative voltage.

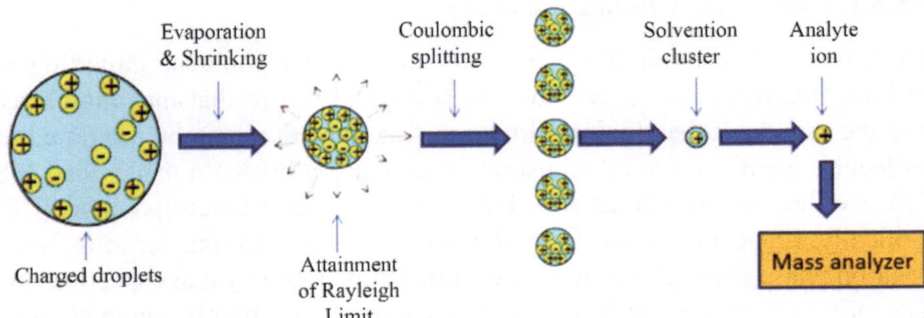

Fig. 7.6: Illustration of ESI

Thus, two basic modes of operation are possible in ESI.

(1) Positive mode for basic molecules (analytes are applied at low pH): addition of small amounts of formic acid often encourages protonation of the sample molecules.

(2) Negative mode for acidic molecules (operation is maintained above the isoelectric point facilitating the deprotonation of the molecule): addition of a small amount of ammonia solution or a volatile amine encourages deprotonation of the sample molecules.

Deprotonated molecules ($[M-H]^-$) or adducts ($[M+CH_3COO]^-$) under negative ion mode and protonated molecules ($[M+H]^+$) or adducts ($[M+NH_4]^+$, $[M+Na]^-$) under positive ion mode are commonly found in ESI.

ESI produces multiply charged ions (distribution of molecular ions with different charge numbers: $[M+H]^+$, $[M+2H]^{2+}$, $[M+3H]^{3+}$, etc. are the multiply charged protonated ions) of large molecular mass substances, which are useful for accurate mass measurement for molecules using software. However, multiply charged species require interpretation and mathematical transformation. ESI faces a major drawback for its signal suppression due to the competition of charge between the solvent and the analyte.The sensitivity of ESI is also decreased in the presence of non-volatile buffers and additives, that necessitates a careful consideration of the use of buffers. Besides, it is not good for uncharged and low-polarity compounds. Very little knowledge on structure elucidation of a molecule may be acquired by ESI-MS due to little fragmentation. However, overcoming this drawback of ESI may be achieved by coupling it with tandem mass spectrometry (ESI-MS/MS).

Micro-electrospray, and nano-electrospray are different types or versions of electrospray depending on the flow rate (30 - 1000 nL/min for nanospray), sample volume (1-4 μL for nanospray) and concentration (1 - 10 pmol/μL for nanospray). In general, electrospray functioning at low flow rates causes comparatively smaller initial droplets, leading to higher ionization efficiency.

7.5.8.2. Atmospheric Pressure Chemical Ionization (APCI)

Atmospheric Pressure Chemical Ionization (APCI), being complementary to ESI, is best suited ionization technique for non-polar or less polar compounds. In APCI, a corona discharge is used to ionize the analyte in the atmospheric pressure region.Vapourised solvent, used as mobile phase in HPLC, performs the function of CI reagent gas for analyte ionization. In the first step, the mobile phase and analyte are nebulized (N_2) and vapourized by heating to 350-550 °C. The solvent vapour is subjected to a corona discharge needle that acts as an electron source, and is ionized. Afterwards, the gas-phase ion transfers proton (when analyte exhibits high proton affinity) or exchanges charge (when analyte shows low ionization energy) to analyte through CI mechanism, leading to ionization of analyte molecule. Then, the charged analytes produced at atmospheric pressure are introduced into the mass analyser (**Figure** 7.7).

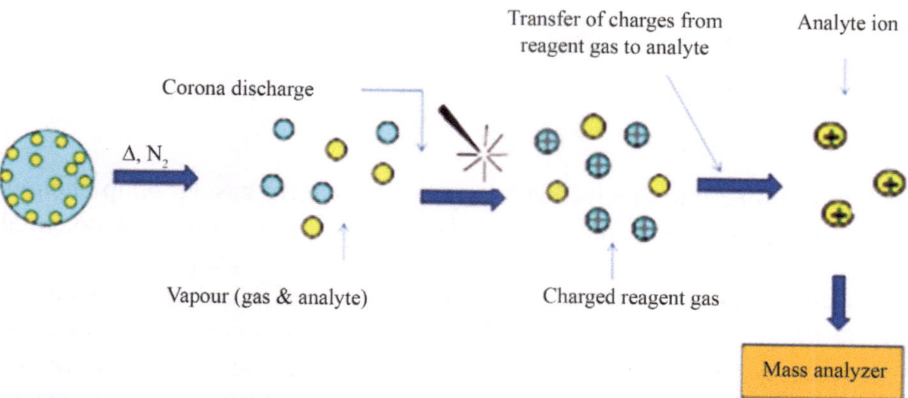

Fig. 7.7: Illustration of APCI

Analysis of thermo-stable compounds is preferred in APCI due to its heat requirement in the vaporization step. Unlike ESI, APCI does not generate multiply charged ions and it performs better at higher flow rate. However, like ESI, APCI may also be performed under both positive and negative ion mode.

7.5.8.3. Atmospheric Pressure Photo Ionization (APPI)

The basic principle of APCI and APPI is quite similar, only differing in the fact that the corona discharge needle (APCI) is substituted by photons (ionization energy of 10 eV & 10.6 eV) of a krypton lamp in APPI. Thus, the APPI source is considered as a modified APCI source. Principally, the photon energy should be higher than the analytes' ionization potentials and lower than that of atmospheric gas (IP nitrogen, 15.58 eV; IP oxygen, 12.07 eV) and solvents commonly used in LC (IP, above 10 eV). As a result the background noise is minimized due to selective ionization of analytes which possess ionizing energy of 7-10 eV. APPI is one of the primary API alternatives in the lab since it extends the ionization range to more non-polar analytes compared to ESI or APCI. The LC eluent, passing through a nebulizer, is made to a fine spray and is vapourized in a ceramic tube. The vapour is then subjected to high energy photons of a krypton lamp. As a result, the sample molecules are ionized either directly or indirectly (aided by dopants). Dopant (D), a solvent modifier, is used to encourage the photoionization process as follows:

Reactions in APPI in absence of dopant:

$M + h\upsilon \rightarrow M^{\cdot +} + e^{-}$ (condition: analyte IP <photon energy)

$M^{\cdot +} + SH \rightarrow [M + H]^{+} + S^{\cdot}$ (CI process)

Reactions in APPI in presence of dopant:

$D + h\upsilon \rightarrow D^{\cdot +} + e^{-}$

$D^{\cdot +} + M \rightarrow [M + H]^{+} + D$

$D^{\cdot +} + M \rightarrow M^{\cdot +} + D$

Dopant is at first photoionized and later acts as reagent gas. Kr lamp photons (10-10.6 eV) photoionize the compounds which bear lower ionization potential (IP). APPI is usually interfaced with HPLC. Therefore, choice of solvent as mobile phase in HPLC is of prime importance. Water (IP, 12.61 eV), acetonitrile (IP, 12.19 eV), methanol (IP, 10.85 eV) are very common HPLC solvents. However, they are not ionized and thus do not encourage ion formation for analytes. Thus, direct photoionization of the compound takes place. However, in this situation dopant [acetone (IP, 9.70 eV), toluene (IP, 8.82 eV), anisole (IP, 8.22)] may be used to aid the ionization of the compound. In the absence of dopant, direct ionization of analytes gives rise to a very weak ionization efficiency due to minimization of available photon number arising due to its absorption by the solvent. Addition of dopant causes many fold increase in ionization efficiency of the analytes.

The unique feature of APPI is its capacity to ionize molecules that are not ionisable by APCI and ESI. Thus, APPI allows the ionization of compounds not detectable in APCI or ESI, mainly non-polar compounds. Intermediate to low polar and thermally stable aromatic compounds (PAHs, PCBs, phthalates, fatty acids, and alcohols) are commonly analysed in APPI technique.

7.5.8.4. Inductively Coupled Plasma (ICP)

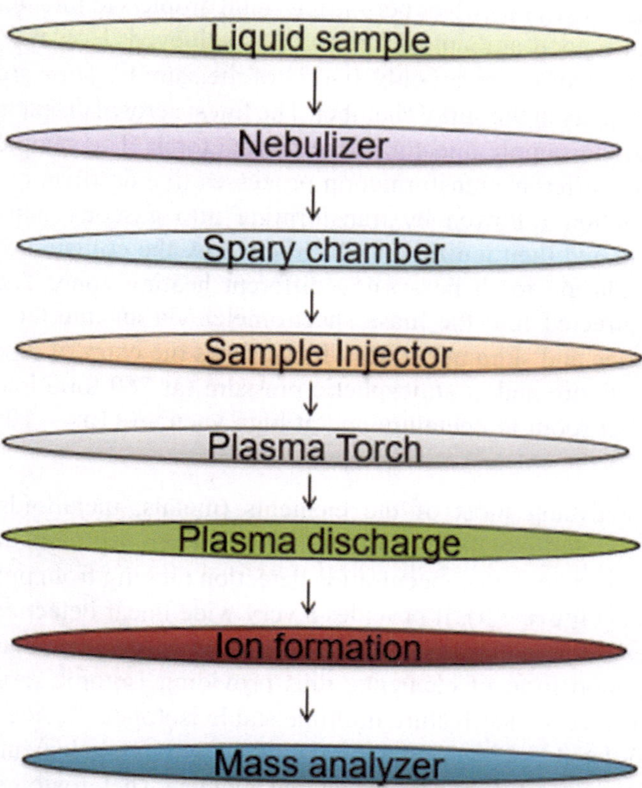

Fig. 7.8: Schematic diagram of ICP-MS

In ICP-MS, positively charged ions are generated using high temperature plasma discharge (**Figure 7.8**). Initially sample in liquid form is made into aerosol in a nebulizer and is introduced into a plasma torch where the sample is dried, vaporized, atomized in the neutral ground state, and finally ionized by stripping the outer electrons from those ground state atoms, using heat from a plasma discharge. Although the ICP is commonly used for positive ion detection, negative ions may also be generated. However, the extraction and transportation of negative ions are quite different from that of the positive ions. The plasma is generated from argon gas and the heating is achieved by a radio-frequency current passed through a coil surrounding the plasma. Most commonly, sample aerosol is formed using argon gas flow at 20-30 psi in the nebulizer which may vary in different designs like concentric, microflow, and crossflow. Formation of fine aerosol by nebulizer and the droplet selection by the spray chamber are the critical points for efficient ionization by plasma discharge. Plasma discharge is not efficient to dissociate large droplets. Therefore, the primary function of the spray chamber, usually of double pass design or cyclonic design, always involves permitting small droplets to interact with the plasma, so that a good amount of ionization is achieved. However, the fine droplets of the aerosol represent only 1 - 2% of the sample. They are separated from larger droplets in the spray chamber. The finest aerosol droplets are then introduced into the sample injector of the plasma torch. The sample aerosol then experiences different transformation processes like desolvation, and solid particle formation followed by transforming into gaseous state, ground state atomization and then ionization which is led by the collision of argon electrons in the plasma torch possessing different heating zone. The generated ion is then directed into the mass spectrometer via an interface, consisting of sampler cone and skimmer cone, which allows the entry of ions generated at high temperature and at atmospheric pressure (at 760 torr) into mass analyser operated at room temperature and at high vacuum (10^{-5} – 10^{-6} torr).

ICP-MS is used for analysing most of the elements (metals, metalloids, and heteroatoms, such as non-metals, semimetals, and halogens) present in environmental samples at a very low concentration detection ranging from ppb (µg/l) to ppt (ng/l) level (**Figure 7.9**). It provides a very wide linear detection range and specificity for accurate detection. This possesses an added advantage of detecting isotope composition of elements, thus providing isotopic ratio information about the elements that feature multiple stable isotopes. Besides, quantification of any molecule, possessing natural covalently bound elements or chemically labelled with an ICP-MS detectable element (heteroatoms, metals, or metalloids), is practically possible by ICP-MS. Thus, detection of complex molecules, such as proteins, nucleic acids or simple and small organic molecules is achieved by ICP-MS. ICP-MS is also applied for pesticidal analysis in environmental samples, as for example element specific detection and determination of phosphorus, sulphur, chlorine, bromine, iodine containing pesticides. It requires a clean room environment for ultra-low detection limits.

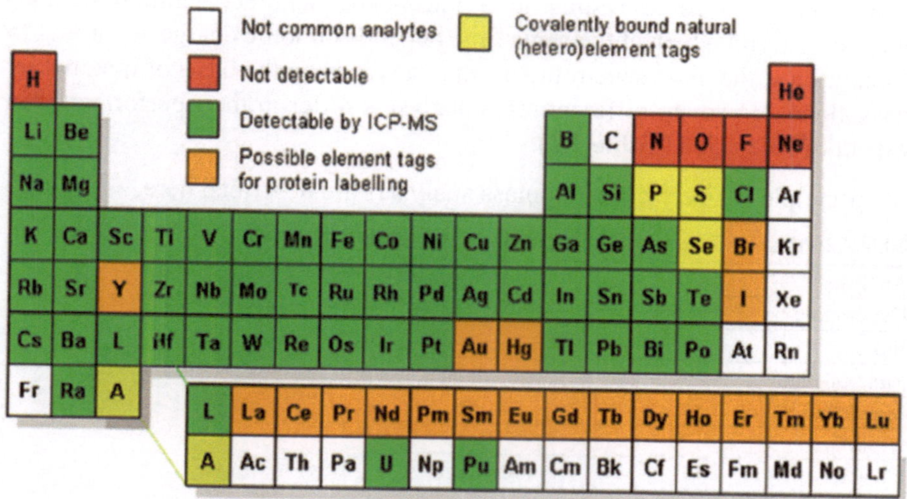

Fig. 7.9: Elements in the periodic table for ICP-MS analysis.

7.6. Mass analyser

This portion of the mass spectrometer separates and resolves ions according to their m/z in different ways. They vary within themselves in the m/z range (the range of *m/z* amenable to analysis by a given analyser), linear dynamic range (the range over which ion signal is linear with analyte concentration), the mass accuracy (the ratio of the *m/z* measurement error to the true m/z in ppm) and the resolution (the measure of the ability to distinguish two peaks of slightly different *m/z*). Before going into detail about the mass analysers, it is better to highlight the reason for considering m/z (not mass alone) while separating ions by mass spectrometer. While electric and magnetic fields are subjected to charged ions to generate a force on them, it is apparent that the force, mass, charge, acceleration, and the applied fields are interrelated and explained by Newton's Second Law (F = ma), and Lorentz Force Law [F = e (E + v . B)].

Where,

F = Force applied,

m = Ion mass,

a = Acceleration,

e = Ionic charge,

E = Electrical field, and

v . B= Vector cross product of ion velocity and magnetic field.

Therefore, since force results in a mass-dependent acceleration of ions (Newton's Law) which at the same time depends on ionic charge according to Lorentz Law, the mass spectrometer separates ions on the basis of m/z, not by mass alone. However, different mass analysers differ in their performance in respect of resolution (**Table 7.2**).

The principles of very common mass analysers are described here.

Table 7.2: Mass analysers with their extent of resolutions

Analysers	Possible resolution
Quadrupole	Low, Medium
Ion trap	Low
Time-of-flight	Medium
Magnetic sector	High
Orbitrap	High
FT-ICR	Very high
MS-MS	Very high

7.6.1. Quadrupole analyser

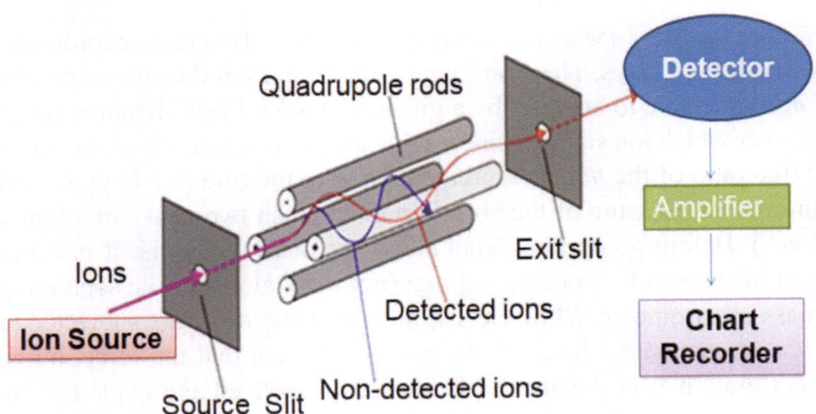

Fig. 7.10: Schematic diagram of Quadrupole mass analyser

Quadrupole type of mass analyser is known for its simplicity, high sensitivity, fast scan rate, easiness to couple with chromatography and moderate vacuum requirement despite its limited ability on mass resolution. The principle of quadrupole mass analyser is based on the separation of ions under their oscillation in an electric field using ac and dc current. It consists of four parallel metal electrodes (**Figure 7.10**). Each set of opposing rods has the same potential, but of different sign and carries a voltage that affects the trajectory of ions during their traversing through the centre of the two pairs of electrodes.

Each diagonally opposing pair is electrically connected and an ac voltage is subjected between the two pairs. A dc voltage is then superimposed on the ac voltage. This helps the ions to lead an oscillatory trajectory, while they traverse. Choosing a given combination of ac and dc voltages, ions with m/z of interest are allowed to go through the electrode to the detector, while other ions with m/z of no interest follow different trajectory paths and are removed by vacuum. By ramping the voltages at a nearly constant ratio on each pair of electrodes, a complete range of ion masses goes through the mass filter to the detector and thus, a mass spectrum is produced.

7.6.2. Time-of-flight (TOF) analyser

There are mainly two parts in a TOF type of analyser: (a) an ion-accelerating region and (b) a flight tube followed by a detector. TOF separates ions taking time factor into consideration. All generated ions are introduced in the ion-accelerating region where ions are accelerated by an electric field (no magnetic field is required in TOF) and are then pulsed into a field free flight tube in short, well defined packets possessing same kinetic energy, same potential difference, and a constant homogeneous electrostatic field application. The ions of the same kinetic energy exhibit different velocities based on their masses that encourage temporal separation of ions before reaching the detector. Thus, masses of the ions dictate their arrival time at the detector and the difference of flight times of different ions to reach the detector is simply exploited to resolve them. The ions reach the detector in increasing order of their masses. Lower is the mass of an ion, faster it reaches the detector. This is very much ideal for pulsed ionization. Majority of MALDI mass spectrometers are equipped with TOF. However, pulse is not applicable to all ions with the same intensity and this phenomenon leads to a kinetic energy distribution and, in turn, also leads to a time-of-flight distribution for each discrete ions. This results in minimizing the resolution. The resolution may be increased theoretically by increasing time of flight and by increasing length of the flight tube because resolution is directly proportional to both of these parameters. But a very long flight tube results in loss of ions by scattering and, thus, decreases the performance of the analyser. Again, time of flight may be increased by decreasing the acceleration voltage. However, decrease in acceleration voltage results in lesser sensitivity. It has been experimentally found that a flight tube of 1-2 m and an acceleration voltage of 20 kV are maintained for high resolution and high sensitivity. All ions, that encourage high transmission efficiency and high S/N level, are detected in TOF. This is important to note that sorting of ions in TOF mass analyser is performed in absence of any magnetic field.

7.6.3. Sector analyser

A sector field mass analyser employs a static electric and/or magnetic field affecting the path and/or velocity of the charged particles.

In a magnetic sector analyser, a magnetic field perpendicular to the direction of ion motion is applied with an aim that the trajectories of the ions are bended in such mass analyser on the basis of their mass-to-charge ratios, following the principle that the velocity of the same m/z remains constant. The degree of ion deflection in a magnetic field is proportional to the square root of their m/z ratio. The radius of curvature of an individual ion beam at a given magnetic field is a function of m/z, thus separating the individual ion beams spatially. It is possible to separate ions of different m/z by manipulating the magnetic field or the accelerating voltage. If the magnitude of the magnetic field and the acceleration voltage remain constant, the ions of the same m/z experience the same trajectory path and pass to the detector. As a result, the ions, which are not selected by the magnitude of magnetic field and the acceleration voltage, collide on the wall of the flight tube, and cannot reach the detector through the slit. Highly charged, faster moving, and lighter ions are deflected too much and heavier ions are deflected too little (**Figure 7.11**). Only ions that match the small mass range reach the detector.

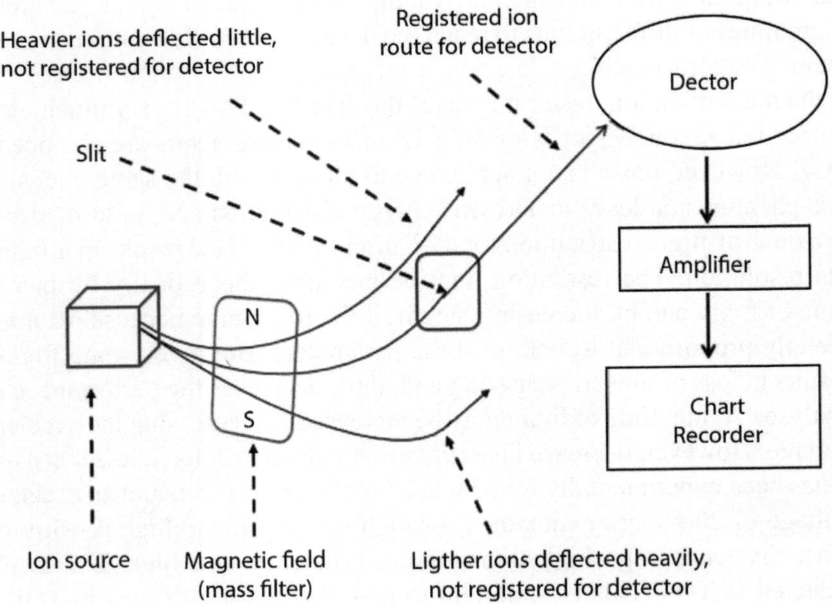

Fig. 7.11: Schematic diagram of magnetic sector mass analyzer

In electrostatic sector analyser, electric fields are employed, perpendicular to the direction of ionic motion, to focus fast moving ion beams in accordance with the kinetic energy of each ion. Under the electric field, ions possessing constant kinetic energy but different mass-to-charge ratios are focused.

Magnetic sector analyser and electrostatic sector analyser could be used individually (**single focussing instrument**) or both together (**double focussing instrument**). In double focussing instrument, the mass focussing of ions of same angular dispersion at magnetic sector as well as mass focussing of ions of same kinetic energy at electrostatic sector improves the resolution of the mass analyser in such a manner that ions differing in their m/z values in ppm level are also separated and analysed. The sector analyser can be used to select a narrow range of m/z. This is helpful for isotope ratio measurements. However, it is not suitable for pulsed ionization methods (MALDI).

7.6.4. Ion-trap analyser

A combination of electrostatic and magnetic field is used to capture ions inside the mass analyser. Ion traps of various configurations are available, viz., 3D ion trap, linear ion trap (2D trap), electrostatic field-based ion trap (orbitrap) and magnetic field-based ion trap (ion cyclotron resonance). Ion trap analyser functions for ions confinement in the first stage and measures the m/z ratio as a mass spectrometer in the second stage.

3D ion trap: This is handled with two possibilities: (a) Ions are generated within the ion trap or (b) ions obtained from an external source are injected into an ion trap. It is equipped with three electrodes. One cylindrical ring electrode is positioned symmetrically between two end-cap electrodes (inverted saucer appearance). All three electrodes possess a hyperboloidal geometry. The geometries of electrodes are unique in nature and are very important to produce an ideal quadrupole field. Simultaneously, a parabolic potential well, created by application of RF voltages to the ring electrodes, is produced for confinement of ions. Ejection of ions from the trap to the detector is accomplished by ramping the RF voltage at the ring electrode. Helium gas is used as a 'bath gas' to help containing the ions in the trap by stabilizing the ion trajectories. The ions, entering through one of the end-cap electrodes, are subjected to an electrical field at a space in between the electrodes. As a result, ions start orbiting. Orbits of heavier mass ions are more stabilized at higher RF voltages, while that of light ions are less stabilized. Thus, the light ions collide on the walls and escape the detection. Therefore, a mass selective ejection of ions to the detector through the aperture of other end-cap electrode is possible in 3D ion traps in the ascending order of mass by gradually increasing the RF voltage.

The ability in storing and in manipulations of ions makes the 3D quadrupole ion trap more sensitive and convenient than a simple quadrupole for full scan mode, since the properties are helpful to acquire both tandem mass spectra and multiple stage tandem mass spectra.

2D ion trap: Two-dimensional ion trap, also known as Linear Ion Trap (LIT), came into existence as an alternative to 3D ion trap. Relative to 3D ion traps with similar mass range, 2D ion traps exhibit improved trapping efficiency and ion capacity; so they have a better response and a wider dynamic range.

Orbitrap mass analyser: Orbitrap delivers high resolution and high mass accuracy determination. It works without any magnetic field, but on an electrostatic field. It involves orbital trapping of ions, consisting of three electrodes: inner or central electrode (spindle type) and two outer electrodes. High vacuum (10^{-8} torr approximately) is subjected to the space between internal and external electrodes. A linear electric field along the axis is generated upon subjection of voltage between outer and central electrodes. The ions, moving around the central electrode, show harmonic oscillation along the horizontal axis. The frequencies of oscillations vary with difference in mass to charge ratio and this phenomenon governs the separation of ions based on their m/z. Mass analysis may be performed by either FT (Fourier Transform) mode where coherent harmonic oscillations in the axial direction for the same mass to charge ratio induce an electric current that is measured by image current detection where the image current from the trapped ions is detected and converted to a mass spectrum using the Fourier transform of the frequency signal, or MSI (Mass Selective Instability) mode which engages in ion ejection and collection in a detector. In MSI, the ions after ejection from the orbitrap collide with a dynode. As a result, secondary electrons are generated, multiplied further and then detected. Orbitrap mass analyser may be operated in a pulsed ion source and also in a continuous ion source. But both types need different designs for the process of ion introduction.

Fourier Transform-Ion Cyclotron Resonance (FT-ICR) mass analyser:

In FT-ICR, frequency of movement of ion is measured very accurately and is then converted to m/z via a Fourier transform giving a more accurate m/z measurement. In this technique, ions are trapped in a strong magnetic field using huge and expensive superconducting magnets, combined with a weak electric field. The ion trap is housed in a magnetic field. The captured ions resonate at their cyclotron frequency in the presence of a magnetic field. A uniform electric field is applied to excite the ions. The detector measures the cyclotron frequency of all the ions in the trap and uses a Fourier transform to convert these frequencies into m/z values. FT-ICR is very suitable for proteomic study, since it provides high mass accuracy, high resolution, and high sensitivity.

7.6.5. Tandem (MS-MS) mass spectrometers

MS-MS studies involve usually with two mass analysers, either of same geometry (Quadrupole-Quadrupole) or of different geometric combination [Hybrid type, e.g., Quadrupole -TOF (Q-TOF), magnetic sector-quadrupole, etc.], combined in tandem. Specificity of tandemness lies in space (Tandem-in-space) and in time (Tandem-in-time). TQ (triple Quadrupole), instrument is a unique example of tandem-in-space because the precursor ion selection, collision-induced dissociation (CID) or fragmentation of precursor ion to generate product ions, and mass analysis of the product ions are carried out in three spatially arranged devices. This is described as tandem-in-space. However, these three steps, involving precursor ion selection to product ions formation, may be carried out in the same device, as seen in the ion-trap instrument. This is known as tandem-in-time. A collision gas cell exists in between the two analysers (MS1 and MS2). Precursor ions selected by MS1 collide with a high-pressure noble gas (usually He, Ar or Xe) in the cell. As a result, the precursor ion gains internal energy by conversion of its kinetic energy (process is known as Collisional Activation) and when these internal energies suffice to break the bonds, the ion undergoes fragmentation to produce product ions. The process is known as collisionally activated dissociation (CAD) or collision-induced dissociation (CID). Besides CID, other techniques are also available for the fragmentation of precursor ions viz., electron capture dissociation (ECD), electron transfer dissociation (ETD), infrared multiphoton dissociation (IRMPD), blackbody infrared radiative dissociation (BIRD), electron-detachment dissociation (EDD) and surface-induced dissociation (SID).

MS/MS may be performed using precursor-ion scan mode, product-ion scan mode, constant neutral-loss scan mode or selected reaction monitoring (SRM) mode.

Tripple Quadrupole (TQ or QqQ) is a mass analyzer which consists of three quadrupoles. First one is used as a filter for specific m/z (precursor ions). The second quadrupole is used as a collision cell where fragmentation of precursor ions and generation of product ions are obtained. Third quadrupole is set to specific m/z (SRM or MRM) or scan mode (product ion scan). In scan mode MS/MS, the third quadrupole allows only one m/z at a defined time to reach the detector.

7.7. Detector and recorder

Ions, hitting a detector, induce a current. The generated current is proportional to its abundance and is measured. Electron multiplier detector, Faraday cups detector and Daly detector are commonly used detectors in MS systems, which are selected on the basis of types of mass analyser.

In an **electron multiplier detector** containing discrete dynodes, an ion strikes the first dynode and many secondary electrons are generated, which eventually hit the second dynode and furthermore electrons are produced. Therefore, using a series of dynodes at a varied potential in the electron multiplier, the greater number of electrons are produced by a cascade effect to produce a current. Electron multiplier detector may also contain a continuous-dynode system. Microchannel plate detector, Channeltron and Microsphere plate detector are examples of continuous-dynode systems.

In **Faraday cup detectors**, ions hit the inner wall of a metal cup and electrons, flowing through the circuit, neutralize the ions at the cup's surface. This event leads to current generation which is then amplified and detected.

A **Daly detector** uses a photomultiplier for amplification of the signal. Ions initially strike a dynode and secondary electrons are emitted in this process. Electrons, in turn, hit the scintillator (phosphor screen) and are converted to photons. Photons then pass through a photomultiplier to be amplified in a cascade manner like an electron multiplier.

However, in FTMS and orbitrap, the mass analyser is the detector where ions orbit at different frequencies in accordance with their m/z values under the influence of a magnetic field and produce RF signals whose frequencies are related to the m/z of the ions. The detector measures the frequency of all the ions in the trap and uses a Fourier transform to convert these frequencies into m/z values.

Cryodetectors and **Focal-plane array detectors** are newly developed detectors used now-a-days.

The signals obtained from the detector are recorded in a system where the components in a sample are obtained by plotting the m/z values against their intensities. Thus, the recorder provides the molecular mass of each component in a sample along with their relative abundance.

7.8. Model questions

i. What is M^+ in mass spectrometry? What are the main criteria on which mass spectrometer is used?

ii. What is the basic principle of mass spectrometry? What are the applications of mass spectrometry?

iii. Why is mass spectrometry important? Why is methane used as the most common reagent gas in CI mass spectrometry?

iv. What is a quadrupole mass spectrometer? What is an "ion source"?

v. What is 'Dopant'? Explain the mode of action of dopant illustrating the reactions in APPI in the presence and absence of dopant.

vi. Name different sources of ionization in mass spectroscopy. Narrate in detail APPI, APCI and ESI.

vii. Mention the differences between FAB and SIMS. How does ICP-MS work?

viii. Illustrate the basic principle of TOF and FT-ICR mass analyser.

ix. Write a short note on MS-MS.

x. What is the basic principle of MALDI-MS? What are the main advantages of using MALDI-MS? Which detector is usually used for MALDI-MS?

7.9. Suggested Readings

(i) Herbert, C.; Johnstone, R. (2003) Mass Spectrometry Basics, CRC Press LLC.

(ii) Skoog, D. A.; Holler, F.J.; Grouch, S. R. (2007) Principles of Instrumental Analysis, Thomson Brooks/Cole.

(iii) de Hoffmann, E.; Stroobant, V (2001) Mass Spectrometry: Principles and Applications, 2nd ed.; Wiley: England.

(iv) Glish, G.L.; Vachet, W. R (2003) The basics of mass spectrometry in the twenty-first century. Nature Reviews, 2: 140-150.

(v) Downard, K. (2004) Mass Spectrometry: A Foundation Course, RSC, London

(vi) Lössl, P.; van de Waterbeemd, M.; Heck, A.J.R. (2016) The diverse and expanding role of mass spectrometry in structural and molecular biology. The EMBO Journal, 2016, 1-24 (DOI 10.15252/embj.201694818).

8

Chromatography

8.1. Introduction

Research related to biochemical analysis has been greatly benefited due to the development of chromatographic technique, since this technique makes it possible to resolve mixtures of closely related substances on a milligram to picogram range and to identify individual components by comparison with reference standard compounds. Chromatography 'gives information where other analytical means have failed' [Tswett, M. (1906) The macro-chemical and micro-chemical detection of carotene. Ber. Deut. Botan.Ges. 24 : 384-393]. David Talbot Day (1859-1925) was the first scientist who introduced chromatography in fractionating crude petroleum oil by passing crude oil through a layer of finely pulverized fuller's earth or limestone [Day, D. T. (1897) A suggestion as to the origin of Pennsylvania petroleum. Proceedings of the American Philosophical Society, 36(154), 112-115.]. The first fraction collected had remarkable similarity to light petroleum fraction obtained by distillation. The next fraction was heavier oils followed by petroleum jelly because of selective adsorption. This concept of chromatography he acquired from his observation while he was engaged in studying the origin of different Pennsylvanian oils and their geographical relationships. He proposed that the differences in color, viscosity and other properties of oil were due to large-scale fractional filtration processes, although they were of common origin. This forced him to apply this very principle in the laboratory. In the meantime, Michael Tswett (1872-1919) also initiated an attempt to separate plant pigments into their constituent parts. It is not known whether Tswett was aware of Day's work. He was the first person to say that there are two chlorophylls in plant leaves. He resolved plant pigments on a column of calcium carbonate using petroleum ether and 1% alcohol to the petroleum ether [Tswett, M. (1906) The macro-chemical and micro-chemical detection of carotene. Ber. Deut. Botan. Ges. 24: 384-393.]. Later, amino acids were separated by developing partition chromatography [Martin, A. J. P.; Synge, R. L. M. (1941) A new form of chromatogram employing two liquid phases: 1. A theory of chromatography.

2. Application to the micro-determination of the higher monoamino-acids in proteins. *Biochem. J.* 35(12), 1358–1368. doi: 10.1042/bj0351358] and gas liquid chromatography was developed to separate fatty acids [James, A. T.; Martin, A. J. P. (1952) Gas-liquid partition chromatography: the separation and micro-estimation of volatile fatty acids from formic acid to dodecanoic acid. *Biochem. J.* 50(5), 679-690. doi: 10.1042/bj0500679]. With gradual improvement in science, chromatographic techniques are also improved. Such an improved chromatographic technique is HPLC (High Performance Liquid Chromatography) with higher performance resulting from technological advances in both instrumentation and column design leading to greater flexibility and broad separation spectrum.

8.2. Definition

Chromatography is a group of techniques, facilitating separation of a mixture of solute into individual components by the differential migration of individual component through a stationary phase under influence of a mobile phase in such a manner that qualitative identification, quantitative estimation, isolation, and purification of the individual components are possible. However, identification and estimation of a component from a mixture through chromatographic technique require a reference standard for comparison. Otherwise, isolated products must be analyzed further by infrared spectroscopy, nuclear magnetic resonance, mass spectrometry or other convenient methods.

8.3. Principle

In any chromatographic technique it is observed that while a mixture of solutes is separated into their individual entity, it interacts with a stationary phase and a mobile phase. A stationary phase may be an immobilized solid, liquid (coated on a solid) or gel and a moving phase may be a liquid or gas. In the separation process, the moving phase with solute flows over the stationary phase and the solute mixture is separated due to the differences in the affinity of individual components toward the stationary and moving phase. This affinity may be characterized on the basis of physical properties like solubility, adsorption, volatility, mass, etc. Usually, two or more of these physical properties are involved in any chromatographic separation. The overall process of chromatography is a differential migration phenomenon based on these physical properties and as a result if the component spends more time in the stationary phase, it moves slowly with the mobile phase. On the other hand, a component that spends less time in the stationary phase and more time in the mobile phase moves very fast. Thus, a complete separation of the two components is possible. Therefore, one can say that chromatography is a physical method that helps components to be separated by distributing them in two phases at different rates. Due to the difference in their distribution or partition between two immiscible phases, separation is possible. An uneven distribution of a compound between two immiscible phases occurs, if it is shaken vigorously with two immiscible solvents. At equilibrium, ratio of concentration of the solute in two phases remains constant and is referred to as partition or distribution coefficient (K_D, Equation 8.1). Under this circumstance, components in a mixture, having quite different distribution coefficients, may thus be separated.

K_D = Concentration of solute in stationary phase at equilibrium / Concentration of solute in moving phase at equilibrium Equation 8.1

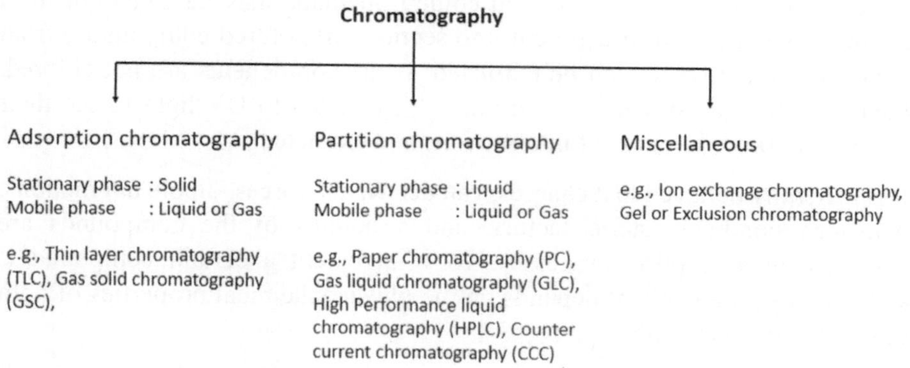

Fig. 8.1: Types of chromatography

8.4. Classification of chromatographic technique

Based on the physical state or nature of stationary and mobile phase used, chromatographic technique is mainly divided into two types: Adsorption and Partition chromatography (**Figure 8.1**).

8.4.1. Adsorption chromatography

A stationary solid phase and a mobile liquid or gas phase are used in adsorption chromatography. The separation is obtained due to the difference in the adsorption behavior of solid stationary phase toward individual solute components, viz., thin layer chromatography (TLC) and Gas solid chromatography (GSC). In the TLC mobile phase is liquid, whereas in GSC the mobile phase is gas.

The principle is based on the fact that a solid adsorbs different individual components of a mixture, which is present in a solution, to a varying degree. When a mixture of solute is poured in a vertically oriented column containing adsorbent, they will be adsorbed on the upper portion of the adsorbent. This is subjected to a free downward movement of solvent. As a result, components which have a greater tendency to be adsorbed on the solid surface, remain on the upper layers of adsorbents and the other components elute downward along with solvent. The less easily adsorbed components are thus removed from the upper part but are readsorbed at the lower part of the column. The adsorption of each solute is virtually restricted to a definite layer in the column, where solute that is strongly adsorbed on the solid remains on the top and the solute adsorbed weakly remains at the bottom of the column. This makes a separation of individual components from a mixture of solutes. Individual components may be collected through continued elution with solvent. One may apply other methods also to collect and identify individual components. On completion of separation on column, adsorbent column in shape may be taken out from column container and directly cut into sections (if colored compounds) from which separated solute can be extracted. If the components are not colored, then the whole adsorbent column may be subjected to UV light to see their fluorescent behavior and cut into sections and extracted.

It is generally believed that charges, van der Waals' forces, dipole interactions, hydrogen bondings, steric factors, and structures of the compounds are operative in adsorption processes. To what extent a certain compound is adsorbed by an adsorbent depends on the physicochemical properties of both adsorbent and adsorbate.

8.4.1.1. Adsorption isotherm

An adsorption isotherm is a graphical presentation of adsorption exerted by an adsorbent on a solution of a given substance at a fixed temperature. From an adsorption isotherm one may get an idea about the amount of substance adsorbed per unit weight of adsorbent from a solution of a particular concentration at equilibrium.

An increased pressure of a gas or vapor or an increase in the concentration of a solute causes increased adsorption. The relationship between the amount adsorbed and the pressure may be expressed by Freundlich Adsorption Isotherm (Equation 8.2 and 8.3).

$x/m = k \cdot P^{1/n}$, or $(x/m)^n = kP$ Equation 8.2

Where,

x = amount adsorbed,

m = weight of adsorbent,

P = pressure, and

K & n = constants for any given combination at any given temperature.

The disadvantage of this equation is that it does not hold good over a wide range of pressure. The same equation holds good to describe the behavior of adsorption from solution also. But in that case, instead of using pressure (P), concentration (c) should be used. Thus, it becomes:

$(x/m)^n = kc$

Taking logarithm of both sides, we get,

$n \log (x/m) = \log k + \log c$ Equation 8.3

i.e. if the logarithm of the amount of solute adsorbed per unit weight of adsorbent is plotted against the logarithm of the final concentration of the substance adsorbed, a straight line is obtained.

Langmuir (1916) derived an equation (Equation 8.4) based on the assumption that the process of adsorption is due to an equilibrium established between adsorbed and non-adsorbed molecules and adsorption is of monolayer thickness. He proposed that the molecules striking on the surface of solid adsorbent adhere to the surface and after an appreciable time it again comes back to the solution phase. The time lag between adherence and leaving the solid surface makes adsorption.

$m = k_1 K_2 c / 1 + k_2 c$ Equation 8.4

Where,

k_1= a measure of the number of active sites per unit weight of adsorbent, and

k_2= a measure of the affinity of solute toward the adsorbent. It is affected by all components of the system.

Hinshelwood suggested another equation (Equation 8.5) which is of more practical meaning as it encounters a number of different sites having different affinities toward the solute on the surface of adsorbent, while Langmuir adsorption isotherm assumed only one type of binding site exhibiting same type of affinity toward the solute.

$m = \sum k_1 . k_2 . c / 1 + k_2 . c$ Equation 8.5

The presence of many different binding sites varying in their affinities toward the solute is clearly seen in the elution profile of the solute having 'tailing' effect (**Figure 8.2**). This 'tailing' effect may be minimized or totally curtailed by using suitable gradients of pH, ionic strength, or polarity in such a manner that a more strongly adsorbed molecule faces a higher concentration of displacing compound.

Principally three different shapes of adsorption isotherms may be obtained: a) straight, b) convex, and c) concave adsorption isotherm (**Figure 8.2**). From the shape of the curve one can get an idea related to the speed of elution of a compound through a column and its concentration in solution. In the first case of straight adsorption isotherm (a) partition coefficient is constant irrespective of its concentration in solution.This means that the solute is distributed within the mobile and stationary phase in the column according to a set ratio of the distribution coefficient, giving rise to a straight-line isotherm, with the slope being the distribution coefficient. Under this situation, an ideal symmetrical shaped (Gaussian) peak is obtained.

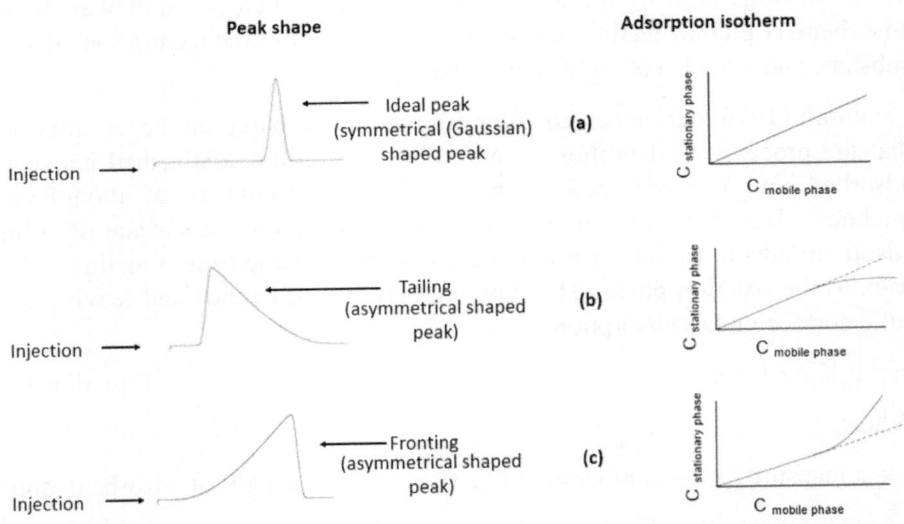

Fig. 8.2: Peak shape showing the elution pattern of solute due to its varying affinity to the different binding sites and their corresponding adsorption isotherm

$C_{\text{stationary phase}}$ and $C_{\text{mobile phase}}$ denote the analyte concentration in stationary and mobile phase, respectively.

In the second case (b), convex adsorption isotherm, the stationary phase is saturated, resulting in overloading the column capacity and a 'tailing' effect in the peak shape (leading to an asymmetrical peak shape) is envisaged. However, the third case (c), concave isotherm, may also be obtained under two situations: 1. When solute is dissolved in the stationary phase and resultant mixed phase increases the solute's solubility. However, this possibility is very remote; and 2. When injected solute concentration is too high, the increased vapor pressure allows its major portion of the solute to be retained in the stationary liquid phase. This may be evidenced in gas liquid chromatography. The concave isotherm leads to an asymmetrical shaped peak with a 'fronting' effect. In convex adsorption isotherm, substance travels faster at higher concentration than at lower concentration, while in concave adsorption isotherm the picture is just the opposite, i.e., at higher concentration in solution the substance moves slower. This is more understandable if we consider the conditions at two concentrations. One concentration is two times more than the other. Under this condition, in the straight adsorption isotherm the amount of adsorbed substance per unit mass of adsorbent is just double at the concentration where it is two times more than the other i.e., partition coefficient is constant at two different concentrations. But in convex adsorption isotherm amount of adsorbed substance per unit mass of adsorbent is less than double at the concentration where it is two times more than the other, and in concave adsorption isotherm the amount of adsorbed substance per unit mass of adsorbent is more than double at the concentration where it is two times more than the other. Straight adsorption isotherms are applicable at very low concentration. Usually in adsorption chromatography convex adsorption isotherm and in partition chromatography straight adsorption isotherm are the most common and predominant form of isotherm. However, they are also encountered in ion exchange chromatography. Chromatographic separation rarely faces concave type of adsorption isotherm. In convex adsorption isotherms, there lies a greater possibility of separation of the substances if the distance between the adsorption isotherms of the substances is more. The same is true for straight adsorption isotherms too.

Adsorption isotherms are determined by vigorous shaking (until equilibrium between adsorbed amount and amount remaining in solution is reached) known volume of the solution at fixed concentration with known amount of adsorbent. In the next step, solutions are separated by centrifugation and concentration of substance in solution are determined. Thus, at equilibrium the amounts of substance adsorbed per unit mass of adsorbent at different concentrations are calculated and plotted against concentrations of substances in solution.

8.4.2. Partition chromatography

Use of a stationary liquid phase (mostly an immobilized liquid film on a solid support) and a mobile liquid or gas phase is under practice in partition chromatography. Differential partitioning of the components between two phases leads to the separation. The extent of separation may be measured from the partition coefficient (Equation 8.1). Under ideal conditions at any given temperature and concentration, the partition coefficient remains constant. It remains unaltered even by the presence of some other substances. Paper chromatography (PC), Gas liquid chromatography (GLC) and High-performance liquid chromatography (HPLC) are the classical examples of partition chromatography. In PC and HPLC, stationary and mobile phase are liquid. In GLC, the stationary phase is liquid and the mobile phase is gas. The difference between PC and HPLC lies only in the mode of mobile phase movement. In PC, mobile phase moves either by capillary action or gravitational force, while mobile phase is forced to move by subjecting a high pressure from external source in HPLC. The principle of partition chromatography may be applicable to some protein separation, obviously with few difficulties. Proteins may denature in organic solvents and at the interfaces. This is why it is suggested to work at temperature below –0.5°C. But again, lower temperature may cause decreased solubility of most of the proteins in organic solvents. Even minimum change in temperature leads to their changing behavior in respect of their affinity to the phases. One of the first applications of the partition chromatography was purification of catalase by partition between ethanol and aqueous ammonium sulfate. Partitioning of ribonuclease in aqueous ammonium sulfate – cellosolve yields two enzymatically active fractions.

8.4.3. Miscellaneous

Additionally, other types of chromatography may also be employed, viz., Ion exchange chromatography, exclusion or gel chromatography, etc. Ion exchange chromatography is involved with an ion-exchanger as stationary phase and liquid electrolyte as mobile phase. In exclusion or gel chromatography, a liquid phase inside pores of a stationary porous structure and a mobile liquid phase are used. The molecules of smaller size are trapped inside the pores and the bigger molecules move with the mobile liquid phase.

8.5. Paper chromatography

8.5.1. Principle

In paper chromatography, support is paper (cellulose) or treated paper. Cellulose is a fibre network of polymeric carbohydrate chains, capable of holding appreciable quantities of water. Water is adsorbed between cellulose fibers and it makes a hull around them. This water hull acts as the stationary hydrophilic liquid phase. A liquid organic phase is used as the mobile phase. The solute mixture is dissolved in an organic volatile liquid and spotted onto the paper at one end and dried. The paper spotted with solute mixture is then housed in a chamber containing the solvent that works as a mobile phase. Solvent flows upward due to capillary action (Ascending paper chromatography) or moves downward by gravity (Descending paper chromatography). As solvent flows along the paper sheet, individual components of the mixture, if at all they show solubility toward the developing solvent, move on the paper at its own characteristic rate. After development, the solvent front is marked and the paper is air-dried. The position of individual components is then visualized by a suitable staining reaction. In the resultant chromatogram, the components are characterized by R_F (*Retention Factor*) value which is defined as the ratio of distance moved by a compound to that by the solvent (Equation 8.6).

R_F = Distance traveled by the compound / Distance traveled by the solvent
Equation 8.6

R_F value measures the relative mobility of a sample solute to that of the developer solvent. It is a unique characteristic property of a compound and remains constant, if all operating parameters, such as solvent, solute concentration, pH, temperature, humidity, etc. remain same. Minute fluctuation in any of the aforementioned operating parameters leads to altered R_F. R_F value of one signifies that the solute has no affinity for the stationary phase and travels with the solvent front, while R_F value of zero indicates that the solute remains in the stationary phase and it does not move with the mobile phase across the stationary phase. Thus, the solute is immobile in the applied chromatographic system. It is important to note that R_F value never exceeds one.

Out of the two types of paper chromatography depending on the direction of solvent flow: ascending and descending paper chromatography, the former is more preferred than the other because of its simplicity of set up and handling. However, solvent flow is faster in descending paper chromatography, since solvent moves by gravity. There is one more added advantage of descending type PC over the ascending type: in descending PC, resolution of the individual components for their separation may be enhanced by allowing solvent to drip out of the edge of the paper. But in that case, R_F measurement for the compounds is not possible.

8.5.2. Support

Chemically, the support medium is cellulose or modified cellulose. Most frequently used support media in paper chromatography are Whatman No. 1 filter paper, Schleicher and Schüll 2043b paper. Whatman No. 4 and 5 are also used, but with a lesser resolution compared to Whatman No. 1. Whatman No. 1 gives more reproducible results. In all papers, loading of sample should be minimum to get good results; overloading of sample on the paper may cause a 'tailing' effect hampering the resolution. For preparative use to separate larger quantities of material, it is suggested to use Whatman 3 MM and 31 ET, Schleicher and Schüll 2040a paper. Cellulose after chemical treatment by acetylation is also often used to avoid 'tailing' and for the separation of hydrophobic substances, such as carotenoids. Ion exchange papers, silica impregnated papers, etc. are now commercially available. To separate lipids and similar hydrophobic materials silica impregnated papers are best suited. For separation of polar substances, papers with increased exchange capacity due to higher carboxyl content are now being used. So-called carboxyl papers are found to separate cations, amines, and amino acids. Cellulose ion exchangers, such as carboxymethylcellulose, aminoethyl cellulose, Diethylaminoethyl (DEAE) cellulose papers give good results in separating acids and inorganic compounds. Cellulose phosphate appears to be an excellent selective exchanger. It is chemically cellulose dihydrogen phosphate and acts as a bifunctional exchanger, since it contains both strong acid and very weak acid groups. Other than cellulose paper, now glass fibre papers are also available. The added advantage of the glass fibre papers over cellulose papers is rapid analysis. Besides, one can use corrosive reagents in glass fibre papers, which cannot be used in cellulose papers.

Whatever papers are used in paper chromatography, they should have uniformity in their thickness; otherwise, non-uniformity in thickness may interfere in the flow rate of the mobile phase and thus, may lead to altered R_F. Papers for paper chromatography are commercially available in different shapes and sizes, which save time and provide more reproducible results.

8.5.3. Experimental procedure

8.5.3.1. Sample preparation

Sample should be dissolved in a suitable organic solvent with low boiling point, e.g., acetone, ethanol, chloroform, etc. Amount of solvent taken to dissolve the substance is such that approximately 10 μl solutions contain nearly 5-15 μg of the test sample. But for preparative use loading may be done up to 1000 μg. Biological samples that contain test material at low concentration are extracted with suitable solvent, or a mixture of solvents followed by a concentration step. Co-extractives or impurities, such as proteins, lipids, inorganic ions, pigments, etc. should be removed from the biological samples; otherwise, they may interfere in the detection step. Moreover, presence of excess impurities results in poor chromatogram with spreading of spots and alteration in R_F values. Proteins are removed by ultrafiltration, ultracentrifugation, using ion exchangers, or by some other methods. Inorganic materials are eliminated by using ion exchangers or electrodialysis. Lipids may be avoided by extraction with hydrophobic solvents. Sometimes it is a practice to develop a chromatogram with lipophilic solvent first, so that lipids are washed off and collected at the solvent front which is then cut off. The material remaining at the origin is subsequently developed with suitable solvent. Derivatization of substances, though it is time consuming, sometimes gives chromatograms with good resolution, where a volatile substance is chromatographed in the form of a non-volatile substance. Derivatization process also facilitates detection, where it is difficult to detect the parent compound.

8.5.3.2. Sample application

First of all, a faint line is drawn at the edge of the paper with a pencil and samples are applied on the line with the help of glass capillary tubes or micro pipettes in the form of a spot or a streak. Mostly for qualitative purposes a sample is applied in the form of a spot, while for semi-quantitative purposes a streak is much preferred. Spots or streaks should be dried by blowing air with a hair-drier before another application to the same spot or streak is made. Now-a-days automatic applicator coupled with a drying system is available, which automatically controls the amount of solution to be applied. This is essentially useful in the case where a large quantity of biological samples (1 or > 1 ml aqueous solution) is to be applied on the paper for semi-quantitative separation purposes. The exact amount of sample applied is critical. There must be a sufficient amount of sample on the spot, so that the amount of sample present on the developed zone is enough to be detected. At the same time care should be taken to avoid overloading which may cause tailing and lack of resolution. Finding the proper sample size is determined by trial-and-error method.

8.5.3.3. Mobile solvent

Chemical nature of the solute (polar or nonpolar) is the prime factor to determine the choice of mobile solvent for its separation by paper chromatography. If the solute is extremely polar in nature, it becomes highly soluble in polar solvents, such as acetone, methanol, etc.; and if these solvents are used as the mobile phase, it moves close to the solvent front, R_F value is then near or equal to unity. On the other side, if hexane, petroleum ether, etc. which are nonpolar in nature, are used as mobile phase, the solute is not sufficiently soluble and does not move. They remain at or near the origin and R_F is nearly zero. The event is not expected for the purpose of separation. Therefore, it is imperative to use a mixture of polar and nonpolar solvent, so that the solutes spread across the length of paper with reasonable R_F value. Very often acetic acid or ammonia is added to the mobile organic phase to retain an acidic or basic environment, since pH also affects the solutes' separation. Adjustment of pH in mobile systems is occasionally done for suppression and / or enhancement of dissociation, and for increasing water solubility in mobile phase. This makes the system more hydrophilic in character.

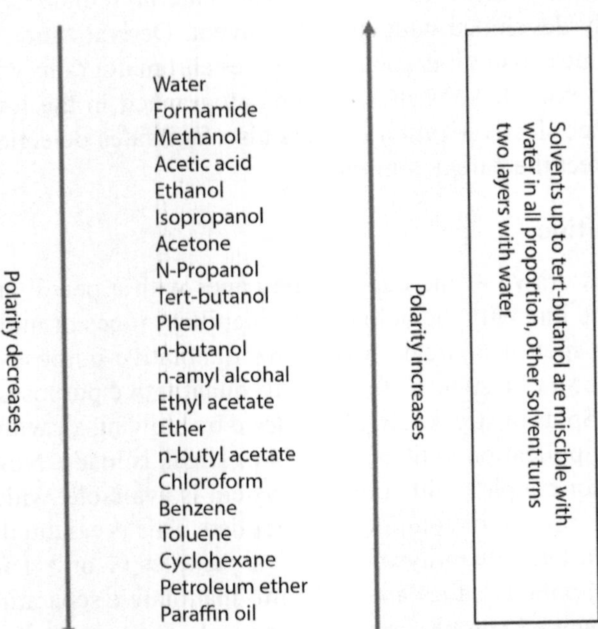

Fig. 8.3: Eluotropic solvent series

To separate polar hydrophilic substances or substances of medium polarity, water is bound to the carrier (cellulose), and a mobile water-immiscible organic solvent flows over it. To separate medium to low polarity substances, it is a general practice to use methanol, formamide or propylene glycol, etc. as stationary phase and a hydrophobic solvent as mobile phase. On some occasions, reverse phase paper chromatography is used successfully. In this case hydrophobic organic solvent is used as the mobile phase. But to accomplish this, cellulose being of hydrophilic character must be modified in such a way that it can hold hydrophobic stationary phase. The modification is done by impregnation of cellulose with silicone oil, or esterification of hydroxyl group of cellulose. This reverse phase technique is primarily employed for separation of hydrophobic substances. It is noteworthy to mention that in this reverse phase chromatography individual components migrate in the reverse order, i.e., nonpolar components move first followed by polar components. A list of solvents analogous to the eluotropic series is given in **Figure 8.3**. At the top of the series solvents are able to form intermolecular hydrogen bridges (hydrophilic or polar solvents or lipophobic) and at the bottom of the series solvents do not form any intermolecular hydrogen bridges (hydrophobic or nonpolar or lipophilic). In between these two types, there is a continuous transition, formed by the solvents of medium polarity.

8.5.3.4. Development

Paper strips spotted with samples should be placed on the developing tank in such a manner that spots are just above the surface of the mobile solvent. The developing tank should be air-tight and saturated with solvent vapor. Filter papers moist with developing solvent are kept on the two sides of the tank to enhance vaporization of the solvent.

Paper strips are developed in different arrangement modes: (1) ascending, (2) descending, (3) ascending-descending hybrid, (4) horizontal, and (5) radial mode or circular development; out of which ascending and descending development are common.

Ascending chromatography is more frequently used with an added advantage that separation may be carried out in two dimensions. However, descending development is associated with faster migration as the flow is governed by both gravity and capillary action. Separation may also be done with better resolution in case of descending development for compounds possessing low and quite similar R_F values, since developing solvent may be allowed to drip off the paper. Although R_F measurement is not possible in this case, it is unique for qualitative separation. However, ascending development is preferred due to its simplicity of set up and ease of handling. In ascending paper chromatography, the solute rises up the paper along with the mobile phase by capillary action against the gravitational force.

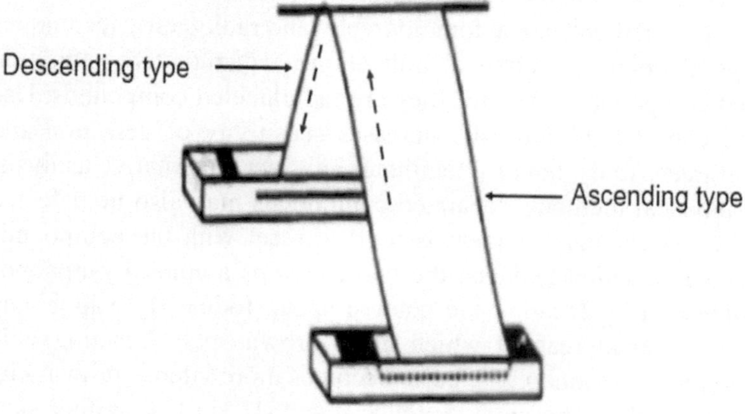

Fig. 8.4: Ascending-descending hybrid paper chromatography

In ascending-descending hybrid development (**Figure 8.4**) ascending is followed by the descending type. Length of separation is increased in this hybrid development of chromatogram.

Horizontal development is not so popular compared to others. In this type of paper chromatography, the development ensures solvent migration in the horizontal direction.

In radial or circular mode, solvent moves from the midpoint of a circular paper toward its periphery. The wick at the center of the paper dips into a mobile phase in a Petri dish by which the solvent drains on to the paper and moves the solute radially. Eventually, the solutes are separated in the form of concentric rings on the paper.

When development solvent reaches near the edge of the other end of the paper strip, development is over. The chromatogram is then taken out of the chamber, air-dried, and treated with suitable chromogenic reagents that give some color reactions with the individual components.

8.5.3.5. Detection and identification of spots

After development, the paper is usually air-dried or dried at an elevated temperature (≈ 80 °C) in order to avoid solvent.There is no problem to detect compounds on the chromatogram if they themselves are colored. Problem lies with colorless compounds to be detected in paper chromatography. Substances that are highly conjugated may be detected by fluorescence under UV-lamp. Large number of organic compounds exhibit absorption of light at 240-260 nm range. Compounds show dark spots against a fluorescent background of the paper. Again, some compounds show characteristic fluorescence under UV-light. Radioisotopic methods like autoradiography and radiometric techniques also provide a good tool for detection, identification and quantitative estimation of the separated compounds, provided they are radiolabeled compounds. Use of radiolabeled compounds not only increases sensitivity of detection and identification of many folds, but also facilitates an exact quantitative analysis. Besides these physical methods, separated compounds may also be detected chemically, where a chemical reagent is used to react with the compounds and the resultant interaction between the two develops a color. Compounds may be then detected by locating the colored spots. Iodine (I_2) vapor may be served as a universal reagent which gives brown spots, reacting with unsaturated organic compounds. The color becomes more intense brown with more unsaturation in the compound. Sulphuric acid (H_2SO_4) is another such universal reagent which churns organic compounds and gives black spots. But this is not applicable to paper chromatography, since it churns paper too.

However, it is a preferred universal reagent used in thin layer chromatography. Besides, there are some specific reagents that are applicable for a particular class of compounds, viz., ninhydrin for amino acids, rhodamine B for lipids, aniline phthalate for carbohydrates, etc. General procedure to apply reagents on compounds is to spray the same with an atomizer on the chromatogram, so that uniform spraying is maintained. Detection of enzymes is frequently carried out using their substrate. Detection of amylase is accomplished by spraying starch solution and, after a suitable incubation period, with an iodine solution. Amylases appear as white spots on the blue background.

After separation and detection in paper chromatography, each individual component in the solute mixture is identified by their relative mobility in a suitable solvent, R_F values (Equation 8.6, **Figure 8.5**). This is based on the fact that a substance in a mixture shows the same R_F value in a certain solvent system as the reference standard substance does.

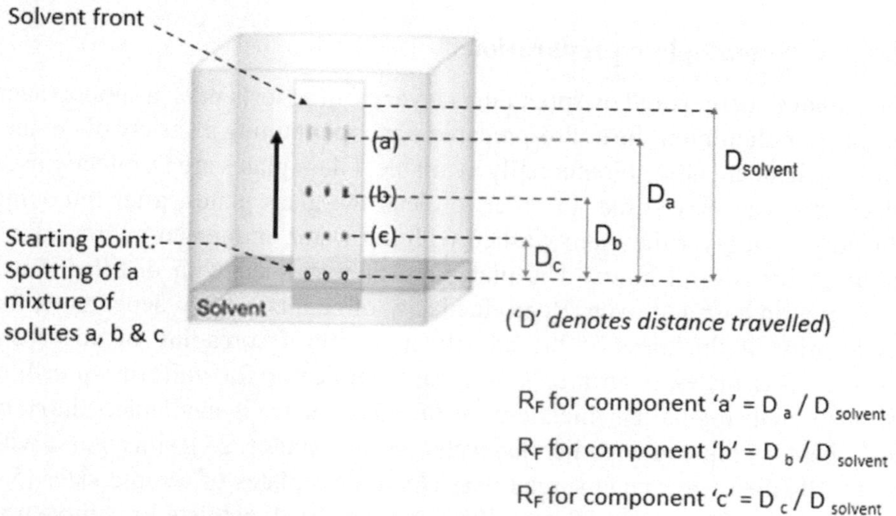

Fig. 8.5: Paper chromatogram development and R_F determination

8.6. Thin-layer chromatography

8.6.1. Principle

Thin-layer chromatography (TLC) is a rapid technique to separate and analyze a small quantity of related substances from microgram to nanogram level. This technique is similar to paper chromatography. But the difference between them is that PC is exclusively partition type chromatography; while depending on the nature of the medium in TLC, separation may be obtained by adsorption, partition, gel filtration or ion exchange phenomenon. However, most commonly used TLC is adsorption type of chromatography, where a solid adsorbent is used as a stationary phase and a liquid as a mobile phase. The advantage of TLC over PC is that a variety of support medium, even a mixture, can be used in TLC, whereas only cellulose or modified cellulose acts as support in PC. A better resolution is achieved in TLC as compared with PC due to compactness of spots. TLC is less time consuming and there is a possibility of choosing a wide spectrum of chromogenic reagents for detection purposes, since in TLC one can use a very corrosive reagent, which is not possible in PC. Moreover, after spraying reagents for color formation, chromatoplates may be subjected to high temperature in TLC, if it is needed. High temperatures might churn the paper on the PC.

Originally, loose layers of adsorbents spread on glass plates are used to separate compounds. Adsorbents are bound to glass plates by suitable binding agents, such as starch, gypsum, etc.

8.6.2. Chromatoplate preparation

TLC plates are prepared by spreading a layer of an adsorbent on a support such as glass or aluminum sheet. Pre-prepared plates containing a variety of choices on adsorbent are now commercially available. Glass plates are frequently used since they can be re-used. In order to clean the glass plates, after removing the adsorbent layer these are washed with tap water, immersed in 10% alkali solution for several hours. The plates are then washed with distilled water and dried in a drying oven. Reproducibility of results greatly depends on the uniformity of thickness of the adsorbent. Uniform spreading of adsorbent on the glass plates is an art. Now-a-days applicator for uniform spreading of adsorbent, regulating thickness on the glass plate, is available. Slurry of adsorbent is prepared with distilled water (as an instance, 25 g silica gel G with 50 ml distilled water) and spread over clean glass plates of desired sizes (5 x 20 cm, 10 x 20 cm, 20 x 20 cm). Plates are air-dried, activated in an oven at 110 °C for a few hours and then kept in a desiccator. Plates must be activated, particularly under humid weather conditions; otherwise, presence of moisture hampers resolving power of the adsorbents.

8.6.3. Adsorbents

Adsorbents should be chosen in such a manner that it should not induce any chemical reaction with compounds to be separated. Thus, choice of adsorbents largely depends on the chemical nature of compounds, whether it is acidic, basic, neutral or amphoteric. Silica gel is most commonly used adsorbent in

TLC for separation of neutral compounds. Besides silica gel, aluminum oxide, Fuller's earth (hydrous magnesium alumino silicates), activated charcoal, calcium carbonate, etc. are used to a great extent as adsorbent. Freshly prepared alumina possesses a high adsorption power but tends to lose the same by adsorption of water from the surrounding atmosphere. It is available in three different forms: acidic (pH 4.0), basic (pH 9.0) and neutral (pH about 7.5). Charcoal acts as a strong adsorbent because it contains carbon-carbon spacing like a graphite-type lattice and a variety of polar and ionic groups. Very often cellulose and acetylated cellulose are also used in TLC with partitions governing the separation as cellulose contains water and the water hull acts as a stationary phase. A list of adsorbents, used for separation of different compounds, is summarized in **Table 8.1**. Usually adsorbents adsorb solutes by H-bonding, ion exchange, van der Waals forces etc. Adsorbed solute molecules appear to cover the whole external surface of the adsorbent.

Table 8.1: Adsorbents used in TLC to separate different compounds.

Sl.No.	Adsorbents	Compounds for separation
1	Silica gel	Sterols of fatty acids and glycerides, azoated carbohydrates, sugar acetates, amino acids
2	Alumina	Almost all organic compounds except saturated aliphatic hydrocarbons
3	Fuller's Earth	Color from petroleum oil, amino acids, pyridines
4	Charcoal	Aromatic substances
5	Cellulose and acetylated cellulose	Amino acids, carbohydrates, naturally occurring hydrophilic compounds

8.6.4. Solvents

Choice of solvents is another important criterion to separate compounds through adsorption or partition type of TLC. Organic solvents analogous to the eluotropic series (**Figure 8.3**) are ideal for the purpose. Besides, it is most frequently observed that a mixture of two or more solvents of different polarity performs better to separate compounds than a single solvent does.

8.6.5. Tank

TLC plates are developed in a rectangular glass tank which should be saturated with a solvent system being used. Solvent saturated environment inside the tank is created by pouring solvent to a depth of 1.5 cm and then filter paper cuts wet with the same solvent system are placed on three sides of tank, dipping a part into solvent. Front side is left without filter paper to monitor the plates during development. The lid of the tank is sealed with petroleum jelly and removed for minimum time necessary to place the plate.

8.6.6. Sample application

The procedure of sample spotting is similar to that of paper chromatography. With a capillary glass tube, a microsyringe or a micropipette, samples are applied as spots, 3-5 mm in diameter, 2-3 cm from the bottom edge of the plate. Usually, a non-polar solvent is used to dissolve the sample. This avoids spreading of the spot and helps to obtain a small area of application. Smaller is the size of the spots, better is the resolution. Dissolving solvent should be significantly volatile so that it can be removed easily before starting development. During application of sample on the plate, care should be taken in such a manner that the adsorbent layer is kept intact without any damage made by the capillary tube. Irregularity in the movement of mobile solvent may occur due to damage of the adsorbent layer resulting in poor resolution. It is always recommended to run standards on the same plate since R_F values may show some variations among plates on different days. To avoid any misinterpretation of chromatograms, the standard should be pure enough.

8.6.7. Development

TLC plates spotted with sample and standard reference compound are left at room temperature for 5 minutes to remove residual solvent. The plate is then developed in an ascending manner till mobile solvent reaches 1 cm from the top of the plate.

8.6.8. Detection

If compounds are colored, they might be visualized in front of ordinary light or daylight. For colorless compounds, like paper chromatography, chromatoplate is sprayed with a suitable chromogenic reagent; so that, after going through reaction, the compounds become colored and are easily visualized. In case of radiolabeled compounds, autoradiography shows the spots as dark areas on X-ray film or the plate may be scanned by radioautograph scanner. This is applicable to paper chromatography also. A list of chromogenic reagents frequently used in TLC is presented in **Table 8.2**.

Table 8.2: Few examples of chromogenic reagents giving color after reacting with compounds.

Sl.No.	Reagents	Color developed	Compounds detected
1.	Daylight	Various colors	Colored compounds
2.	UV-light	Fluorescence spot on dark background	Many organic compounds
3.	Sulfuric acid and heating	Black spot	Mostly all organic compounds
4.	Iodine vapor	Brown spot on yellow background	Mostly all unsaturated organic compounds
5.	0.25% Ninhydrin in acetone and heating	Pink to purple spot on white background	Amino acids and amino group containing substances
6.	Ammoniacal silver nitrate	Black spot	Chlorinated hydrocarbons
7.	50% $SbCl_3$ in glacial acetic acid	Characteristic colors	Steroids, carotenoids, vitamins, etc.
8.	Molybdenum blue reagent (MoO_3 + Mo-powder + 25 N H_2SO_4)	Dark blue spot on a white or light blue grey background	Phospholipids
9.	α-naphthol in methanol-water + 50% H_2SO_4 and heating	Purple spot	Glycolipids
10.	Potassium iodide + bismuth subnitrate in acetic acid	Orange red spot	Chlorine containing phospholipids (Dragendorff's test)
11.	Radiolabeled compounds	Dark spot on x-ray film through autoradiography	Exclusively for all compounds

8.6.9. Preparative-TLC and –PC

Thin layer and paper chromatography are not restricted only to qualitative analysis, they can also be extended for quantitative purposes. When the analysis is being made quantitatively with TLC or PC, the technique is referred to as preparative-TLC or preparative-PC. Basic principle of analytical and preparative chromatography is the same. Only difference between them is the thickness of the stationary phase used. For preparative purposes, a thick solid support cum stationary phase is used to increase sample capacity. In TLC for analytical separations, the layer is in the order of 0.25 mm thick and for preparative separations, it may go up to 2 mm. For preparative-PC, thick chromatographic paper cardboards like Schleicher and Schüll 2071 are most suitable. Up to 200 mg of mixture may be applied at one spot. Edrol 202 S, Schleicher and Schüll 2040 b GI, 2043 GI, 2045 a GI, Whatman 31 ET, Whatman 3 MM papers are also used in preparative-PC. It is easy to handle these papers in ascending paper chromatography. In the descending method, a strip of ordinary filter paper is sewed on above the start, permitting a slower flow rate and a good separation. The sample is applied as a band or a streak across the plate or paper cardboard rather than as a single spot. In preparative-PC, spots may be cut with scissors and paper containing the compound is extracted with a suitable solvent. In case of preparative-TLC, solid support from a spotted region is scrapped with a knife edge or a razor blade and then is extracted with a suitable solvent. Quantitative estimation is then carried out by various methods, such as gravimetry, spectrophotometry, GLC / HPLC, etc. Take an example of separation of phospholipids by preparative-TLC. Once phospholipids are separated, the phospholipids spots are scrapped from TLC plate and transferred carefully to a good quality glass tube. This is done immediately after visualizing phospholipids by iodine vapor. In order to quantify phospholipids, its phosphorus is estimated after converting it to inorganic phosphate, which is achieved by hydrolyzing phospholipids by an anhydrous acid (perchloric acid, 60 %). Following the release of phosphorus, it is measured by any method of inorganic phosphate estimation method (e.g., ammonium molybdate method).

8.6.10. Two-dimensional chromatography

This is especially applicable for difficult separation in TLC and PC. Basically, this is accomplished in two separate mobile solvent systems, where K_D values differ each time. The mixture should be spotted at a corner on the solid support and developed with a volatile solvent system in one direction. After drying, the plate or the paper should be developed with another solvent system in a direction at right angle to the first development. For turning the plate 90°clockwise, primary spots then lie along the bottom (**Figure 8.6**). Since development is undertaken in two separate directions, it is known as two-dimensional chromatography. The chromatogram, thus obtained, can be compared with a chromatogram of known compounds developed under the same conditions.

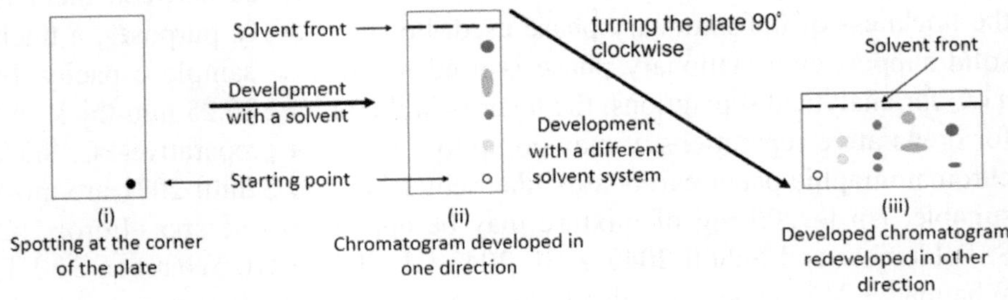

Fig. 8.6: Two-dimensional chromatography

8.7. Column chromatography

8.7.1. Introduction

Column chromatography is mostly adsorption type of chromatography, where a solid adsorbent is used as stationary phase and a liquid solvent as mobile phase. In biochemical work and natural product separation, column chromatography is widely used. In this technique, the column of adsorbent is considered as the heart of the system. It is possible to separate a large number of substances in significant quantities, using a column of adsorbent rather than a thin layer. A glass tube of about 3 feet long and 0.5-inch internal diameter is frequently used, although it may vary according to experimental requirements. In general, long columns give good resolution, but restrict flow rate of the solvent. Therefore, length of the column should be chosen as a compromise between resolution and flow rate. Chromatographic tubes of metal or plastic are also used in place of glass tubes due to their greater strength, more particularly during application of suction or pressure. However, one should take care in selecting chromatographic tubes made up of materials other than glass, so that the materials used in chromatographic tubes should be resistant toward solvents.

When the sample in liquid form is placed on the top of the adsorbent column, compounds distribute themselves between stationary and mobile phase. Different components in the sample mixture have different affinity toward the stationary and the mobile phase and this difference allows them to be separated in two phases. As a result, components move through columns at different rates and times.

8.7.2. Column packing

There are two techniques for column packing with solid adsorbent: (a) dry and (b) wet packing technique. Initially the column is plugged with cotton or glass wool at the bottom of the column to avoid the adsorbent being washed out, but it allows free passage of solvent. It is also possible to use a porous or perforated disk as support for the packing material. In dry packing technique, the column is packed with dry adsorbent with the help of a packer or tamping rod and is then rinsed with solvent thoroughly so that no air-bubbles get trapped inside the column. In the case of wet packing technique, which is more conveniently used, one-third of the column is at first filled with the initial developing solvent and then a slurry of adsorbent made with the same solvent is poured into the column. During pouring the slurry, solvent is allowed to flow out slowly so as to enable settling of adsorbent at the bottom without trapping air. This can be achieved more easily with gentle tapping the column from outside. The slurry may be prepared with different solvent. But in this case, the column should be thoroughly rinsed with the solvent system that is to be used initially after packing is over. Pouring of slurry is continued until packed adsorbent reaches its desired height. In wet packing technique compared to dry packing technique, presence of trapped air is negligible, and it provides a homogeneous column. Sometimes it is a usual practice to pack a column under pressure or suction, but care should be taken when the system is subjected to an external force. Prolonged or high pressure may lead to crack formation in the column inducing air pockets that affect the separation. In general, packing of a column under pressure generates a tightly packed bed that yields more reproducible results. After rinsing the column properly with the solvent, solvent level is kept just above the adsorbent layer to prevent drying of the column. A filter paper disk or glass wool is kept over the adsorbent to avoid disturbing the surface while pouring the sample. It is better to keep some amount of sodium sulfate at the bottom of the column over the glass wool and above the adsorbent surface below the glass wool or filter paper disk. This will trap water, if at all present in the sample. Care should be taken to minimize the space below the packing material (dead space) to avoid the mixing of effluent.

8.7.3. Adsorbents

In column chromatography, a variety of adsorbents of different mesh sizes are used. Alumina (Al_2O_3), silica gel (SiO_2), MgO, resins, calcium phosphates, charcoal, sucrose, florisil (magnesium silicate), hydroxyapatite, etc. of 100-200 mesh size are best suitable column adsorbents for most biochemical analysis. An adsorbent of 100 mesh size means that it passes through a 100-mesh sieve which has 100 small openings per square inch. The adsorption properties of adsorbents may be modified as per the experimental requirement by treating them with acid, water or by heating. Treatment of adsorbent with acid or by heating increases the activity of adsorbent and thus, the phenomenon is known as the 'activation' while treatment with water (e.g., alumina) decreases its activity leading to 'deactivation'. It is found that hydroxy apatite may be used for isolation of polynucleotides, nucleic acids and proteins; silica gel for amino acids, lipids and carbohydrates; florisil for neutral lipids; although there are scopes to select effective adsorbents by trial-and-error method.

8.7.4. Solvents and elution

Usually, organic solvents analogous to eluotropic series are used in column chromatography as eluting solvents. More polar is the solvent, greater is the eluting power; obviously depending on the chemical nature of the substances. It is also a fact that polar compounds are adsorbed strongly on the solid stationary phase. However, the choice of solvent should be made in such a manner that eluting solvent should be inert toward compounds and show difference in solubility of different components in order to allow good separation. All components in a solute mixture may be eluted using a single solvent system. This type of development is known as '**continual elution**'. It may be the case of '**stepwise elution**', using different solvent serially in the same column because a particular group of compounds may be eluted with a particular type of solvent keeping other group of compounds undissolved on the top of the column, which can be eluted by other type of solvents. Take an instance of a complex biological sample which can be first eluted by n-hexane that removes paraffin. Once paraffin is removed from the system, benzene may be used as the second solvent which can remove the aromatics. In the next round, benzene may be replaced by ethyl acetate followed by methanol. In each set of elutions, a particular group of organic molecules are separated. Such procedure makes it possible to separate a solute mixture into components with common functional groups. In general, gradually polarity of solvents is increased to separate components with increasing polarity. Even, different molecules are separated from columns by changing pH of the solvent in a stepwise fashion. Separation of molecules may also be achieved by changing the composition of the eluting solvent by mixing two or more solvents continuously at different ratios that can be gradually changed so as to create a gradient. This type of development is referred to as '**gradient elution**'.

8.7.5. Sample loading

Concentrated sample initially dissolved in the developing solvent (2-5 ml) is poured on to the column with a pipette very carefully, so that the surface of the adsorbent is not disturbed. Sample is allowed to percolate into the adsorbent. When the level of the sample meniscus just touches the top surface of the adsorbent, a few milliliters of solvent is further added to wash the sample into the column material. After that, solvent is added to the column and a steady and constant head of solvent is maintained enabling a steady flow rate. Depending on their affinity toward the stationary and the mobile phase, components in the mixture flow down the column at different rates and thus, are separated.

8.7.6. Eluent collection and detection

Collection of eluents from the column may be accomplished manually or by automatic fraction collector which collects a specified volume of eluent over a specified period. The eluent fractions, after collection, are concentrated to a minimum volume, applied on paper or TLC plates, and processed as discussed in Section 8.5 and 8.6. The fractions showing the same chromatogram are pulled together in order to increase concentration of a particular component. Finally, components may be further characterized by spectrophotometry, mass spectrometry (MS), GLC, HPLC, nuclear magnetic resonance spectroscopy (NMR), etc. Sometimes, eluent is directly attached to a photocell that measures absorption or fluorescence characteristics of the component. If a recorder is attached to a computer, one may get an idea on the eluting components through the graphical representation.

8.8. Ion-exchange chromatography

8.8.1. Introduction

The concept of ion-exchange chromatography has come into existence after knowing the ability of solid substances to exchange ions of the same charge with a solution. Clays are recognized as good ion-exchangers removing metal ions in their vicinity. It is a common practice to use zeolite (sodium aluminum silicates) in order to produce soft water, removing Ca^{+2} and Mg^{+2} from water by Na^{+} ions of zeolite. The ion-exchange phenomenon may also be exhibited by glass surface which is known to adsorb heavy metal ions releasing its Na^{+} ion in solution. Therefore, ion-exchange phenomenon may be utilized to separate ionic species i.e., for separation of cations from anions or vice versa, and ionic from non-ionic species. This technique is successfully employed in Biochemistry and allied fields for separation and purification of amino acids, peptides, proteins, nucleic acids, etc.

8.8.2. Principle

Principle governing the ion-exchange chromatography involves in separation of molecules according to their differences in molecular charge, electrical properties, differential affinity of charged ions or molecules in solution for immobile charged substances (resins) and a reversible interchange of ions of like signs between a solution and an insoluble solid. In this technique, an ion which is electrostatically bound to a chemically inert resin matrix is replaced by an ion present in solution and the process is reversible.

$$Res^{-}X^{+} + Y^{+}_{(solution)} \rightleftharpoons Res^{-}Y^{+} + X^{+}_{(solution)} \quad \text{Equation 8.7}$$

$$Res^{+}A^{-} + B^{-}_{(solution)} \rightleftharpoons Res^{+}B^{-} + A^{-}_{(solution)} \quad \text{Equation 8.8}$$

In Equation 8.7, $Res^{-}X^{+}$ denotes resin matrix containing X cation and Y^{+} represents cations in solution, while in Equation 8, $Res^{+}A^{-}$ is the resin matrix with A^{-} anion and B^{-} is for anions in solution. When solid and solution interact with each other, reversible exchange of ions takes place in two systems. As a result, cations which are bound to the resin matrix come to the solution phase and cations previously in solution become a part of the resin. The sample ions that interact strongly with the resin matrix are retained on the matrix most and in case of weak interaction of sample ions with the resin matrix, sample ions are retained least and thus eluted quickly enhancing a separation. The elution rate of an ionizable compound through a resin column depends on the degree of ionization, ionic strength, pH of the buffer solution and relative affinity of different ions present in the solution toward the charged site on the resin. It is possible to obtain a desired separation of electrostatically held ions on the resin surface by manipulating ionic strength and pH of the eluting buffer solution. Thus, compounds that bind tightly to the ion-exchanger may be eluted by a stepwise elution with buffer solutions of increasing salt concentration or of different pH values. In this case, the rate of ion-exchange is controlled by the law of mass action.

There are substances (e.g., protein molecules) which exhibit both positive and negative charges. Therefore, they can bind to either cationic or anionic exchangers depending on their pH dependent net charge. While purifying one protein from other proteins or polyelectrolytes, it is a usual practice to select pH and ionic strength of the buffer solution in such a manner that the desired protein is immobilized by the specific ion-exchanger with higher affinity and other polyelectrolytes as impurities, which do not bind or bind to the ion-exchanger with relatively low affinity, move through the column. If a mixture of amino acids needs to be separated, ion exchange chromatography may be applied using a cation exchanger, to which an acid solution of amino acid mixture at pH 3.0 is added facilitating cationic form of amino acids with net positive charge. As a consequence, cationic amino acids displace cations of the cationic exchanger and electrostatically bind to the resin matrix. The basic amino acids, such as arginine, histidine and lysine are bound more strongly than the acidic amino acids, such as aspartic and glutamic acid. At this point when the pH and the ionic strength of eluting buffer solution is increased, amino acids move down the column at different rates depending on their chemical nature and differences in degree of ionization. The most anionic amino acid elutes at the beginning, while the most cationic amino acid elutes at the end.

8.8.3. Ion-exchanger

In general, an ion-exchanger, which must be highly permeable and ionic in nature, is composed of two parts: a three-dimensional backbone matrix and a functional group. The functional groups on dissociation give rise to mobile ions that take part in exchange reactions with ions of same charge in the solution. The functional groups attached to the backbone determine the nature and strength of the ion-exchanger. Therefore, depending on the nature of the functional groups and whether it exchanges cations or anions, ion-exchangers may be classified as cationic exchanger and anionic exchanger. Cationic exchangers are those that have a backbone chemically bonded with negatively charged functional groups which, in turn, are associated with positively charged mobile or exchangeable ions. Therefore, the cation exchangers exchange, positive ions. The opposite is true for anion exchangers that exchange negative ions (Equation 8.9 and 8.10).

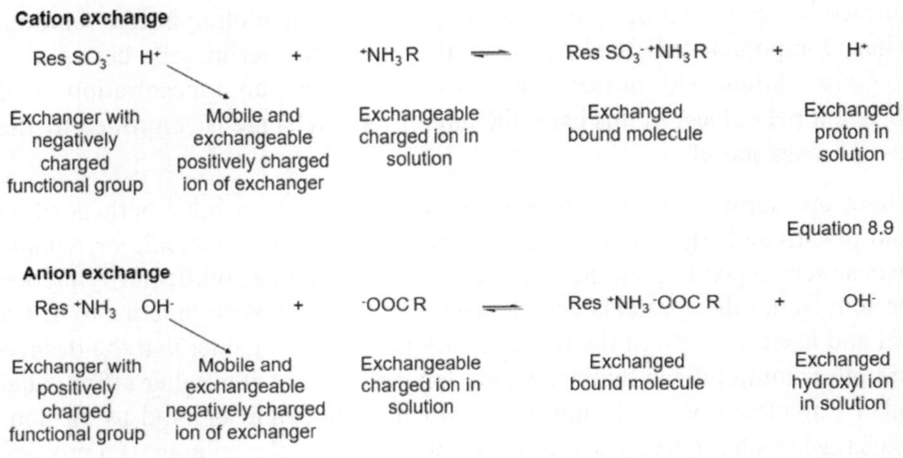

It is the charge carried by mobile or exchangeable ions that defines whether the exchanger is cationic or anionic, but not the charge carried on the backbone. Again, based on the ionizing strength of functional groups present on the backbone, each type of ion-exchanger is further grouped as a strong or weak exchanger (**Figure 8.7**). Strong ion-exchangers are totally ionized at all normal working pH values (i.e., their action is largely independent of pH), whereas weak ion-exchangers are ionized over a narrow range of pH values. For example, exchangers containing carboxyl group and amino group show their maximum activity above pH 6.0 and below pH 6.0, respectively.

There are different kinds of ion-exchangers available today, based on aluminosilicates, hydrous oxides of iron and aluminum, cross-linked synthetic resins (usually of polystyrene type formed by co-polymerization of styrene and divinylbenzene), cross-linked polysaccharide backbone ion exchangers, etc. Synthetic aluminosilicates, prepared by mixing aluminum sulfate and sodium silicate, act as cationic exchanger exchanging its Na^+ by the cations present in the solution. But it is of limited use as it is decomposed by acids and alkalis. Precipitated hydrous oxides of iron and aluminum act as anion exchanger. However, their ion-exchange capacity is small and they exist in an undesirable physical form for use in ion-exchange chromatography. Aluminosilicates and precipitated hydrous oxides, being inorganic ion-exchangers, impart greater selectivity toward simple inorganic ions. They have limited scope for analytical uses in the field of biochemistry.

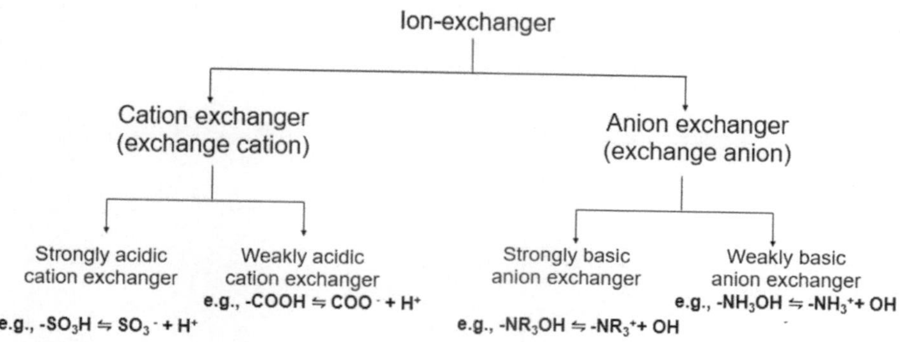

Fig. 8.7: Classification of ion-exchanger

In the field of Biochemistry, widely used ion-exchangers are polystyrene-based resins and polysaccharide backbone ion-exchangers. Polystyrene-based resins are frequently used to separate small ionic molecules, such as peptides, sugar phosphates, etc. They are not suitable for protein separation since their hydrophobic character denatures protein. For protein and nucleic acid like larger molecules, polysaccharide backbone ion-exchanger is most suitable. Cross-linked polystyrene resin is prepared in bead form by co-polymerizing styrene and divinylbenzene (**Figure 8.8**), and the product is then treated for functional group incorporation. Sulfonation leads to the production of a strongly acidic cation exchanger, whereas chloride form of strongly basic anion exchanger is obtained by chloromethylation, followed by treatment with a tertiary amine. Similarly, it is possible to construct a resin with weak acid group [-COOH, $-PO_3H$] or weak basic group [$-NH_3^+$, $-CH_2NH(CH_3)_2$, $-CH_2NH_2CH_3^+$] by derivatization of polystyrene. Usually, polystyrene resin is constructed with varying degree of cross-linkage that depends on the amount of divinyl benzene present in the resin. A polystyrene backbone of 2% cross-linkage refers to resin containing 2% divinyl benzene and 98% styrene. Cross-linkage varies from 1-16% in commercial resins with 4-8% being the best for general use. Low cross-linking permits relatively larger ions to enter into resin. But in this situation swollen resin volume changes significantly when the surrounding solution is changed. High cross-linkage retards penetration of molecules into resin granules. However, high degree cross-linkage retards expansion and contraction of resin during salt changes, which is an advantage in biochemical work.

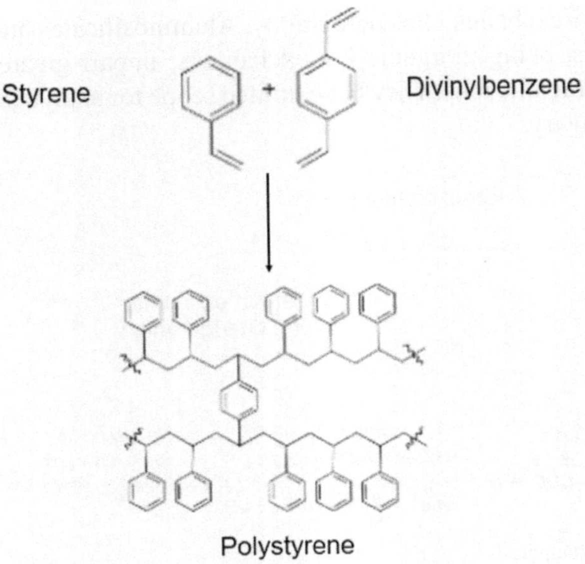

Styrene + Divinylbenzene

Polystyrene

Fig. 8.8: Polystyrene resin as a backbone for ion-exchanging

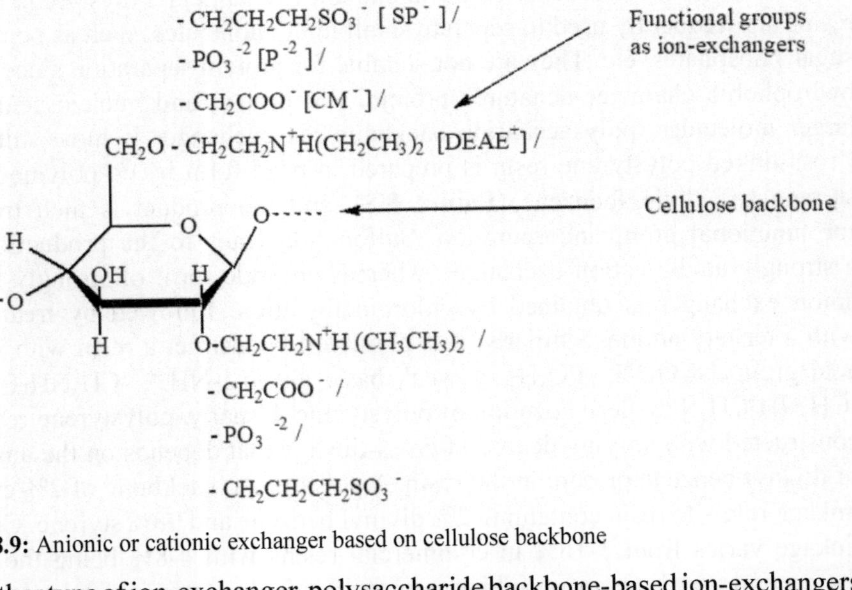

- $CH_2CH_2CH_2SO_3^-$ [SP^-]/
- PO_3^{-2} [P^{-2}]/
- CH_2COO^- [CM^-]/

CH_2O - $CH_2CH_2N^+H(CH_2CH_3)_2$ [$DEAE^+$] /

Functional groups as ion-exchangers

Cellulose backbone

H OH H O O H

O-$CH_2CH_2N^+H$ $(CH_3CH_3)_2$ /

- CH_2COO^- /
- PO_3^{-2} /
- $CH_2CH_2CH_2SO_3^-$

Fig. 8.9: Anionic or cationic exchanger based on cellulose backbone

Another type of ion-exchanger, polysaccharide backbone-based ion-exchangers are made from cross-linked cellulose (Sephacel), cross-linked agarose (Sepharose-CL) and cross-linked dextran (Sephadex G-25 or G-50). These types of ion-exchangers are generated by chemical modification of pre-existing hydroxyl groups on the backbone with charged groups like diethylaminoethyl ($DEAE^+$), quaternary aminoethyl (QAE^+), carboxymethyl (CM^-), sulphopropyl

(SP^-) or phospho (P^-) [**Figure 8.9**]. Sometimes these ion-exchangers are further treated to obtain benzoylated products (e.g., Benzoylated DEAE) that enhance both ion-exchanging and adsorptive capacity. Some commercially available ion-exchangers with their characteristics are listed in **Table 8.3**.

Table 8.3: Few commercially available ion-exchangers

Commercial name	Matrix	Functional groups	Type
		Cation exchanger	
CM-Sephadex	Dextran	- CH_2COO^- (carboxymethyl)	Weakly acidic
Bio-Rex 70	Acrylic	- COO^- (carboxy)	Weakly acidic
AG-50	Polystyrene	- SO_3^- (sulfo)	Strongly acidic
SP-Sephadex	Dextran	- $CH_2CH_2CH_2SO_3^-$ (sulfopropyl)	Strongly acidic
P-Cellulose	Cellulose	- $PO_4H_2^-$ (phosphate)	Intermediate
Bio-Rex 40	Acrylic	-$CH_2SO_3^-$ (sulfomethyl)	Intermediate
Amberlite IR 120	Polystyrene	- SO_3^- (sulfo)	Strongly acidic
Amberlite IRC 50	Acrylic	- COO^- (carboxy)	Weakly acidic
BioGel-CM 100	Acrylic	- CH_2COO^- (carboxymethyl)	Weakly acidic
		Anion exchanger	
DEAE-Sephadex	Dextran	$CH_2CH_2N^+H(CH_2CH_2)_2$ [diethylaminoethyl]	Weakly basic
DEAE-Cellulose	Cellulose	$CH_2CH_2N^+H(CH_2CH_2)_2$ [diethylaminoethyl]	Weakly basic
AG-1	Polystyrene	-$CH_2N^+(CH_3)_3$ (tetramethylammonium)	Strongly basic
QAE-Sephadex	Dextran	-$CH_2CH_2N^+(CH_2CH_3)_2$ with CH_2CHCH_3 on N, OH on CH (diethyl-2-hydroxypropylaminoethyl)	Strongly basic
Dowex-1	Polystyrene	-$CH_2N^+(CH_3)_3$ (tetramethylammonium)	Strongly basic

8.8.3.1. Essential common properties of ion-exchangers

Ion-exchangers possess the following essential common properties:

a) They are of complex nature and polymeric.

b) They must be ionic in nature.

c) They should be highly permeable.

d) Ion-exchange resins must be cross-linked to exhibit only a negligible solubility. They are almost insoluble in water and organic solvents.

e) They must be chemically stable.

f) They should contain an adequate number of accessible ionic exchange groups.

8.8.3.2. Available forms of ion-exchangers

Acidic cation exchangers are commercially available in the sodium and hydrogen forms, and basic anion exchangers in the hydroxide or chloride forms. The chloride form of basic anion exchanger is used widely because in case of hydroxide form there is a possibility of its partial conversion into carbonate form, absorbing atmospheric carbon dioxide.

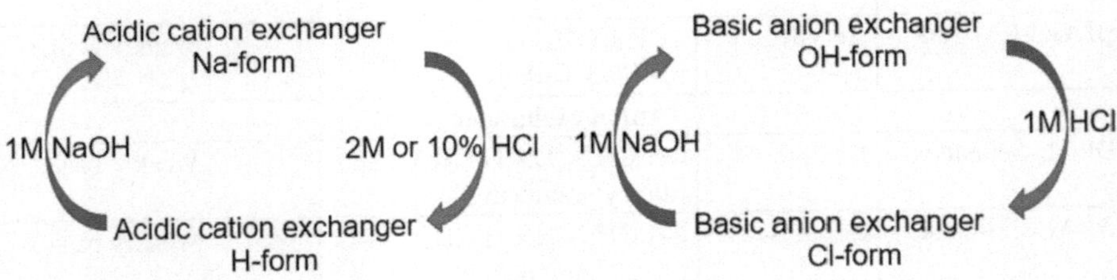

Fig. 8.10: Changing the ionic forms of resins

Changing the ionic form of ion-exchangers is possible, taking their characters into consideration – strong or weak. A very general outline to convert cation exchangers in their respective forms is cited in **Figure 8.10**. During conversion of the resin into its desired form, in each case, it should be sufficiently rinsed with deionized water to remove excess ions.

8.8.4. Ion-exchange capacity

It is a measure of performance of the ion-exchanger. Two specific terms are applied to express ion-exchange capacity: total capacity and available capacity. The total capacity of an ion-exchanger is the total number of ionizable and potentially ionizable groups per unit weight (in dry state) or volume (in wet state) of material. It is expressed as milli-equivalents of ionizable groups per gram of dry resin or milli-equivalents of ionizable groups per cubic cm of wet resin. The ion-exchange capacity is a function of resin porosity. Due to the cross-linking nature of resins, sometimes large molecules do not get access to the ionized functional groups that are present within the matrix. The large molecules do not fit into the pores. As a result, the ionized functional groups that are present deep inside the matrix are not readily available to exchange reactions; only the charged groups on the surface are utilized for the purpose. Thus, in this case available capacity is practically a more suitable term to be used than total capacity. Available capacity may be defined as the amount of a given molecule that can bind under defined experimental conditions. Available capacity may remain equal to total capacity for small molecules, but for large molecules it will be less than the total capacity. It is also important to note that although pH does not affect the ion exchange capacity for strongly acidic and strongly basic ion-exchanger as the strong ion-exchangers are totally ionized at all normal working pH values; but for the weakly acidic and weakly basic ion-exchanger, ion-exchange capacities are functions of pH. Ion-exchange capacity for weakly acidic cation exchanger becomes constant at pH value above about 8.0 and for weakly basic anion exchanger at pH below 5.0. The idea of ion exchange capacity is a helpful guide and very useful in estimating the quantity of resin required for a particular determination.

8.8.5. Choice of ion-exchanger

If the sample material is of unknown nature, the type of ion-exchanger, cationic or anionic, to be taken for separation purpose may be chosen by a trial-and-error method that may be accomplished by a very simple and easy process. A small quantity of solute mixture in the buffer is mixed with cationic and anionic exchanger separately and left for equilibration for 30 minutes followed by centrifugation. Amount of solute is then measured in the supernatant, monitoring the absorbance at 280 nm for proteins, 260 nm for nucleic acids, catalytic activity for enzymes or some other ways depending on the nature of solute. The ion-exchanger exhibiting the lowest level of added solute in the supernatant binds more solute and may be selected for use. Again, in the reverse way, it may also be utilized to determine pH and ionic strength of the eluent to release the solute molecules from the ion-exchanger. However, polystyrene based synthetic resins for small molecules that do not denature on hydrophobic forces and fibrous cellulose and low-percent cross-linked dextran-based ion-exchangers for larger molecules are most suitable. For amphoteric molecules both cationic and anionic exchangers may be used, obviously considering pH range of their stability in which they are not denatured.

8.8.6. Choice of buffer

The composition, pH, and ionic strength of buffer solution must be taken into consideration to select a proper buffer. There is a possibility that the buffer ions carrying the same charge may compete with the solute for binding sites on the ion exchangers reducing its total capacity. To avoid this problem, in general, cationic buffers, such as Tris, pyridine, and alkyl amines with anionic exchangers and anionic buffers such as acetate, barbiturate and phosphate with cationic exchangers are used. Buffer pH should be selected within the range of stability so that the biomolecules are not denatured. Initial buffer pH and ionic strength should be such that the desired solute molecules bind to the ion-exchanger and the undesired molecules remain unbound. At a later stage, desired bound molecules are made unbound by changing the buffer pH. In case of protein molecules, as the pH reaches its isoelectric point, they lose their net charge and become unbound. Thus, a good separation is achieved while other charged protein molecules remain bound on the column.

8.8.7. Operation

Prior to use, resins must be pretreated to remove traces of colloidal matter or 'fines', to transform to the desired ionic state and to remove trace impurities. Trace metals as impurities may be removed by refluxing with HCl followed by thorough rinsing of resins with deionized water. To remove the 'fines', ion-exchangers must be suspended in a large volume of water. When 80-90% of ion-exchangers settle down, the cloudy supernatant is decanted. One should carry on the process until the supernatant is completely clear. It is essential to remove the 'fines' from the exchanger, otherwise, the 'fines' clog the ion-exchange column affecting the flow rate of eluent. For all resins, final treatment should be with a solution leading to the resin acquiring desired ionic form.

Final separation of charged solute molecules through ion-exchangers is accomplished by column method or by equilibration. However, the ion-exchange column is the most extensively used technique where the column is operated by elution development. The performance of gradient elution is better than isocratic elution. Initially the solute mixture to be separated is absorbed on the column and subsequently the mixture is separated manipulating pH and ionic strength of buffer eluent. The sample volume should be 1-10% of the exchange capacity of the column. Elution is carried out @ 0.01-0.1 ml of eluate / ml of resin / minute. An alternative method is batch separation; where the solute mixture is mixed directly with an ion-exchanger, allowed to equilibrate for 1 hour, and then filtered. The desired solute is in the filtrate, if it remains unbound in solution. If the case is opposite i.e., desired solutes bind to the exchanger; for its subsequent removal, the exchanger should be suspended in a buffer of higher ionic strength or different pH. In the batch equilibration process, it is possible to measure the affinity of solute molecules to be separated over a range of concentration and pH, and with various eluents.

Finally, the eluates, collected at certain volumes, are monitored by measuring radioactivity, UV-absorbance, catalytic activity for enzymes, refractive index, etc. to understand the progress of separation.

8.8.8. Ion-exchange paper chromatography

It is a special type of chromatographic technique combining the advantages of paper chromatography and ion exchange chromatography. In this case, a dual-action chromatographic material is yielded by inducing ion exchange capacity in cellulose paper through introducing ion exchanging functional groups like sulfonic, carboxylic, etc. in it.

8.9. Gel-filtration chromatography

8.9.1. Introduction

Gel-filtration chromatography is a separation technique which separates molecules of different sizes utilizing molecular sieve properties of porous materials in the form of gel. In usual and conventional chromatography, partition and adsorption chromatography, molecules are separated based on the difference in their physical properties, such as solubility, adsorption, and volatility. But in gel-filtration chromatography separation of molecules is achieved exploiting their differences in molecular size. Gel-filtration chromatography is also known as **size exclusion chromatography** (since the gel matrix excludes larger molecules and retains smaller molecules), **molecular sieve chromatography** (since sieving of larger molecules from smaller molecules with the help of porous beads is accomplished) and **gel permeation chromatography** (since larger molecules pass through the interstices of the gel beads). Interestingly, largest molecules elute first and smallest molecules last from the gel-filtration column. This technique is utilized to purify and also to determine molecular weights of polysaccharides, enzymes, proteins, nucleic acids, and other biologically important organic molecules. It may act as an alternative to dialysis.

8.9.2. Principle

Differential migration of dissolved solutes differing in their molecular sizes through a particulate matrix containing semi-uniform pores is the basis of separation. Here, the stationary phase is inert porous beads containing controlled pore size and the mobile phase is a liquid, usually aqueous in nature. Organic solvents may also be used as mobile phase; but for this, special type of gel matrix (Sephadex LH, Sepharose CL, Styragel, etc.) that swells in a particular organic solvent (dimethylsulfoxide, dimethylformamide, acetonitrile, ethanol, acetone, tetrahydrofuran, etc.) is to be used.

When a solute mixture in solution is allowed to pass through a column packed with sponge like gel matrix that is made up of spherical beads containing relatively semi-uniform pores, under a continuous solvent flow solute of larger molecular size than the pores cannot diffuse into the pores. They remain in the interstices of the gel i.e., void volume of the column and thus, they are eluted from the column easily without any resistance. On the other hand, molecules that possess smaller size than the pore size enter inside the pores and are removed from the mainstream of the mobile phase. The small molecules are capable of diffusing in and out of the pores. Thus, larger volume of the column is available to the small molecules, and they take more time to be eluted from the column. The elution rate of the molecules of intermediate sizes falls in between those for large and small molecules because they penetrate only some pores and are retained to a lesser degree than the small ones. As a result, a complete separation of solutes based on their molecular size is possible. The molecule's size and pore's diameter are two important key factors that govern the molecule's degree of exclusion from the column.

In a gel-filtration column, one can expect two types of water volume: outer or void volume (V_0) of water retained in the interstices of the gel beads and inner volume of water retained inside the pore (V_i). Therefore, a total volume of column (V_T) represents:

$$V_T = V_0 + V_i + V_g \qquad \text{Equation 8.11}$$

Where,

V_g = volume of the gel matrix

Since the dry porous granules on soaking with water swell and create inner volume (V_i), there exists a direct relationship of V_i with dry weight (m) of the beads and its water regain (W_r), which may be represented as:

$$V_i = m \times W_r \qquad \text{Equation 8.12}$$

In general, small V_i leads to a tighter cross-linking network. On the other hand, large V_i depicts less cross-linking resulting in a loose network. Thus, swelling or water regain (W_r) is a direct measure of cross-linking.

The elution volume (V_e) of a solute from a gel column may be calculated from a relationship related to void volume (V_0), inner volume (V_i) of solvent present inside gel particles and the distribution coefficient (K_D), since a distribution equilibrium is established between solute's concentration in inner and outer volume when a solute in solution passes through the gel. K_D is a constant, characteristic for each solute.

$V_e = V_0 + K_D \times V_i$ Equation 8.13

$K_D = (V_e - V_0) / V_i$ Equation 8.14

$= (V_e - V_0) / m \times W_r$ Equation 8.15

When the solute's molecular size is larger than that of the pore, it does not diffuse inside the pore. It remains only in the void space and is completely excluded from gel. In this case K_D shows zero value, since $V_e = V_0$ (Equation 14). But when solvent present in the column is equally available to the solute i.e., a molecule has complete accessibility to the gel, then $V_e = V_0 + V_i$ and thus, K_D becomes unity. Hence, K_D value varies between 0 and 1. Since K_D varies with solutes of different molecular sizes, it is possible to fractionate them on gel filtration chromatography. Sometimes larger K_D values may be obtained due to interaction (e.g., adsorption) between the gel and the solute. Larger differences in molecular sizes and K_D values among the solutes lead to a better separation.

When a solution with two solutes ($V_e = V_0 + K_D \times V_i$ for one solute and $V_e' = V_0 + K_D' \times V_i'$ for the other) passes through a gel column, the difference in elution volume of solutes represents the upper limit of the sample volume to be applied for a complete separation.

$V_e - V'_e = V_i (K_D - K'_D)$ Equation 8.16

In practice, it is observed that sample volume should not exceed a volume of 0.4 V_i which corresponds to a volume of nearly 25-30% of total bed volume. Otherwise, zone spreading may occur.

8.9.3. Gel-chromatographic materials

Basically, gel materials are dextran, agarose, and polyacrylamide. Mixed-type gel materials are also available which are derived from dextran and polyacrylamide (sephacryl gel) or agarose and dextran (superdex gel). Mixed-type gels provide a good gel rigidity and allow a high degree of sieving property. Dextran, a polymer of glucose, is a natural polysaccharide of high molecular weight and is produced by the bacterium *Leuconostoc mesenteroids*. Dextran is soluble in water. An alkaline solution of epichlorohydrin reacts with two polymer strands of Dextran and brings about the cross-linkage between two polymers of dextran (**Figure 8.11**). Agarose is a linear polymer of alternating D-galactose and 3,6-anhydro-L-galactose. It is obtained from red algae. Polyacrylamide gel is derived by copolymerization of acrylamide and N, N/-methylenebisacrylamide. These are particularly suited for separation at low molecular weight range. This is discussed in detail in Chapter 5: Electrophoresis. Some commonly used gel filtration materials are listed in **Table 8.4**. An ideal gel matrix should be chemically inert (there should be no chemical interaction between gel matrix and solute), mechanically rigid, stable over a wide pH range and temperature, and it should possess a narrow particle size distribution to provide a good elution flow rate.

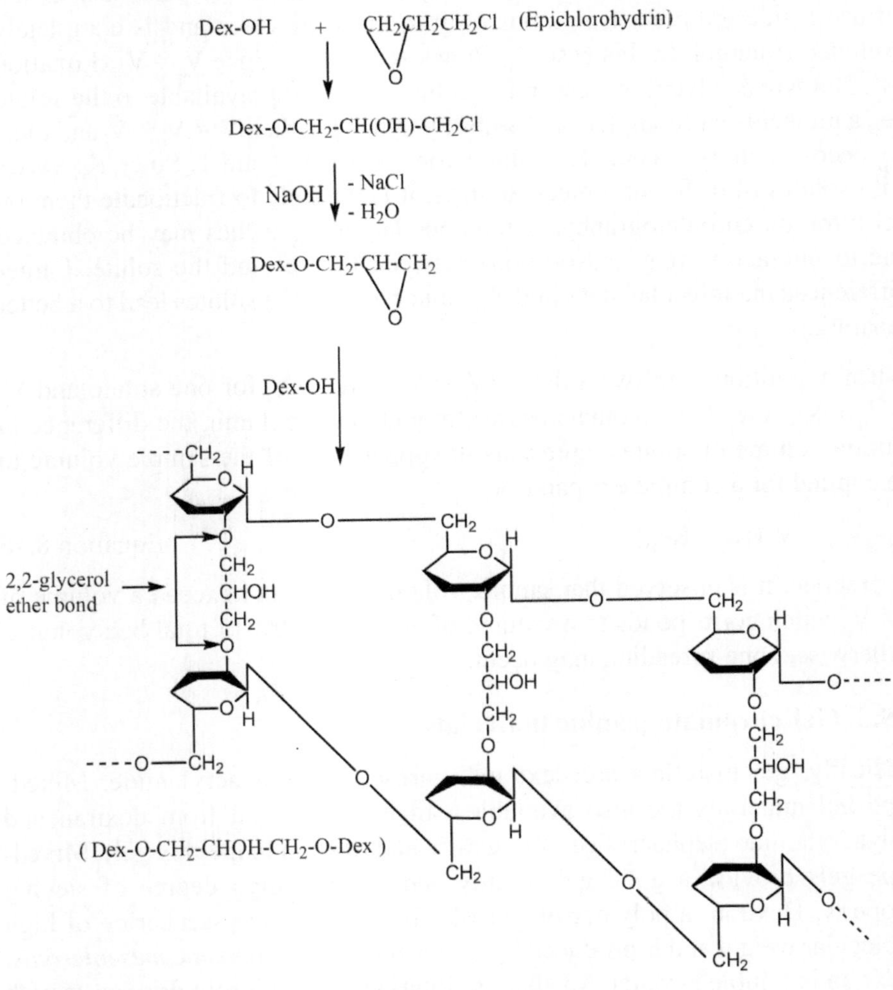

Dex-OH + $CH_2CH_2CH_2Cl$ (Epichlorohydrin)

Dex-O-CH_2-CH(OH)-CH_2Cl

NaOH, - NaCl, - H_2O

Dex-O-CH_2-CH-CH_2

Dex-OH

2,2-glycerol ether bond

(Dex-O-CH_2-CHOH-CH_2-O-Dex)

Fig. 11: Structure of cross-linked dextran polymer

Table 8.4: Gel-filtration materials with their fractionation range.

Trade name	Type	pH stability	Fractionation range for proteins (KD)
Bio-Gel P-2	Polyacrylamide	2.0 – 10.0	0.1 – 1.8
Bio-Gel P-4	Polyacrylamide	2.0 – 10.0	0.8 – 4.0
Bio-Gel P-6	Polyacrylamide	2.0 – 10.0	1.0 – 6.0
Bio-Gel P-10	Polyacrylamide	2.0 – 10.0	1.5 – 20.0
Bio-Gel P-30	Polyacrylamide	2.0 – 10.0	2.5 – 40.0
Bio-Gel P-100	Polyacrylamide	2.0 – 10.0	5.0 – 100.0
Bio-Gel P-200	Polyacrylamide	2.0 – 10.0	30.0 – 200.0
Bio-Gel P-300	Polyacrylamide	2.0 – 10.0	60.0 – 400.0
Sephadex G-10	Dextran	2.0 – 10.0	up to 0.7
Sephadex G-15	Dextran	2.0 – 10.0	up to 1.5
Sephadex G-25	Dextran	2.0 – 10.0	1.0 – 5.0
Sephadex G-50	Dextran	2.0 – 10.0	1.5 – 30.0
Sephadex G-100	Dextran	2.0 – 10.0	4.0 – 150.0
Sephadex G-150	Dextran	2.0 – 10.0	5.0 – 300.0
Sephadex G-200	Dextran	2.0 – 10.0	5.0 – 600.0
Sepharose-6B	Agarose	3.0 – 13.0	10.0 – 4000.0
Sepharose-4B	Agarose	3.0 – 13.0	60.0 – 20000.0
Sepharose-2B	Agarose	3.0 – 13.0	70.0 – 40000.0
Bio-Gel – A-50	Agarose	3.0 – 13.0	100.0 – 50000.0
Bio-Gel – A-150	Agarose	3.0 – 13.0	1000.0 – 150000.0
Sephacryl-S-200 HR	Dextran-polyacrylamide	3.0 – 11.0	5.0 – 250.0
Sephacryl-S-300 HR	Dextran-polyacrylamide	3.0 – 11.0	10.0 – 1500.0
Sephacryl-S-400 HR	Dextran-polyacrylamide	3.0 – 11.0	20.0 – 8000.0
Superdex 30	Agarose-dextran	3.0 – 12.0	0 – 10
Superdex 75	Agarose-dextran	3.0 – 12.0	3.0 – 70.0
Superdex 200	Agarose-dextran	3.0 – 12.0	10.0 – 600.0

8.9.4. Applications

Apart from its main application to purify biological macromolecules, gel-filtration chromatography may also be employed to determine molecular masses, to concentrate solutions, to remove salts and buffer ions from protein solutions as an alternative to dialysis.

8.9.4.1. Molecular mass determination

It has been observed that over a considerable molecular mass range a linear relationship exists between logarithm of molecular mass of a substance and its relative elution volume (**Figure 8.12**). This enables us to estimate the molecular weight of an unknown substance on the basis of its elution characteristics on a gel column from a plot using macromolecules of known molecular masses.

The method assumes that the standard reference compounds and the unknown compounds must possess the identical shape. Another limitation of this method to be considered is the choice of gel which would be such that the molecular mass of the unknown substance must be within the linear section of the curve. Estimation of molecular weight by this method is very simple. It requires very simple equipment and thus, it is inexpensive. It is applied to determine molecular masses of highly purified or crude samples.

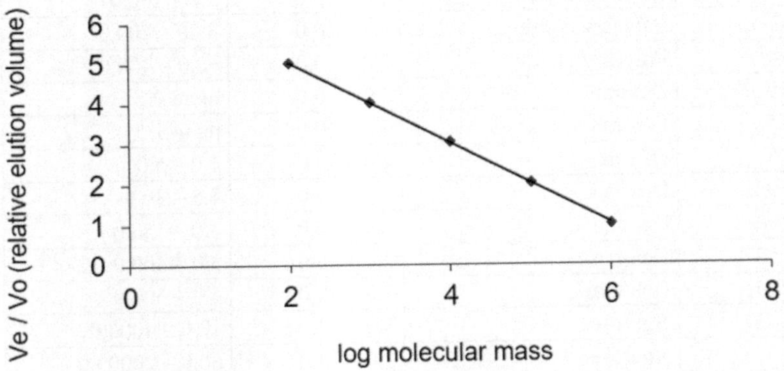

Fig. 8.12: A linear relationship of relative elution volume versus log molecular mass

8.9.4.2. Desalting

A compound may be desalted very easily using a column of gel with an exclusion limit less than the molecular mass of the compound. Thus, a protein from ammonium sulfate-precipitated protein form may be separated. This is faster and more efficient than dialysis. Protein is totally excluded and salt ions are totally included in the gel-filtration granules.

8.9.4.3. Making concentrated solution

To make a concentrated solution of high molecular mass, it is suggested to add a dry gel matrix to the solution. As a result, the gel matrix swells; water and compounds of low molecular masses are absorbed by the gel and the compounds of high molecular mass remain in solution. The mixture, after keeping for a considerable period, is centrifuged and the supernatant is decanted. A concentrated solution of high molecular masses is thus obtained.

8.10. Affinity chromatography

8.10.1. Principle

Unlike most chromatographic methods that depend on the differences in physicochemical properties of the solutes to be separated, affinity chromatography exploits unique biochemical properties. Since the technique is based on natural specificities of the interacting molecules and the interaction of biochemical origin, it exhibits high selectivity that may not be applicable to other chromatographic methods. The physicochemical properties, such as net charge, size, mass, polarity, adsorption, volatility, solubility, etc. that govern basically the chromatographic methods other than affinity chromatography, are not capable enough to separate and isolate biomolecules with higher selectivity. In affinity chromatography, high affinity of a biomolecule toward the other (ligand) is the basis of separation of that biomolecule from a solute mixture. It is a type of adsorption chromatography where adsorption is caused by the biological property of the molecule. The molecule, to be separated from a mixture, must be specifically, non-covalently and reversibly adsorbed by the ligand, a complementary binding substance that is immobilized on a solid or gel matrix. For example, attaching a specific inhibitor or substrate or coenzyme covalently to a matrix (agarose gel) may be used to isolate and purify a specific enzyme. Likewise, avidin may act as a ligand to isolate biotin containing enzymes. Affinity chromatography is principally concerned with three major events: a) immobilizing ligand to a solid or gel matrix, b) adsorption of molecule (to be separated) to the ligand, and c) finally, desorption of bound molecule from ligand. The whole process is summarized in **Figure 8.13**. Affinity chromatography is virtually used to purify biomolecules, such as proteins, enzymes, nucleotides, nucleic acids, antibodies, hormone receptors, membranes, and whole cells. In clinical laboratories, affinity chromatography has been employed for preparative separation of proteins and antibodies.

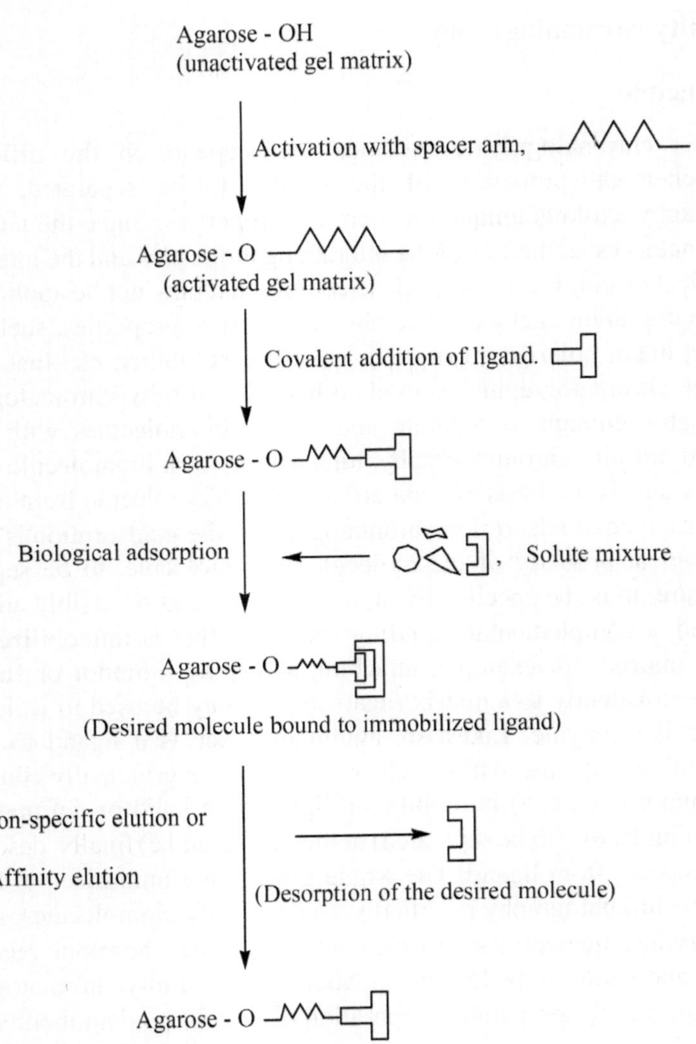

Fig. 13: Separation of desired molecule by affinity chromatography

8.10.2. Matrix

An ideal matrix used in affinity chromatography must be of high porosity and macro-porous so that it exhibits satisfactory flow characteristics and permits macromolecules to get access to the binding sites. It should be physically and chemically stable during the entire process of separation. It must possess a large number of suitable reactive functional groups. These functional groups are utilized to form covalent linkages to ligands. While doing so, it must show its stability. Likewise, it must be stable enough when desired molecules bind to the ligand. It should exhibit extremely low non-specific adsorption to avoid other undesired biomolecules. Usually cross-linked agarose (Sepharose, Bio-Gel A), polyacrylamide (Bio-Gel P), polystyrene (Bio-Beads S), dextrans (Sephacryl S), controlled-porosity glass (CPG) beads are used as solid / gel matrix in affinity chromatography; out of which agarose is widely used, since it qualifies all criteria of an ideal matrix.

8.10.3. Ligand

As per the choice of the experimenter, selection of ligand may be made. For example, when one employs affinity chromatography to isolate a specific enzyme; one may use its substrate, reversible inhibitor or allosteric activator as a ligand. A special care must be taken to avoid catalytic transformation of the bound substrate by modifying the column conditions when the substrate itself acts as a ligand. Hormone acts as a ligand when hormone receptor protein is to be isolated. Likewise, antigen may be used for antibody isolation. Whichever ligand is used, one must keep it into consideration that the ligand must possess specific and reversible binding sites for the molecule to be purified and must contain suitable functional groups, such as $-NH_2$, -COOH, -SH and alcoholic or phenolic –OH. These functional groups help the ligand to bind with the gel matrix. Ligands may be of two types: (i) specific ligand to bind exclusively a specific macromolecule, i.e., substance specific and (ii) ligand showing group selectivity, e.g., 5′-AMP may act as a ligand for many NAD^+ dependent dehydrogenases. Ligands commonly used in affinity chromatography are listed in **Table 8.5**.

Table 8.5: Examples of some ligands with their specifications

Ligand	Specificity
Avidin	Biotin containing enzyme
Cibacron Blue F 3G-A	Enzymes with nucleotide cofactors (dehydrogenases, kinases, DNA polymerases), serum albumin, coagulation factors
Heparin	Nucleic acids binding proteins, restriction endonucleases, lipases, lipoprotein, steroid receptor proteins
5′-AMP	NAD^+ dependent dehydrogenases, ATP dependent kinases
2′, 5′-ADP	$NADP^+$ dependent dehydrogenases
Antibody	Antigen, virus, cell
Benzamidine	Serine, proteases
Concanavalin A	Glucose, glycoproteins, glycolipids
Calmodulin	Calmodulin binding enzymes
Insulin	Insulin receptor proteins
Iminodiacetate	Serum proteins, interferons, proteins with affinity for metal ions
Lectin	Glycoproteins, polysaccharides, cell surface receptor, cell

Ligand	Specificity
Lysine	Nucleic acids (tRNA)
Octyl ligand	Weakly hydrophobic proteins
Phenyl ligand	Strongly hydrophobic proteins
Phenyl boronate	Glycoproteins, ribonucleotides
Poly (A)	Nucleic acid containing Poly (U) sequences, m-RNA specific proteins
Poly (U)	Nucleic acid containing Poly (A) sequences
Protein A	Immunoglobulins (IgG-type antibodies)
Fatty acids	Fatty acid binding proteins
Disulfide 2′-pyridyl ligand	Thiol (-SH) containing proteins

8.10.4. Gel matrix activation and introduction of a spacer arm

It is observed that many macromolecules are unable to bind to ligand, when ligand is directly coupled with the gel matrix, due to steric hindrance of the matrix to the macromolecule. This is because of the presence of active sites deep in the matrix exhibiting low capacity of adsorption. This problem may be overcome through activation of the gel matrix, where a spacer arm may be introduced in between the matrix and the ligand. Thus, to increase the efficiency of separation, a ligand may be attached to the matrix by two steps: (i) Firstly, functional groups of the gel must be activated and ii) ligand is then attached to the activated gel matrix. Insertion of a spacer arm in between the gel matrix and the ligand helps macromolecules to gain access to the binding sites. But it is also essential to be careful about the length of the spacer arm. Optimum length of the spacer arm is three to ten carbon atoms. Too long spacer arm may even increase the risk of undesired non-specific adsorption, reducing selectivity of the separation. If it is too short, the arm is ineffective, and the ligand fails to bind desired macromolecules. The activation process involves the reaction of hydroxyl groups on agarose matrices.

8.10.4.1. Cyanogen bromide activated gel matrix

Cyanogen bromide activated agarose is commercially available (CNBr activated Sepharose 4B, Pharmacia) and widely used. It helps to attach ligands containing amino groups. Synthesis is outlined in **Figure 8.14**. However, since small ligands attach very closely to the matrix surface, many proteins are unable to bind the cyanogen bromide coupled ligand due to steric interference with the agarose matrix.

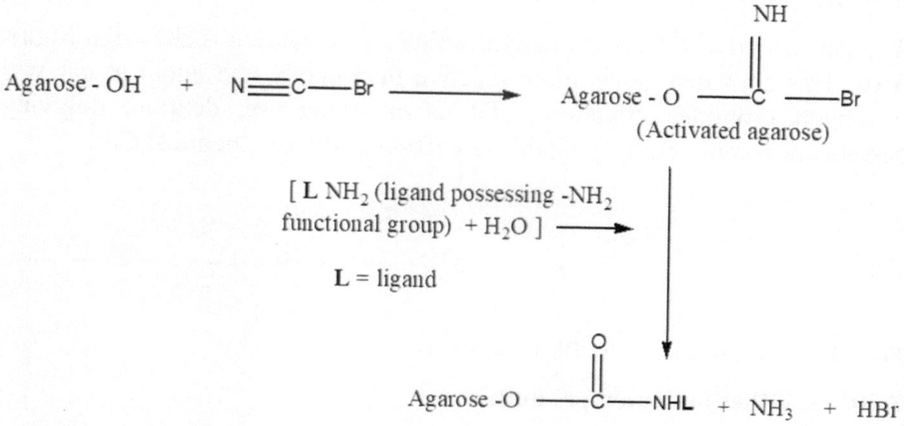

$$\text{Agarose-OH} + N{\equiv}C{-}Br \longrightarrow \underset{\text{(Activated agarose)}}{\text{Agarose-O}{-}\overset{\overset{NH}{\|}}{C}{-}Br}$$

[L NH_2 (ligand possessing -NH_2 functional group) + H_2O] ⟶

L = ligand

$$\text{Agarose-O}{-}\overset{\overset{O}{\|}}{C}{-}NHL + NH_3 + HBr$$

Fig. 8.14: Synthesis of cyanogen bromide activated agarose

8.10.4.2. Epoxy activated gel matrix

Epoxy activated gel matrix (epoxy activated Sepharose 6B, Pharmacia) is more convenient and efficient compared to cyanogen bromide activated matrix. It provides active epoxy groups as the active electrophiles and ligands containing nucleophilic groups, such as amino, hydroxyl, thiol, or carboxyl groups may be covalently attached to epoxy activated gel matrix via reaction with its epoxide groups. The epoxy-activated agarose coupled with ligands containing different nucleophilic groups are shown in **Figure 8.15**.

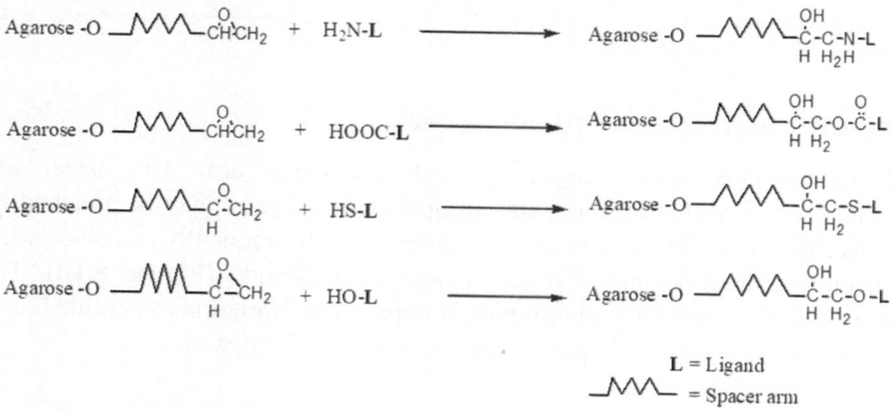

Agarose -O —ᴧᴧᴧ— CH(O)CH₂ + H₂N-L ⟶ Agarose -O —ᴧᴧᴧ— C(OH)H-CH₂-NH-L

Agarose -O —ᴧᴧᴧ— CH(O)CH₂ + HOOC-L ⟶ Agarose -O —ᴧᴧᴧ— C(OH)H-CH₂-O-C(O)-L

Agarose -O —ᴧᴧᴧ— CH(O)CH₂ + HS-L ⟶ Agarose -O —ᴧᴧᴧ— C(OH)H-CH₂-S-L

Agarose -O —ᴧᴧᴧ— CH(O)CH₂ + HO-L ⟶ Agarose -O —ᴧᴧᴧ— C(OH)H-CH₂-O-L

L = Ligand

—ᴧᴧᴧ— = Spacer arm

Fig. 8.15: Ligands coupled with epoxy activated agarose

8.10.4.3. N, N′-carbonyldiimidazole (CDI) activated matrix

The gel structure of N, N′-carbonyldiimidazole activation is shown in **Figure 8.16**. This activated gel is more reactive than epoxy and about at par with cyanogen bromide activation. CDI-activated agarose, dextran, polyvinyl acetate are commercially available (Reacti-Gel, Pierce Chemical Co.).

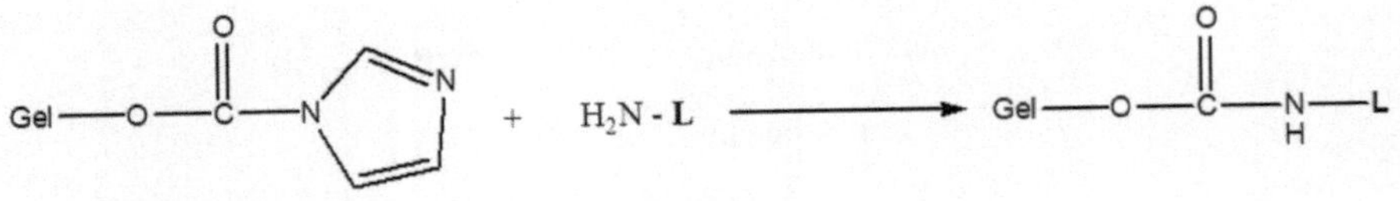

$$\text{Gel}-O-\overset{O}{\overset{\|}{C}}-N(\text{imidazole}) + H_2N-\mathbf{L} \longrightarrow \text{Gel}-O-\overset{O}{\overset{\|}{C}}-\underset{H}{N}-\mathbf{L}$$

Fig. 8.16: Ligand coupled with CDI activated gel

8.10.4.4. Other activated gel matrix

The steric interference problem may be overcome very conveniently by interposing a spacer arm between matrix and ligand. 6-aminohexanoic acid – agarose (CH-Sepharose 4B), and 1,6-diaminohexane – agarose (AH-Sepharose 4B), that can be covalently attached to free primary amino and free carboxyl groups of ligand, respectively, are the typical examples of this group (**Figure 8.17**).

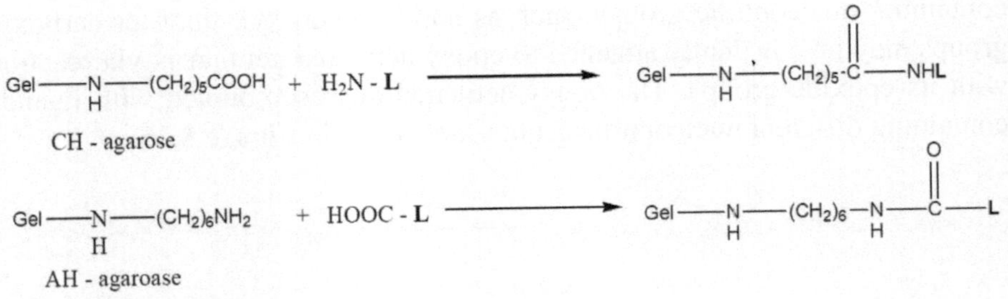

$$\underset{\text{CH - agarose}}{\text{Gel}-\underset{H}{N}-(CH_2)_5COOH} + H_2N-\mathbf{L} \longrightarrow \text{Gel}-\underset{H}{N}-(CH_2)_5-\overset{O}{\overset{\|}{C}}-NHL$$

$$\underset{\text{AH - agaroase}}{\text{Gel}-\underset{H}{N}-(CH_2)_6NH_2} + HOOC-\mathbf{L} \longrightarrow \text{Gel}-\underset{H}{N}-(CH_2)_6-\underset{H}{N}-\overset{O}{\overset{\|}{C}}-L$$

Fig. 8.17: Ligand coupled with CH / AH - agarose

Diaminobutane is also frequently used as a spacer arm. The spacer arm must possess a second functional group to which ligand may be attached by conventional organosynthetic procedures, which frequently involve using succinic anhydride and a water soluble carbodiimide (**Figure 8.18**). The function of the soluble carbodiimide is to promote formation of peptide bonds between – COOH of the spacer arm and – NH_2 of the ligand.

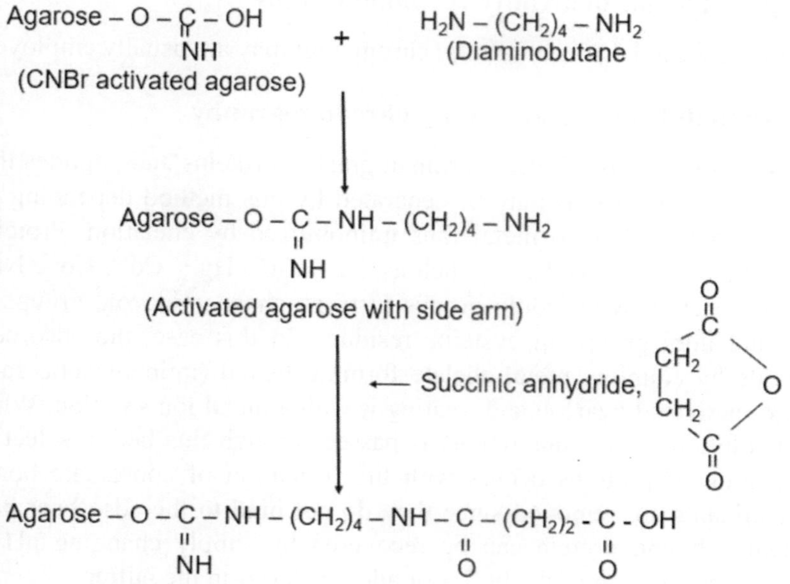

Agarose – O – C(=NH) – OH (CNBr activated agarose) + $H_2N – (CH_2)_4 – NH_2$ (Diaminobutane

↓

Agarose – O – C(=NH) – NH – $(CH_2)_4$ – NH_2

(Activated agarose with side arm)

↓ ← Succinic anhydride

Agarose – O – C(=NH) – NH – $(CH_2)_4$ – NH – C(=O) – $(CH_2)_2$ - C(=O) - OH

Fig. 8.18: Succinylation to the spacer arm activated gel

Dichlorotriazines are also frequently used to activate gel matrix (**Figure 8.19**).

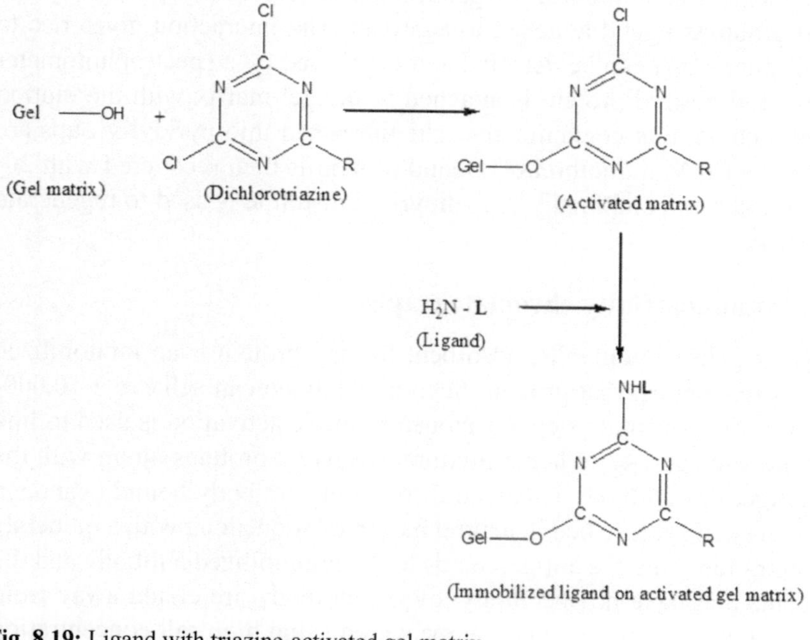

Gel —OH (Gel matrix) + (Dichlorotriazine) → Gel—O— (Activated matrix)

H_2N - L (Ligand) →

(Immobilized ligand on activated gel matrix)

Fig. 8.19: Ligand with triazine-activated gel matrix

8.10.5. Special forms of affinity chromatography

The following special forms of affinity chromatography are usually employed.

8.10.5.1. Immobilized metal affinity chromatography

It is also known as metal chelate chromatography. Proteins and peptides that have affinity for metal ions may be separated by this method depending on their differential binding to metal ions immobilized by chelation. Proteins selectively bind to the metal ions, such as Cu^{+2}, Zn^{+2}, Hg^{+2}, Cd^{+2}, Co^{+2}, Ni^{+2}, or Mn^{+2} by reaction with indole groups in tryptophan, imidazole groups in histidine and thiol groups in cysteine residues. In this case, the adsorbent bed is made by coupling metal chelate forming ligand (iminodiacetic acid) with epoxy-activated agarose and treating it with a metal ion solution. When a protein solution with contaminants is passed through this bed, a selective immobilization of proteins occurs with the formation of coordinate bonds and contaminants are removed since they do not bind to the adsorbent bed. Subsequently, bound protein can be recovered by simply changing pH or increasing ionic strength of the buffer or adding EDTA in the buffer.

8.10.5.2. Covalent chromatography

This is particularly applicable to the separation of thiol containing proteins which interact with an immobilized ligand, containing disulfide group (disulfide 2′-pyridyl group as ligand attached to agarose). The interaction gives rise to pyridine-2-thione that can be detected and monitored by a spectrophotometer at 343 nm, and desired protein is attached to the gel matrix with the elution of non-thiol-containing contaminants. The unreacted thio pyridyl groups are removed with 4 mM dithiothreitol. Bound protein is then recovered with 20-50 mM dithiothreitol and finally 2, 2′-dipyridyl disulfide is used to regenerate the gel matrix.

8.10.5.3. Immunoaffinity chromatography

In this type of chromatography, adsorbent for any protein is an immobilized antibody raised against that protein. Such a technique can achieve a 10,000-fold purification in a single step. Cyanogen bromide activation is used to link the antibodies to agarose. When a mixture of several proteins along with the protein antigen for antibody is passed through the antibody bound cyanogen bromide activated agarose bed in neutral buffer solution along with a moderate salt concentration, only the antigen binds to the immobilized antibody and the contaminants that have no specificity toward antibody are eluted away from the column. Later on, protein may be recovered using high salt concentration with use of urea, SDS or guanidine hydrochloride. But these may cause denaturation. This is why chaotropic agents, such as thiocyanate, trifluoracetate and perchlorate or lowering pH at about 3.0 are preferred to recover proteins from antibody binding.

8.10.5.4. Lectin affinity chromatography

Lectin bound agarose gel matrix is commercially available, where lectins are immobilized on the gel matrix by conventional methods. Since lectins, obtained from leguminous plants, can bind specific saccharides; lectin affinity chromatography is employed to separate and isolate glycoprotein, member receptor protein and cells. Specific glycoprotein after binding to specific lectin may be recovered in many ways. Elution with a borate buffer helps to recover glycoprotein forming a complex. Lectin's hydrophobic interaction with glycoprotein is reduced with addition of ethylene glycol. Ligand complex remains no longer stable by changes in pH (change within pH 3.0 – 10.0).

8.10.5.5. Hydrophobic interaction chromatography (HIC)

This technique separates native proteins based on their hydrophobic characters due to the presence of hydrophobic amino acid side chains. Octyl- and phenyl-Sepharose which have group specificity toward weakly and strongly hydrophobic proteins, respectively are used as gel materials in this chromatographic technique. Since hydrophobic interactions are strengthened by increased ionic strength, the eluent in HIC, whose gradient must progressively reduce these hydrophobic interactions, is aqueous buffer with decreasing salt concentration, increasing concentration of detergents, or increasing pH.

8.10.5.6. Dye-ligand chromatography

In dye-ligand chromatography, vinyl sulphone reactive dyes and different dyes of triazinyl chloride type are immobilized in gel matrix. In general, dyes are immobilized on agarose by treating agarose with 0.2% dye solution at pH 10.0 in the presence of 2% NaCl and proteins are attached to immobilized dye at pH 7.0 – 8.5. Later on, proteins are recovered using a salt gradient or affinity elution. Dye selection for particular protein purification is empirical and is made on a trial- and error- basis.

8.10.6. Operating conditions

Like ion-exchange chromatography, affinity chromatography is also concerned with two operations: Adsorption of substance (to be separated) on the ligand bound to a gel matrix and subsequent recovery of that substance through desorption. The second step, desorption via elution method is usually manipulated exploiting the nature and properties of ligand and substance. The choice of elution method depends on types of forces responsible for complex formation and stability of ligand matrix and substance. The whole process of affinity chromatography may be carried out either in column or in batch, out of which column operation is widely used. Once the sample has been bound to an affinity chromatographic column, the column is washed to remove impurities that are non-specifically bound to the column. Later, the desired substance is released from the column by either specific or non-specific elution.

8.10.6.1. Specific elution

This is also known as affinity elution. In this method either a selective and competing binding substance or a ligand is used in the eluting buffer. Since the competing substance in the eluting buffer possesses more affinity toward the ligand, the desired substance is desorbed. Similarly, a ligand of more affinity to the adsorbed substance may be used in the eluting buffer to desorb it. Monosaccharide or its derivative is used as competing binding substance to elute column of immobilized lectin which is used to fractionate carbohydrates and glycoprotein. A high concentration of substrate may be used in eluting buffer to release enzyme from the immobilized ligand.

8.10.6.2. Non-specific elution

Release of bound macromolecules from the column by non-specific elution method is accomplished altering the solution condition, using deforming or denaturing agents and hydrolyzing the matrix or spacer arm.

8.10.6.2.1. Alteration of solution condition

Solution condition is altered in such a way that the macromolecule-ligand complex is no longer stable. If the complex (macromolecule-ligand) is formed by ionic interaction, a change in ionic strength or pH leads to desorption of the macromolecule by changing the extent of ionization of ligand or macromolecule. A gradual decrease in pH or increase in ionic strength may even fractionate mixtures of adsorbed substances. An increase in temperature also is responsible to decrease the extent of affinity, thereby releasing the macromolecules. However, care should be taken that alteration of the solution condition is not made so drastic that the macromolecules to be separated from the ligand are subjected to irreversible damage. If protein is released by changing pH of the buffer solution, pH of the collected fraction must be readjusted to avoid its denaturation.

8.10.6.2.2. Elution with deforming or denaturing (chaotropic) agents

This method is well practiced to obtain desorption of macromolecules, where gentle affinity elution is not sufficient to release the bound macromolecules. The deforming and chaotropic agents being used are urea, guanidine hydrochloride, thiocyanate (CNS^-), and trichloroacetate (CCl_3COO^-). The conformation of the bound macromolecules is changed by these substances; thereby, their affinity for ligands is drastically reduced. The desorbed and deformed substance is then reconstituted by removing deforming agents. This method gives a good result in immunoaffinity chromatography to dissociate antibody-antigen complexes.

8.10.6.2.3. Hydrolysis of matrix or spacer-arm

When the macromolecule is not easily released due to its very tight binding to the ligand and its instability in presence of chaotropic agent, hydrolysis method may be used. The matrix is digested with suitable hydrolytic enzymes and the macromolecules, particularly viable cells whose recovery is critically dependent on the composition of eluent, are released. Sephadex can be digested and solubilized with dextranase enzymes.

8.11. High performance liquid chromatography

8.11.1. Introduction

High performance liquid chromatography (HPLC) is an improved chromatographic technique possessing high-sensitivity and high-speed. It is so called because it is designed to obtain good resolution in the least amount of time while separating a mixture. It is also known as high pressure liquid chromatography, since separation of a solute mixture is accomplished under the influence of a liquid mobile phase flown under high pressure. This chromatographic technique is coupled with an improved technological advance both in instrumentation and column design providing a good flexibility and broad separation spectrum. One may handle this technique with a lot of choices leading to versatility in its use. HPLC is a widely used and popular technique practised almost in all biochemical laboratories due to its high efficiency. Now-a-days its popularity in clinical analysis of biological materials at picogram level has increased several folds due to its reliable, highly automated and rapid data producing capacity. However, it is true that gas-liquid chromatography (GLC) is also an improved technique that provides rapid and reliable quantitative estimation of various biological components. But its major limitation is that the components must be thermally stable and volatile, otherwise it fails to separate and estimate. It is interesting to note that many biological components are neither volatile nor thermally stable.

Therefore, they must be derivatized to volatile and thermally stable derivatives for their estimation by GLC. But it is time consuming and limits rapidity in estimation. This problem may be overcome easily using HPLC that enables it to separate and estimate non-volatile and thermally unstable compounds directly (without derivatization step), since it can be operated at ambient temperatures. Biologically active molecules, such as carbohydrates, nucleotides, nucleic acids, amino acids, peptides, proteins, lipids, fatty acids, vitamins, hormones, steroids, pigments, drugs, pesticides and their metabolites are easily separated and identified by HPLC.

8.11.2. Principle

It is a separation process where the mobile phase used is liquid and the stationary phase is specific functional groups present on micro-particulate solid particles. The individual components in a solute mixture are separated due to their differential interaction and equilibration between liquid phase and stationary functional groups. The separation is based on adsorption, partition (normal phase and reverse phase), ion-exchange, size exclusion and hydrophobic interaction. It is the reason why its use is so versatile in nature. The principle of each of these interactions has been discussed earlier separately. The same may be applied for HPLC too. However, since in HPLC better resolution is achieved by reducing particle size (5-20 μ) of the stationary matrix (increasing the number of theoretical plates), the liquid mobile phase faces a greater resistance leading to a backpressure in the column. This backpressure that may damage stationary matrix structure, reduce flow rate and finally affect resolution; is overcome by introducing a high pressure of up to 5000 psi. Therefore, the major requirement is to prepare suitable column materials which can withstand high pressure.

8.11.2.1. Major chromatographic modes in HPLC

8.11.2.1.1. Normal-phase mode

In normal phase HPLC the packing material, usually of unmodified silica gel (**Figure 8.20**), is more polar than the mobile phase, such as hexane, heptane, chloroform, dichloromethane, ethyl acetate, methylene chloride, diethyl ether or their mixtures.

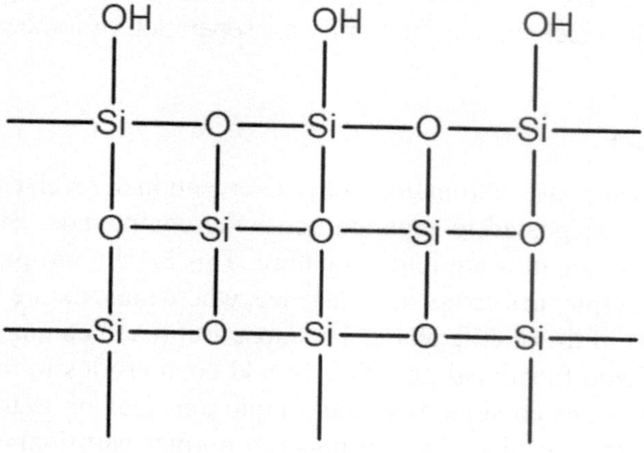

Fig. 8.20: Silica gel

Separation is based on the differences in polar functionalities of different components in a mixture. The polar molecules are retained in the column and the non-polar components are eluted from the column with non-polar mobile phase. However, a gradient elution of a mobile phase with a gradual increase in polarity, which may be achieved by use of methanol or dioxane, helps to elute the polar analytes.

8.11.2.1.2. Reversed-phase mode

Polar character of the stationary and mobile phase as it exists in normal-phase HPLC is reversed in reversed-phase HPLC. The stationary phase consists of a non-polar liquid immobilized on an inert solid and the mobile phase is a relatively polar liquid. Most commonly used stationary phase is the bonded-phase type in which alkyl silane groups, such as butyl (C_4), octyl (C_8), and octadecyl (C_{18}) silane groups are chemically bonded to the silica. C_8 and C_{18} chains are most popular for peptides and C_4 and C_{18} chains for protein analysis. Commonly used mobile phases are water, aqueous buffer, methanol, acetonitrile, tetrahydrofuran, and/or their mixture. In this chromatography, polar compounds elute first followed by non-polar ones. The non-polar analytes retained in the column may be eluted using a gradient solvent system of increasing non-polarity.

8.11.2.1.3. Ion-exchange mode

The column packing is either cation- or anion-exchangers. The cation-exchangers have either sulfonic or carboxylic acid functional groups, while anion-exchangers have quaternary ammonium functional groups. Resin-based and silica-based columns are available. High-performance-ion-exchange chromatography is a flexible technique used mainly for separation of ionic or easily ionizable species.

8.11.2.1.4. Ion-pair mode

It is a version of reversed-phase chromatography. Conventional reversed-phase chromatography is designated to separate non-ionic compounds. But most biologically active compounds are ionic in nature. This is why ion-pair reversed-phase chromatography has come into existence, where ionic nature is suppressed by altering pH of the mobile phase. The most useful and common method to suppress the ionic functional groups is to add counter-ions to the mobile phase. The counter-ions combine with the sample ions leading to the formation of neutral ion-pairs which can then undergo normal partitioning. Quaternary alkyl amine, such as tetrabutylammonium or tetradecylammonium may be used as counter ion to facilitate the separation of strong acidic compounds and alkyl sulfonates, such as sodium lauryl sulfonate, sodium heptane sulfonate or other sulfonic acids for strong basic compounds. Usually water-acetonitrile or water-methanol is passed through the octyl silane or octadecyl silane-bonded stationary phase in ion-pair HPLC.

8.11.2.1.5. Size-exclusion mode

Size-exclusion HPLC shows an added advantage of giving high resolution and sensitivity due to application of high pressure. Although the principle of conventional size-exclusion chromatography is similar to that of size-exclusion HPLC, their size-exclusion stationary matrices are not same. The structure of the gel matrix used in conventional size-exclusion chromatography is susceptible to collapsing under high pressure due to their softness. Semi-rigid cross-linked polymers (styrene-divinylbenzene, polyvinyl chloride, polyacrylamide, vinyl acetate copolymers, etc.) or rigid controlled-pore glass or silica which can withstand high pressure, are available for size-exclusion HPLC. It requires a single mobile phase and isocratic elution. In this chromatography, larger molecules that are excluded from the pores move with mobile phase and elute first.

8.11.2.1.6. Hydrophobic interaction mode

Hydrophobic interaction chromatography (HIC) and reversed-phase chromatography (RPC) are based on hydrophobic interaction between solute molecule and stationary phase. However, in RPC the hydrophobic character of the stationary phase is very strong compared to HIC and as a result protein may get denatured. But in HIC, it is a hydrophilic substance (agarose matrix) substituted lightly with hydrophobic groups (hexyl-, octyl- or phenyl-) and exhibits weak hydrophobic characters. Hence, the protein exists in native state in HIC. Hydrophobic interaction is facilitated by increasing ionic strength. Therefore, the eluting mobile phase in HIC must be with a gradient of decreasing salt concentration, increasing detergent concentration, or increasing pH. Increased pH increases the hydrophobicity of protein molecules. Detergents, such as Tween-20, Triton X-100, etc. displace the bound protein molecules because they have a stronger affinity for the stationary phase than that of the protein molecules.

8.11.2.2. Few important terminologies

8.11.2.2.1. Retention time (R_t)

A component in a mixture is identified by its R_t value. The time required for the elution of each component from the column is called its retention time (R_t). Alternatively, it may be defined as the length of time that each component spends in the column. This is actually the time lapse between injection of the sample and the maximum signal obtained from the recorder. This is also applicable in gas-liquid chromatography, and it is analogous to the R_F value in TLC and PC. R_t remains constant for a specific compound at specific and defined chromatographic conditions.

8.11.2.2.2. Retention volume (R_v)

The volume of mobile phase required to elute a compound from a column is defined as its retention volume (R_v). This is also referred to as elution volume. Retention time and retention volume are related with the flow rate (F) of the mobile phase (Equation 8.17).

$$R_v = F \times R_t \qquad \text{Equation 8.17}$$

Retention time of a particular compound may be increased or decreased by decreasing or increasing the flow rate of the solvent, respectively. The flow rate, in turn, is dependent on internal diameter and length of the column, size, shape and porosity of the stationary phase, and viscosity of the mobile phase.

8.11.2.2.3. Void volume or dead space (V_m)

This is the volume taken up by the liquid mobile phase around and within the packed stationary phase and it is represented by the solvent peak in an HPLC chromatogram. There exists a distinct relationship (Equation 8.18) among retention volume (R_v), distribution coefficient (K_D) of the solute between stationary and mobile phase, void volume (V_m) and volume of the stationary phase (V_S).

$R_v = V_m + K_D \times V_S$ Equation 8.18

It is noteworthy to mention that this relationship is applicable to adsorption column chromatography only when V_S is replaced by surface area (A_S) of the adsorbent.

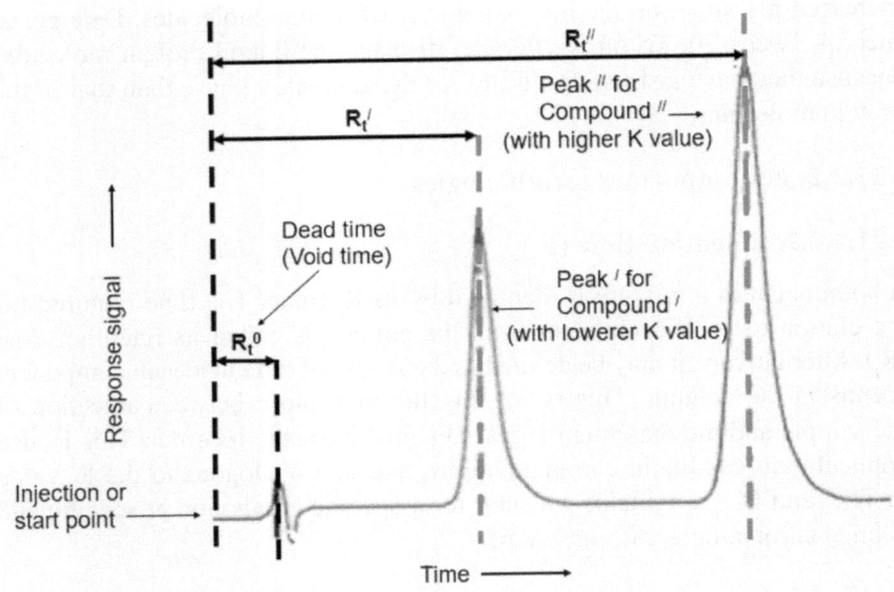

Fig. 8.21: HPLC chromatogram showing retention time of two compounds and void time

8.11.2.2.4. Capacity factor (*k*)

It is a measure of the column's ability to retain the compound. Alternatively, it is a measure of the time spent by the compound in the stationary phase relative to the time spent in the mobile phase. This may be expressed in terms of the elution volume for the compound and the void volume of the column. As depicted in **Figure 8.21**, capacity factor (k') for peak 1 becomes

$k' = (R_V' - V_m) / V_m = (R_t' - R_t^0) / R_t^0$ Equation 8.19

Where,

R_V' = retention volume of peak 1,

V_m = void volume,

R_t' = retention time of peak 1, and

R_t^0 = retention time of unretained solute.

Smaller $k^{/}$ value represents a low degree of association of solute molecules with the column packing resulting in sharp and early eluting peaks and larger $k^{/}$ value for a high degree of association of the compound with the stationary phase representing higher retention of the compound in the column. If there is no partitioning of an analyte into the stationary phase (i.e., for an unretained one), the $k^{/}$ becomes zero. Capacity factor may be influenced by changing polarity of the eluting solvent, temperature and surface area of the solid stationary phase or amount of liquid stationary phase. Usually $k^{/}$ value ranging between 1-10 gives optimum resolution, separation time and sensitivity. Band spreading due to large elution volume decreases peak height at $k^{/}$ value of >10.

8.11.2.2.5. Column selectivity (α)

Column selectivity (α) is a measure of relative separation of two compounds, taking peak centers into consideration. It is expressed as follows:

$$\alpha = (R_t^{//} - R_t^{0}) / (R_t^{/} - R_t^{0}) = K_D^{//} / K_D^{/} \qquad \text{Equation 8.20}$$

Where,

$R_t^{/}$ and $R_t^{//}$ = the retention time for two compounds (Figure 8.21, peak 1 and 2),

R_t^{0} = the retention time for unretained compound (solvent front), and

$K_D^{/}$ and $K_D^{//}$ = the distribution coefficient for compounds in peak 1 and 2, respectively.

When $K_D^{/}$ and $K_D^{//}$ are equal, α becomes unity and represents no separation of the two compounds. This value may be increased, and in turn, column selectivity may be improved by changing the mobile phase composition, its pH and ionic strength or by changing stationary phase or by changing the sample chemistry through derivatization or by changing the physical condition of separation, such as temperature. When the peak of interest is fused with the interfering peaks, this problem is overcome by changing α. However, an increase in retention time causes a decrease in peak height and thus loss in sensitivity. Therefore, it is suggested that α will be changed in such a manner that retention time for the peak of interest decreases compared to interfering peaks with an increase in peak height improving sensitivity.

8.11.2.2.6. Column efficiency

Column efficiency is a measure of band spreading for a given analyte. It is increased by decreasing the band spreading. It is expressed as the number of theoretical plates (N). It is imagined that columns contain a number of theoretical plates in each of which analytes obtain complete equilibration between the stationary and mobile phase. The theoretical plates are expressed from a relationship mentioned below (Equation 8.21):

$N = 16(R_t^{/} / W^{/})^2 = 5.5(R_t^{/} / W^{/}_{1/2})^2 = (R_t^{/} / \sigma)^2$ Equation 8.21

Where,

N = number of theoretical plates for peak 1 (**Figure 8.21**),

$R_t^{/}$ = retention time for peak 1,

$W^{/}$ = peak width of peak 1 at the base line,

$W^{/}_{1/2}$ = peak width of peak1 at ½ peak height, and

σ = ¼ of the peak width of peak 1 at the base line.

Higher is the N value, more is the column efficiency. Number of theoretical plates may be increased by increasing the column length (L). However, there must be a limit of increasing column length. The retention time and peak width increase with excessive increase in column length, causing a decrease in peak height and thereby affecting the sensitivity of the column. Therefore, to ascertain column efficiency it is more useful and practical to take help of height equivalent to a theoretical plate (HETP) because this term takes the column length into consideration.

HETP = L / N Equation 8.22

Where,

L and N are column length and number of theoretical plates, respectively.

For a given column length, N may be increased by decreasing particle size of the stationary phase. It has been found that 20-30 cm columns packed with 5-10 μ particles provide 1000 or more plates.

8.11.2.2.7. Peak capacity (η)

Peak capacity denotes the maximum possible number of peaks separated by a specific chromatographic system. It may be increased by gradient elution and temperature programming in liquid chromatography and gas-liquid chromatography, respectively. It is expressed as

$\eta = 1 + \sqrt{N}/16 (\ln R_V^{L}/R_V^{F})$ Equation 8.23

Where,

R_V^L and $/R_V^F$ are retention volumes of last and first peak, respectively.

8.11.2.2.8. Resolution (R_S)

The separation of two analytes depends not only on the distance between their peak centers (difference between the retention times) but also on the widths of the peaks. Resolution is the measure of relative separation of two compounds taking their both retention time and peak width into consideration. It is expressed as:

$R_S = (R_t^{\prime\prime} - R_t^{\prime}) / (W^{\prime} + W^{\prime\prime}) \times 1/2$ Equation 8.24

Where, $R_t^{\prime}$ and $R_t^{\prime\prime}$ are retention time and $W^{\prime}$ and $W^{\prime\prime}$ are base widths of peak 1 & 2, respectively.

Since resolution is a function of both retention time and peak width spreading, it can be manipulated by increasing or decreasing the flow rate of the mobile phase. A more useful and practical expression of resolution may be achieved by considering selectivity, capacity, and efficiency factors.

$R_S = \frac{1}{4} [(\alpha - 1)/ \alpha] \times [k^{\prime} / (k^{\prime}+ 1)] \times \sqrt{N}$ Equation 8.25

Thus, a good resolution may be obtained for a favorable separation by manipulating selectivity, capacity, and efficiency. It is observed that R_S value of 1.0 corresponds to 98% separation and is adequate for qualitative analysis. R_S value of 1.5 signifies 99.7% separation of two peaks (compounds). Inadequate separation usually occurs when the R_S value is below 0.8.

8.11.3. Instrumentation

Instrumentation for HPLC does not differ too much from that of the conventional column chromatography. To achieve a good resolution, a stationary matrix of small-sized particles (5-10 μ) possessing large surface area is essentially required and to keep pace with the flow rate (to overcome the back pressure of the solvent), an applied pressure is necessary for HPLC. This is why a pressure generating pump and a specially designed column are present in HPLC in addition to other accessories usually used in conventional column chromatography. The basic components of HPLC consist of a solvent reservoir, one or more pumps, a sample injector, a guard column, a column, a frit filter, a detector, sample collector and data recorder. A schematic representation of an HPLC system is shown in **Figure 8.22**.

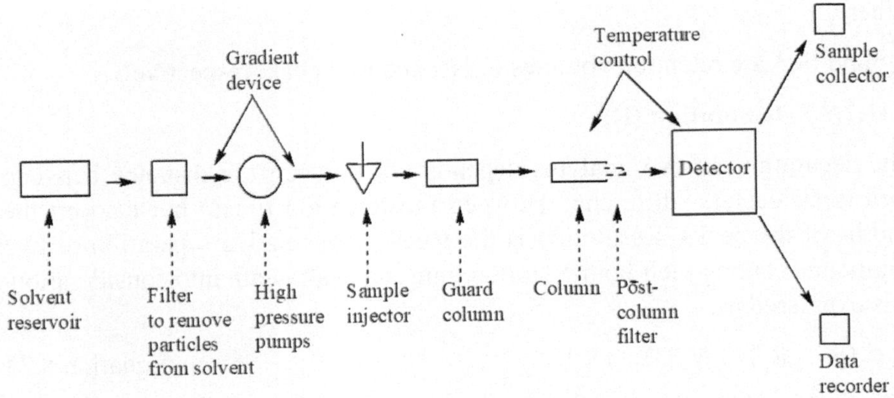

Fig. 8.22: Schematic representation of HPLC

8.11.3.1. Solvent reservoir

Solvent reservoir, usually glass bottle(s) or flask(s), stores mobile phase. It may be one in number for isocratic and two or more for gradient elution. It must be of sufficient volume to allow for several analyses (at least of 500 ml capacity). The mobile phase to be used must be degassed by ultrasonic vibration, purging with an inert gas like helium, filtration through a vacuum filter or refluxing; otherwise, unwanted gas may affect the column and interfere with detection process for the compound of interest. The solvent reservoir is followed by a micro-filter which removes the particulate matter from the solvent.

8.11.3.2. Pump

To obtain a constant flow rate of mobile phase through the column under pressure, a pumping system is necessary for an HPLC. An ideal pump should provide a pressure up to 5000 psi with minimal pulsations (cyclical variations in pressure) at low and high pressure. The pumping device should be capable of delivering a mobile phase at a speed up to 1 ml/minute for analytical purposes and up to 10 ml/minute for preparative purposes. For isocratic elution, where a single solvent or a mixture of two or more solvents premixed in fixed proportion, a single pump is needed. However, for a gradient elution separate pumps are needed to deliver two or more solvents in proportion predetermined by a gradient programmer. Several types of pumps are now available in the market. The most common one is a reciprocating pump that delivers a fixed volume of solvent into the column by a piston driven by a motor. In most advanced pumps, electronic pulse compensations (pulse dampeners) are used to give accurate flow rate minimizing the pulsing effect. A unique characteristic of the most modern pumps used in HPLC now is that they have in-built safety cut-out systems. This safety cut-out system helps to inactivate the pump automatically when the pressure within the column changes from a pre-set limit.

8.11.3.3. Gradient device

With a gradient device it is possible to provide a gradient elution of mobile phase facilitating the manipulation of capacity factor (*k*-value). The gradient may be generated by changing polarity, pH, and ionic strength. The gradient device is placed either on the low- or high-pressure side of the pump. On the low-pressure side of the pump this device offers the gradient at atmospheric pressure, while on the high-pressure side two or more pumps are employed with programmable electronic devices.

8.11.3.4. Sample injector

A sample is introduced in a column through injection mainly in three ways: at high pressure, reduced pressure (stop flow condition) and at high-pressure type but by using a loop. For introducing samples at high pressure, injection is made through a system of neoprene or Teflon septum with a microliter syringe that can withstand high backpressure. To eliminate back-pressure the solvent flow is stopped, sample is introduced and then again flow is resumed to carry the sample to the column. This is the state of the art of injection under stop flow conditions. The most widely used injector is loop type. There is a fixed volume loop that is coupled with valves. Rotating the valves helps the loop to be on-line with the waste port once and with the column next time. Sample is loaded in the loop first at atmospheric pressure and excess solvent goes to the waste port. Next time by rotating, the valve position of the loop becomes in between the pump and the column, so that a mobile phase flashes the loop and a sample is introduced into the column. Therefore, with a loop system there is no need to stop the flow of the mobile phase, as evidenced in the stop-flow system, to avoid the backpressure.

8.11.3.5. Guard column

A guard-column is a pre-column placed between the injector and the main column, used to safe-guard the main column from contaminating materials and particulate matters. In addition, the guard-column saturates the main mobile phase with the stationary phase so that losses of the stationary phase from the main analytical column are minimized. It is of small size ($\approx$ 2 cm in length) and has an internal diameter similar to the main column. The stationary phase of the guard column and the main column remains the same. The guard column retains the contaminating materials and the compounds of interest enter into the main column for separation. Thus, use of a guard column increases the longevity of the main column, while contaminated guard-column may be replaced at regular intervals.

8.11.3.6. Column

Column is the heart of the HPLC system. The column and its fittings are designed to minimize dead volume and provide uniform flow. Usually, the columns are made of stainless steel that can withstand a pressure of about 5000 psi or more. Straight columns of 1-4 mm in internal diameter are generally used for analytical purposes. Commercially available preparative columns are of up to 25 mm in internal diameter. Though column length ranges from 5 cm to 100 cm, a column of 10-25 cm in length is very common. It is a common practice to operate the column at ambient temperature, but in some cases elevation of column temperature above room temperature gives a better result, particularly in liquid-partition and ion-exchange mode. For this, the column is placed in a special heating device (which is optional) for warming-up. Elevated temperature is used to reduce retention volume and solvent viscosity.

8.11.3.6.1. Column packing material

The column packing material is composed of a rigid solid material of silica, alumina, charcoal, organic polymers, etc. However, silica is the most commonly used support. HPLC-supports usually possess particle diameters of about 5-10 μm. There are three forms of packing materials: macro particulate porous support (≈ 40 μm diameter), micro-particulate porous support (5-10 μm diameter) and pellicular superficially porous support (20-60 μm diameter). In pellicular support there is a thin, porous pellicle of stationary phase coated on to an inert solid core, such as glass bead. Micro-particulate support is most widely used because of its high efficiency and sample loading capacity.

Table 8.6: Commonly used stationary phases in HPLC

Mode	Commercial name	Nature	Type of support
Adsorption	Corasil	Silica	Pellicular
	MicroPak Al	Alumina	Microporous
	Partisil	Silica	Micro porous
	Partisil C_8	Octylsilane	Porous
	Pellumina	Alumina	Pellicular
Exclusion	Bio-Glas	Glass	Rigid solid
	Fractogel TSK	Polyvinyl chloride	Semi-rigid gel
	Superose	Agarose	Soft gel
	Styragel	Polystyrene-divinylbenzene	Semi-rigid gel
Ion-exchange	AS Pellionex-SAX	Strong base	Pellicular

Mode	Commercial name	Nature	Type of support
	MicroPak-NH_2	Weak base	Porous
	Partisil-SAX	Strong base	Porous
	Partisil-SCX	Strong acid	Porous
	Perisorb-KAT	Strong acid	Pellicular
	Zipak-WAX	Weak base	Pellicular
Partition	ULTRA pak TSK ODS	Octadecylsilane	Porous
	ULTRA pak TSK ODS	Alkylamine	Porous
	MBondapak-NH_2	Alkylamine	Porous

8.11.3.6.2. Stationary phase

Some commonly used stationary phases are listed in **Table 8.6**. Usually, the stationary phase is coated on to the inert support. But the disadvantage of this system is that the stationary phase is gradually washed off from the column by the mobile phase. This is overcome with the advent of bonded phase packing materials where the stationary phase is chemically bonded to the surface of the solid particles through a silica ester or preferably a silicone polymeric linkage. The most useful bonded-phase coatings are siloxanes. Siloxanes are generated by reacting the hydrolyzed surface of silica with an organochlorosilane (**Figure 8.23**). The silica surface is hydrolyzed by heating it with 0.1M HCl for a day or two to give silanol groups.

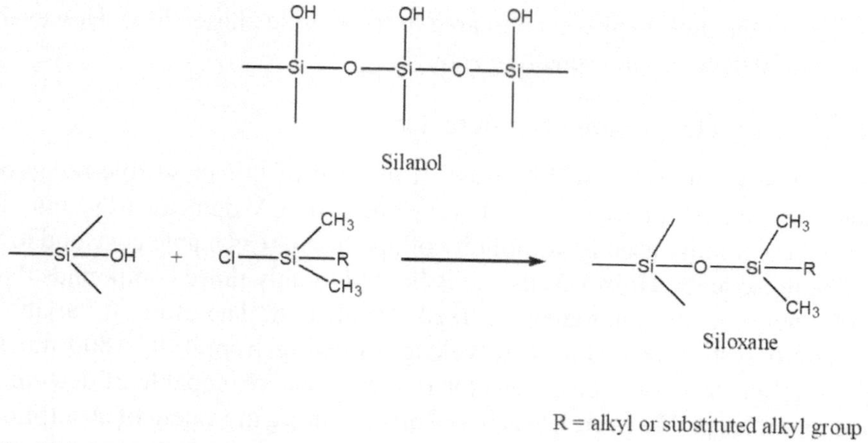

Silanol

Siloxane

R = alkyl or substituted alkyl group

Fig. 8.23: Preparation of siloxane

Bonded-phase packing may be of reversed-phase (containing non-polar functional groups) or normal-phase in nature. In reversed-phase type bonded packing, most commonly n-octyl (C_8 chain) or n-octadecyl (C_{18} chain) is used as R-group of siloxane and relatively polar solvents like water, methanol, acetonitrile, etc. are used for elution. In normal-phase bonded packing, R-group represents polar functional groups, such as dimethyl amino [-$C_3H_6N(CH_3)_2$], amino (–$C_3H_6NH_2$), diol (-$C_3H_6OCH_2CHOHCH_2OH$) and cyano (-C_2H_4CN); where amino type is most polar and diol type is intermediate in polarity.

8.11.3.7. Post-column filter

A filter (2-5 μ), porous plug of stainless steel or Teflon, is attached at the end of the column; so that column material should not come out under high pressure.

8.11.3.8. Detector

There are many detectors available now-a-days for HPLC to detect compounds as they elute from the chromatographic column. But most commonly used are RI (refractive index), spectrophotometer, fluorescence and electrochemical detectors. To obtain increased sensitivity and a better choice of analysis, the main unit of HPLC may be coupled to a mass spectrometer, nuclear magnetic resonance (NMR) spectroscope, photodiode array and fourier transformed infrared (FTIR) detectors.

8.11.3.8.1. Refractive index (RI) detector

The main advantage of this type of detector is its wide versatility because it is considered universal in its response to organics. It is widely used for carbohydrate and lipid analysis. It measures the difference in refractive index of the solute in the mobile phase, compared to the mobile phase alone. However, this type of detector is not sensitive enough.

8.11.3.8.2. Spectrophotometric detector

This is most popular and widely used. It may be of UV or visible-range or in the combination of two. A fixed wavelength of UV-detector (254 nm) is also very commonly used for simplicity of operation, reasonable cost, and low baseline noise level. However, its use is limited to only those compounds that absorb energy at the photometer's fixed wavelength. Therefore, a variable-wavelength system providing a wavelength ranging from 190 – 800 nm is preferred. Spectrophotometric detector is very sensitive, capable of detecting at nanogram level. The principle is based on measuring the extent of absorption of UV or visible radiation by a sample.

8.11.3.8.3. Fluorescence detector

Fluorescence detectors are extremely sensitive, perhaps 100 times more sensitive than UV-detectors. Fluorescence detectors with variable–wavelength (190–600 nm) excitation are available. This detecting method is applied to those compounds that naturally fluoresce or can be made to fluoresce through chemical derivatization. Aflatoxins, opiates, estrogen, etc. can be detected with fluorescence detectors. Amino acids or primary amines and secondary phenols are often derivatized with dansyl chloride (5-amino napthalene-1-sulfonyl chloride), prior to HPLC separation, to give fluorescent compounds that facilitate identification by fluorescence detector after their separation. Dansyl chloride can be used for the detection of amino acids in protein hydrolysates.

8.11.3.8.4. Electrochemical detector

This detecting system comprises two electrodes: a stable counter electrode which may be of Ag / AgCl or calomel and a highly polarizable working electrode. The compound of interest can be either oxidized or reduced at the working electrode surface under a constant potential resulting in a current flow between two electrodes. This detection method is suitable for those compounds that are electrochemically active at the available potentials. However, it can be equally useful to those electrochemically inactive compounds when an electrochemically active tag like bromine is added to them (e.g., unsaturated fatty acids or prostaglandins). Examples of compounds that are capable of undergoing oxidation are quinolones, catecholamines, azines, phenols, aromatic hydroxyls, amines, amides, hydrocarbons, etc., while those capable of undergoing reduction are azo-compounds, nitro-compounds, esters, ethers, olefins, aldehydes, ketones, diazo compounds, halogens, etc.

8.11.3.8.5. Evaporative light scattering detector (ELSD)

This is a recently developed detector and is now commercially available for HPLC. This detector, at first, converts the column effluent into a fine mist with the help of a stream of air or nitrogen in a nebulizer. The mist particles are then taken into a controlled temperature drift tube that causes evaporation of mobile phase and formation of fine particles of analyte. A laser beam helps the cloud of analyte to scatter radiation which is detected at right angle to the flow by a silicon photodiode. This detector is very sensitive (detection limit, 5 ng / 25 μl) and applicable to non-volatile solutes.

8.11.3.9. Sample collector

After separation in a chromatographic column and detection in a detector, the individual compounds come out from the chromatographic system one by one through an outlet port. These compounds can be collected individually and automatically following the nature of the chromatogram. This is most useful in case of preparative-HPLC

8.11.3.10. Data recorder and calculation

Quantitation of a compound is based on the response of the detector, as the detector response is proportional to the amount of compound present in the column effluent. A chromatogram is generated as a series of peaks, each denoting a specific compound, by a strip chart recorder. Quantitation of compounds from each peak can be made possible by measuring peak height or area, taking reference standard compounds as comparison. Peak-area can be measured using the method of triangulation, where peak is assumed to be a triangle, with an equation (Equation 8.26):

Peak area = Height x Width of peak at its one-half height Equation 8. 26

Per cent composition (% X) of a compound in a chromatogram can be estimated as

$$\% X = (\text{Area}_x / \text{Area}_a + \text{Area}_b + \text{Area}_x) . 100 \quad \text{Equation 8.27}$$

Where,

x = compound of interest, and

a & b = other compounds in the sample

Besides, peak area and % composition can be measured automatically using a programmed electronic integrator. With the advent of science, now-a-days, most HPLC instruments are coupled with an electronic integrator and a computer for data handling and processing.

8.11.4. Applications

Separation and quantitative determination of organic compounds, such as drugs, pesticide residues, biogenic amines, steroids, proteins, amino acids, lipids, carbohydrates, vitamins, organic acids, mycotoxins, polycyclic aromatic hydrocarbons, and many other compounds can be achieved in a very short time by HPLC.

8.12. Gas chromatography

8.12.1. Introduction

Gas chromatography is a sophisticated and powerful instrumental technique that helps to resolve a volatile solute mixture into individual components. Although A.J. Jones and A.J.P. Martin introduced this technique experimentally for separation of volatile organic compounds in 1952, the concept of gas chromatography was first enunciated in 1941 by A.J.P.

Martin and R. L. M. Synge. Since then it has become a powerful analytical tool in many branches of science, such as biochemistry, organic chemistry, agricultural chemistry, food chemistry, environmental science, forensic science, pharmacology, clinical science, etc. This is due to its high resolving power, extreme sensitivity, and speedy qualitative and quantitative analysis of many components simultaneously. In general, gas chromatography is a technique by which a mixture is separated into its constituents by a moving phase (gas) passing over a stationary phase. Based on the type of stationary phase being used, gas chromatography is classified as two types: gas-liquid chromatography (stationary phase, liquid; mobile phase, gas) and gas-solid chromatography (stationary phase, solid; mobile phase, gas). However, gas-liquid chromatography (GLC) is more popular than gas-solid chromatography (GSC) and in the present text GLC has been emphasized. GSC has some limitations in its application due to the tailing effect on its elution peaks. GSC is practically employed for analysis of low-molecular weight gaseous species.

Like other chromatographic techniques, in gas chromatography also separation of components is based on their varied time spent in the mobile phase or the stationary phase. When a mixture of two different compounds is introduced in a gas chromatographic column, the compound that spends more time in the stationary phase (thus, lesser time in mobile phase) compared to the other, moves slowly down the column resulting in a clear separation of the two compounds.

8.12.2. Principle

Under a continuous flow of a mobile gas phase through a liquid stationary phase in a column when a sample mixture in a vapor state is introduced into the mobile phase, partitioning of the different constituents in the sample between the mobile and the stationary phase occurs. Due to the difference in partition coefficients (K_D) of each component, they show a difference in their choice to stay with the mobile or stationary phase and thus, pass through the column at a different rate facilitating a complete separation. The mobile gas phase used, that is inert in nature, is not retained by the stationary phase. It is known as carrier gas as it carries the individual components through the column.

The mathematical relationships (related to retention volume, capacity factor, resolution, column selectivity, column efficiency, HETP, etc.), described in Section 8.11.2.2, are also applicable to gas chromatography. However, compressibility of gaseous mobile phase must be considered.

Retention volume is the product of retention time and average flow rate of gas (volume of gas per unit time) within the column. Thus,

$R_V = F \times R_t$ (for solute peak) Equation 8.28

$R_V^u = F \times R_t^u$ (for air peak) Equation 8.29

Where,

R_V & R_V^u= retention volume for retained and unretained (air) species, respectively,

R_t & R_t^u= retention time for retained and unretained (air) species, respectively and

F = average flow rate of gas.

So, the adjusted retention volume ($R_V^/$) for the retained species, which is corrected for the dead space becomes

$R_V^/ = R_V - R_V^u = F.R_t - F.R_t^u$ Equation 8.30

Due to compressibility, the rate of movement of gas is different at the exit and inlet of the column. Therefore, R_V and R_V^u depend on the average pressure inside the column. This average pressure is intermediate between the inlet pressure (P_i) and the outlet pressure (P_o). Hence, a pressure-gradient correction factor or compressibility factor (j) should be added to the adjusted retention volume ($R_V^/$) to give the net retention volume (R_V^n).

$R_V^n = j \,.\, R_V^/$ Equation 8.31

Where,

j can be calculated from the following relationship.

$j = 3[(P_i / P_o)^2 - 1] / 2[(P_i / P_o)^3 - 1]$ Equation 8.32

Again, average flow rate of gas within the column is not directly measurable, but it can be obtained using measured flow rate with a correction factor, considering column temperature and outlet pressure. The average flow rate (F) becomes

$F = F_m \,.\, T_c / [T \,.(P - P_{H_2O})] / P$ Equation 8.33

Where,

F_m = measured flow rate using a soap-bubble meter,

T_c = column temperature in Kelvin,

T = temperature at the meter,

P = gas pressure at the exit of column, and

P_{H_2O} = vapor pressure of water. Water vapor pressure must be considered because gas becomes saturated with H_2O in the soap-bubble meter.

In a soap-bubble meter which is placed at the end of the column, a thin film of soap or detergent solution is allowed to pass through the path of gas. Time taken by the film to move between two graduations on the burette is monitored and volumetric flow rate is measured.

Another terminology, specific retention volume (V_g) comes into existence when the amount of stationary liquid phase (measured at the time of column preparation) is considered because retention volume is changed by changing the amount of stationary liquid phase. The quantity of liquid may be changed by differing the column length or by using a thick or thin layer of liquid. The specific retention volume is expressed as

$$V_g = R_V^n / W_L \times 273/T_c \qquad \text{Equation 8.34}$$

Where,

R_V^n = retention volume,

W_L = mass of stationary liquid phase, and

T_c = column temperature in Kelvin.

Martin and Synge proposed a 'theoretical plate' concept which is entirely hypothetical but very useful to measure the column efficiency for separation. The column is considered to be divided into a number of equal units called theoretical units. The larger is the plate number, the more efficient is the column system and narrower is the peak. The theoretical plate concept assumes that solute is placed on first theoretical plate, equilibrium of the solute between the two phases is instantaneous and complete within each theoretical plate, and the partition coefficients of all components in each theoretical plate remain constant. However, height equivalent to theoretical plate (HETP) is a better expression for column efficiency than plate numbers because it takes column length into account. It seems that the number of plates is increased by increasing the column length and, in turn, column efficiency is increased to separate components. But practically, it does not happen so because due to column extra length the components are likely to be diffused more in the column leading to an overlap between each other. Therefore, there must be a limit in increasing column length.

Band spreading or broadening of the solute constituents inside the column occurs primarily due to two causes: (i) extra column cause and (ii) intra column cause. There are three main ways that direct band spreading for extra column cause: (a) too large sample volume, (b) too slow sample injection, taking more time, and (c) too much dead volume between injection port and column entrance, and between column exit and detector entrance. Intra column causes include eddy diffusion, longitudinal diffusion, mass transport, geometry of the packing material and carrier gas flow rate.

It is not likely that inside a column, molecules always maintain a uniform flow with the carrier gas. There are numerous paths between the particles from the column inlet to the column outlet. Some molecules meeting the stationary granules move around it and follow a zigzag and tortuous path, while some molecules will follow a shortest path. There must be a distribution of velocities around an average value, thus resulting in a band spreading. This phenomenon is known as Eddy diffusion. Bands of two components having similar separation properties may be overlapped due to eddy diffusion. In a poorly packed column, the separation is poor due to higher eddy diffusion. Eddy diffusion is proportional to the average particle diameter of the support. Thus, in an open-tubular column eddy diffusion is zero.

In longitudinal diffusion, the molecules can diffuse in the mobile phase along the direction of flow. At a low flow rate of mobile phase, backward longitudinal diffusion is also possible along with forward diffusion. However, at high flow rate longitudinal diffusion is less important but eddy diffusion becomes more important because turbulence in the column increases with increase in the flow rate. The opposite is true at low flow rate conditions, where eddy diffusion becomes less important due to less turbulence in the column and longitudinal diffusion both at backward and forward direction of flow is more important.

The molecules which diffuse deep inside the stationary phase face a resistance to mass transfer to the gas phase. This resistance leads to non-equilibration of the molecules between the two phases. This unequal distribution results in the fact that some solute molecules stay in the gas phase for a longer time and thus, exerts a high rate of movement. Some molecules which are slow in diffusion from stationary to mobile phase remain in the stationary phase for a longer time and move slower than the average solute molecules. Thus, the resistance to mass transfer is also a very important factor in exerting a peak broadening effect.

Factors like eddy diffusion (A), longitudinal diffusion (B) and resistance to mass transfer of the molecules in the two phases (C) have been utilized to derive a relationship through an equation between the flow rate (V) of the mobile phase and the theoretical plate height (H), which expresses the column efficiency. This is known as Van Deempter equation (Equation 8.35).

$$H = A + B / V + C.V \qquad \text{Equation 8.35}$$

A, B, & C are the characteristics of any particular column and vary from one column to another. According to the Van Deempter equation, the theoretical plate height changes with a change in the velocity of the mobile phase. When H is minimum, the column is most efficient. Optimum value of V for minimum H should be known and may be determined experimentally which can be performed by measuring HETP at different flow rates.

8.12.3. Instrumentation

A gas-liquid chromatograph essentially consists of five parts: (a) carrier gas assembly, (b) injection system, (c) column, (d) detector, and (e) recording system. Injection port, column and detector are housed separately in thermostated compartments to regulate their temperature. A schematic representation of a gas chromatograph is shown in **Figure 8.24**. Now-a-days GLC instruments are entirely computer based. Flow rate, column temperature, injector temperature, detector temperature, data collection and processing are all controlled by computers.

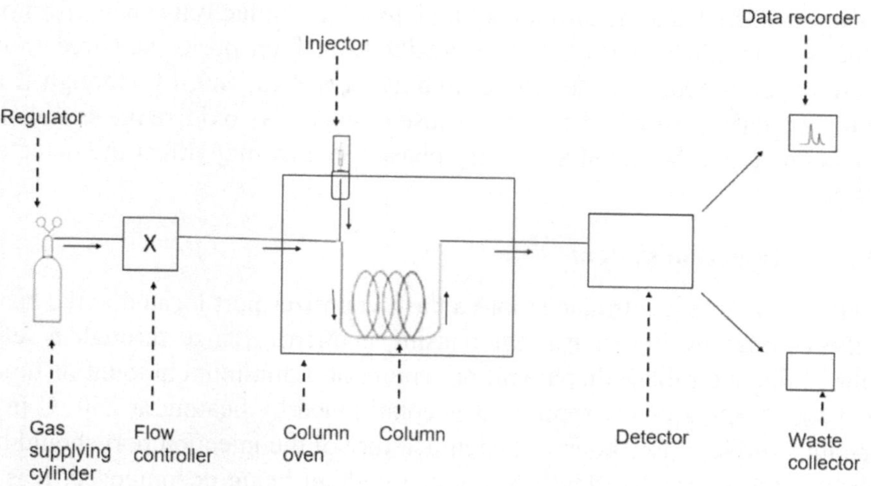

Fig. 8.24: Schematic representation of GLC

8.12.3.1. Carrier gas assembly

Carrier gas assembly is a high-pressure cylinder which supplies carrier gas, accompanied with a pressure regulator, pressure gauge, molecular sieve, flow meter. Pressure regulator, pressure gauge and flow meter are to control and monitor the gas flow. To obtain a reproducible result it is essential to maintain a constant flow rate of carrier gas. Molecular sieve is used to remove moisture, carbon dioxide, hydrocarbons, and other impurities from the carrier gas.

Molecular sieve can be regenerated by heating to 300 ^{0}C for 8 hours with a slow flow of nitrogen. In some cases, oxisorb can be used as a special filter to remove oxygen from the carrier gas, otherwise oxygen chemically affects the liquid stationary phase as well as the compounds to be separated.

8.12.3.1.1. Carrier gas

The most commonly used carrier gases are helium, argon, nitrogen, and hydrogen. However, choice of gas is often dictated by the nature of the detector used in the system. As for example, electron capture detector requires nitrogen or argon; for flame ionization detector any inert gas may be used, however, nitrogen is usually used; greatest sensitivity may be achieved by helium in thermal conductivity detector because helium possesses high thermal conductivity relative to that of the vapors of most organic compounds and so on. In general, an ideal carrier gas must be inert in nature, thermally stable and of low viscosity so that high flow rate can be obtained at a relative low pressure drop and of low diffusivity when high efficiencies are required because low diffusivity leads to a low B value of Van Deempter equation (Equation 8.35). It should also impart high thermal conductivity and ionization property. The carrier gas should be available in high purity and free from hazards. Compressed air cannot be used as carrier gas in GLC, though it is readily available and inexpensive, because oxygen may oxidize the sample to be separated and the liquid stationary phase too, and may affect the detector system.

8.12.3.2. Injection system

Usually, a sample is introduced into a flash vaporizer port located at the head of the column by injection method using a micro syringe through a self-sealing, silicone-rubber diaphragm or septum at a minimum amount of time. To avoid bleeding of the septum, it is conditioned by heating at 250 ^{0}C in a vacuum oven for a few hours. The temperature of the injection port should be such that the sample is rapidly vaporized without being decomposed. It is a common practice that injector temperature is maintained 10-50 ^{0}C above the column temperature to ensure rapid vaporization of the sample upon injection. A suitable sample size, normally 1-10 μl, is used as a 'plug' of vapor with a quick injection method to obtain enough column efficiency. Slow injection leads to diffusion of samples resulting in peak broadening and poor resolution. In some cases, particularly in capillary columns, sample splitter is used so that a small fraction (one part) of sample is introduced into the column and a major fraction (99 parts) goes to the waste, taking an injection volume of 0.5 μl or below of it. Reproducible results are obtained using a rotary valve-like (see section 8.11.3.4.) automated system. Quantity of sample introduced, and the manner of introduction must be reproducible with a high degree of precision. It is important to note that sample overloading causes production of asymmetric peaks due to fronting or tailing (Figure 8.2).

8.12.3.3. Column

This is the most important part of the gas chromatograph and is considered to be the 'heart' of the instrument. The columns are made of glass, fused silica, Teflon and metals, such as stainless steel, aluminum, copper, etc. in a spiral coil- or U-shaped or to a shape that fits into the limited oven space in the instrument. Generally, two types of analytical columns are available: packed and capillary or open tubular columns. For routine analytical work with an internal diameter of 2-4 mm and a length of nearly 2-4 m, packed columns are filled with finely divided inert solid support coated with a non-volatile liquid stationary phase (normally thin layer of 0.05-1 μm size, or filling may be an adsorbent or molecular sieve in gas-solid chromatography). Capillary or open tubular column with an internal diameter of 0.25-0.50 mm and a length of up to 100 m is more sophisticated in nature and efficient in separation of complex mixture than the packed column and is gaining more popularity over packed column, despite some of its disadvantages, such as small sample capacity, fragileness of the column, etc. The basic difference between these two columns is that the gas path through the column is unrestricted in open tubular column leading to quick analysis as the stationary liquid phase is present on the inner wall of the column as a thin film, while movement of the mobile gas phase faces a restriction due to the presence of solid support coated with liquid phase in the packed column. Basically, there are two types of capillary column: (a) wall-coated open tubular column (WCOT) and (b) support-coated open tubular column (SCOT). WCOT is simply a capillary column with its inner wall coated with stationary liquid phase. It offers a very small sample capacity. To overcome this problem, the SCOT column comes into play. The SCOT column contains an inner lining of finely divided support on the inner wall with an objective to hold a larger amount of the liquid stationary phase. Thus, this increases the surface area of the column and, in turn, increases sample capacity. Occasionally, rough surfaces in the inner wall of the WCOT column are generated by passing gaseous hydrochloric acid or strong aqueous hydrochloric acid or potassium hydrogen fluoride, so that the rough surface holds liquid stationary phase more tightly. Fused-silica open tubular column (FSOT), a new type of WCOT column, has become popular for the last few decades. This column contains a minimum amount of metal oxides and due to this reason, the column shows practically no reactivity toward the samples.

The fused-silica column is additionally coated with polyimide on the outside surface providing flexibility and physical strength. Most columns used today are FSOT columns.

In general, a sample splitter device is essential for using capillary columns because of its low sample capacity. The main advantage of using a capillary column is low pressure drop that makes it possible to use very long columns. It is observed that a decrease in column diameter gives a proportional increase in HETP. Additionally, compared to the packed column, very long capillary columns are used. As a result, a narrow column diameter with a greater length in the capillary column helps to achieve higher efficiency in separating complex mixture than the packed column.

8.12.3.3.1. Column preparation

Correct amount of stationary liquid phase is dissolved in a suitable solvent and a weighed quantity of size-graded solid support is added to it. The mixture is well-stirred to ensure uniform distribution or coating of the liquid phase onto the solid support. The solvent is removed preferably in a rotary evaporator. The packing material is re-sieved to avoid any fines produced during preparation. The solid support should remain free flowing after being coated with liquid phase. The coated support is packed into the column with great care so that packing materials are evenly distributed and no voids are left. Any left-over gap inside the column may cause occurrence of eddy diffusion (a value of Van Deempter equation) minimizing the column efficiency. While filling, the column is vertically held, and the packing material is added slowly to the top of the column with occasional shaking. Ultrasonic vibrator is used for this purpose. After packing, porous plugs of stainless steel or Teflon are used at the ends of the column to retain the packing materials and also to prevent their free movement during use. At this stage, the column is spiraled to the required shape. However, glass columns are firstly spiraled and then packed. It is essential to condition the packed column by heating it in a stream of carrier gas for about 24 hours at the upper limit of temperature allowed for the stationary phase.

8.12.3.3.2. Solid support

In gas-liquid chromatography, solid materials are present in the column for holding the liquid stationary phase. This is why it is known as 'support' to the stationary phase. However, in gas-solid chromatography the solid material itself acts as the stationary phase in the form of an adsorbent or molecular sieve. An ideal solid support should be chemically inert in nature with no catalytic sites and should be of uniform granular size with many accessible pores. The support must contain a large surface area (1-2 m^2/g). It should possess enough mechanical stability so that the particles are not likely to break down during handling or other processing. It must exhibit a remarkable thermal stability because in a gas chromatograph the column is supposed to be activated by heating up to 400 ^{0}C while passing carrier gas through it. There is a precise relationship between the particle size and the internal diameter of the column. It has been observed that 10-120 and 80-100 mesh-sized support particles are best suited for 2 and 4 mm i.d. column, respectively to obtain an effective packing, abiding by the general rule that column's i.d. should be 8 times more than the diameter of the support materials. In general, HETP is proportional to the average particle diameter. If the particle size increases, HETP also increases leading to low column efficiency. This is why usually the support size is kept small enough to obtain minimum HETP so as to achieve the maximum column efficiency. However, minimized particle size results in more applied gas pressure to get a desired level of gas flow. To compromise with this problem 80-100 mesh size granules are widely chosen for 4 mm i.d. column.

A very common and popular porous support material used today is 'celite', a brand of diatomaceous earth, which is made up of skeletons of diatoms, single-celled plants. Usually, crushed diatomaceous earth is calcined above 990 ^{0}C (Pink support) or mixed with a small quantity of calcium carbonate flux and calcined above 900 ^{0}C (White support). Chromosorb, Gaschrom, etc. are the available commercial products based on diatomaceous earth, other than celite. They are available with or without pretreatment to reduce their reactivity sites. Another commonly used solid support is ground fire bricks which are also porous in nature like celite. Non-porous small glass beads are also in some cases used successfully as solid support, particularly where it is used to hold non-polar solute during separation of polar solutes. Teflon is used as special support for analysis of corrosive samples. Presence of active sites (-OH groups) on the solid surface causes physical adsorption of polar and polarizable analytes resulting in distorted peaks and tailing effects. Therefore, it is essential to deactivate the solid support. The deactivation is done by silanization with dimethyldichlorosilane (DMDCS) [**Figure 8.25**].

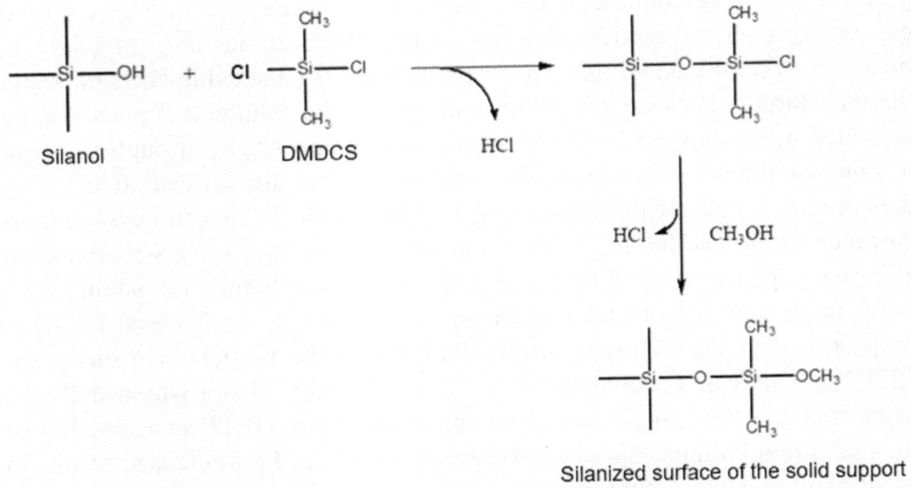

Fig. 8.25: Deactivation of active sites in solid support by silanization

Adsorption may also occur by the presence of metal oxides impurities. Hence, it is suggested to remove metal oxide impurities by acid washing prior to silanization. It is important to note that the FSOT column is free from this type of impurity. This is the reason why the adsorption problem is not encountered with the FSOT column. Silver-plating of the solid support by treating it with silver nitrate solution and a reducing agent leads to a considerable reduction of the tailing effect in many cases.

Porous polymer beads (commercially known as Porapak), made of styrene or divinylbenzene and of large surface area and uniform size, are preferentially applicable to gas-solid chromatography. These porous polymers show minimum affinity toward polar compounds. Polymer beads are strong and can be easily packed.

8.12.3.3.3. Stationary liquid phase

Selection of stationary liquid phase depends on the nature of compounds to be separated and governs elution pattern of solutes. Proper choice of stationary phase is the most crucial and vital factor that decides the success or failure of an operation in separating solute mixture into individual components in gas chromatography. All liquids used as stationary phase in GLC must possess some common properties that include: (a) non-volatility at high temperature (i.e., low vapor pressure), (b) should have selectivity for the components, (c) thermal conductivity, (d) chemical inertness so that it does not react with sample or carrier gas even at high temperature, (e) as a solvent it must show differential partitioning, thus different distribution constants for different solutes, and (f) should have wettability on the solid support surface. A large number of stationary phases are now in use (**Table 8.7**). Based on the chemical nature, a broad classification of liquid stationary phase can be made: (i) non-polar hydrocarbon type e.g., squalane, apiezon L grease, methyl silicone gum, hexadecane, silicon gum rubber, paraffin oil (Nujol), etc.; (ii) liquids of intermediate polarity e.g., didecyl phthalate, dinonyl phthalate, benzyl biphenyl, etc. (a polar group attached to a long non-polar skeleton); (iii) polar liquids, e.g., polyglycols, dimethyl sulfolane, etc.; and (iv) liquids capable of making H-bond e.g., polypropylene glycol, diglycerol, hydroxyl acids, tetrahydroxy ethyl ethylene diamine (THEED), etc. To obtain a good resolution, it is necessary that each solute should show solubility to the stationary liquid phase so that they should have different reasonable residence time in the column. This is why one should follow the 'like dissolves like' rule while choosing a liquid stationary phase. Therefore, polar and nonpolar liquid phases are best for polar and nonpolar samples, respectively. It is simply a fact that unless the sample dissolves in the liquid phase, no separation occurs.

Columns of bonded and/or cross-linked stationary phases are now available. The main objective of their use is to provide a long-lasting stationary phase inhibiting 'bleeding'. 'Bleeding' is a process by which the stationary phase is gradually lost during the elution process. A chemical reaction is involved in preparing a bonded stationary phase; for example, carbowax 400 and β, β′ -oxypropionitrile are bonded to the porous silica surface to produce Si – O – C linking (Durapak series). Zipax series of liquid phase is obtained through Si – O – Si linking with controlled surface –porosity glass beads. Cross-linking of the stationary phase is generated by gamma radiation or by incorporating peroxide. Peroxide helps a cross-linking through C – C bonds by a free radical mechanism.

Sample capacity and its residence time in the column are influenced by liquid thickness which normally ranges from 0.1-0.5 μm. Stationary liquid film thickness is an important factor to reduce or increase C value (mass transfer content) of Van Deempter equation (Equation 8.35). Reduced thickness reduces C and encourages high-speed analysis, high flow rate and low retention volume. One might utilize the film thickness parameter while separating high volatile or low volatile components. If the solutes are highly volatile in nature, it is better to use a reasonably thick film so that it retains the analytes for a longer time providing a sufficient time to separate and the reverse is true for low-volatile compounds.

Table 8.7: Some most widely used stationary liquid phases in GLC

Stationary phase	Trade name	Polarity	Maximum temperature (⁰C)	Applications
Squalane (2, 6, 10, 15, 19, 23-hexamethyl tetracosane	-	Non-polar	200	C_1-C_7 aliphatic and C_6-C_8 aromatic hydrocarbons
Apiezon L grease (high molecular weight aliphatic hydrocarbon)	-	Non-polar	250	Methyl ester of fatty acids, boranes, aromatic hydrocarbons, etc.
Diethylene glycol succinate	DEGS or LAC-2-R-446 or LAC-3-R-728	Polar	200	Methyl ester of fatty acids, essential oils, etc.
Polydimethylsiloxane	OV 1, SE 30	Non-polar	350	High boiling aliphatic and aromatic hydrocarbons, drugs, organometallics, PCBs, alkaloids, pesticides, steroids, etc.
Poly (phenyl ethyl dimethyl) siloxane (10% phenyl)	SE 52, OV 3, DC 550	Intermediate	350	Alkaloids, pesticides, methyl esters of fatty acids, halogenated compounds, aromatic acids as methyl esters, etc.
Poly (dicyano allyl dimethyl) siloxane	OV-275	Polar	240	Free acids, alcohols, polyunsaturated fatty acids, etc.
Polyethylene glycol (average molecular weight 1300-1600)	Carbowax-1540	Polar	150	Esters, ketones, alcohols, aldehydes, amines, etc.
Polyethylene glycol (average molecular weight 15000-20000)	Carbowax-20M	Polar	250	Essential oils, alcohols, ethers, ketones, phenols, sulfur compounds, free acids and glycols
Polypropylene glycol	UCON 550X	Polar	200	Essential oils, aldehydes, ketones, etc

Stationary phase	Trade name	Polarity	Maximum temperature (^{0}C)	Applications
Polyphenyl ether (6 rings)	OS 138	Intermediate	225	Polynuclear aromatics, fatty acid esters, amino acid esters, etc.
Poly (phenyl methyl) siloxane (50% phenyl)	OV 17	Intermediate	250	Phosphorous compounds, sugars (as trimethylsilyl ethers), steroids, pesticides, alkaloids, etc.
Poly (trifluoropropyl dimethyl) siloxane	OV 210	Intermediate	200	Chlorinated and nitro aromatics, alkyl-substituted benzene, alkaloids, steroids, etc.
Fluorinated silicone oil	QF 1	Intermediate	225	Insecticides, steroids, alkaloids, alcohols, ketones, etc.
Di-(2-ethylhexyl) sebacate	-	Intermediate	150	Hydrocarbons, esters, alcohols, halogen compounds, etc.

8.12.3.3.4. Column temperature

Column temperature is one of the many other critical factors (such as carrier gas flow rate, nature of stationary phase, etc.) that influence the resolving ability of the column. Before choosing a column temperature one should keep in mind that it should be within the working range of the stationary liquid phase, otherwise selected column temperature exceeding the working range may cause excessive 'bleeding' and thus minimize its longevity. It is suggested to get a prior knowledge on boiling point of major components or average boiling point of components of a mixture so that an initial choice on column temperature can be made by selecting a temperature that is few degrees lower than the average boiling point to maintain vapor pressure of each component reasonably high. If selected column temperature coincides with boiling points of components, compounds boil and spend most of the time in the gas phase, not in the liquid phase.This leads to fast elution of the compounds from the column without separating.

Column temperature is maintained in two ways: isothermally (constant temperature mode) and temperature programming (within a range, temperature is increased gradually. The advantage of temperature programming over isothermal conditions is that it is possible to obtain a quick separation with reasonable resolution of a mixture of components differing widely in their polarity and molecular weight in a single run.

When a sample mixture contains components varying widely in their boiling points, a selection of single temperature is not sufficient enough to separate the components, maintaining a compromise between their resolutions and retention times. If a low temperature is selected, low-boiling compounds are almost completely immobilized and condensed in liquid phase, resulting in a much extended retention time. Again, if the column temperature is too high, high-boiling compounds are separated with reasonable resolution, but low-boiling compounds without equilibration with the liquid phase emerge from the column so rapidly that they are not separated from each other. This technical drawback of selecting a single temperature is ruled out using temperature programming. In a programmed-temperature mode, sample is initially kept at a low column temperature and then the column temperature is increased either continuously or in steps as the separation proceeds. Initially low boiling compounds will emerge from the column, followed by high-boiling compounds as their vapor pressure increases with gradual increase of temperature. In programmed-temperature gas chromatography, peak height is mostly used for qualitative analysis because the peak height is sharp and uniform in shape.

8.12.3.4. Detector

There are a large number of choices for detectors that can be used in gas chromatography. Obviously, the choice depends on the physico-chemical nature of the compounds and their concentrations to be detected. The most widely used detectors are thermal conductivity detector (TCD), flame ionization detector (FID), electron capture detector (ECD), etc. Sometimes, GLC is coupled with a mass spectrometer (GC-MS) or infra-red spectrophotometer (GC-IR) or nuclear magnetic resonance spectroscopy (CC-NMR) to elucidate the chemical structure of the compounds after their separation.

8.12.3.4.1. Thermal conductivity detector (TCD)

It is also known as Katharometer. Though it is of low sensitivity (10^{-8} g), it is a widely used detector in gas chromatography because of its simplicity, response to almost all organic and inorganic compounds and non-destructive nature. Due to its non-destructive nature, gas chromatography can be adopted on a preparative scale when it is coupled with a TCD detector. After detection and quantitative estimation, it is possible to trap the separated solutes which may then be used for other analytical estimation. Due to its low sensitivity, TCD is not suited for capillary columns, since sample loading capacity of capillary columns is remarkably low.

Suppose a fine wire is connected to an electric circuit. Upon application of current the wire becomes hot. Now, if there is a steady flow of a gas possessing thermal conductivity over the wire, there is a constant rate of heat loss from the wire maintaining its constant temperature and thus, a constant resistance. Under this condition, if the prevailing gas flow is changed by another gas of different thermal conductivity, the temperature and the resistance of the wire are changed. This change in resistance is utilized to detect the presence of the new gas. This is the basic principle of TCD. TCD consists of a metal block containing two chambers. Each chamber possesses a pair of resistance wire or filament. The two pairs of filaments are connected to a Wheatstone bridge circuit. The filament in one chamber (reference cell) is exposed directly to carrier gas only, while in the other chamber (sample cell) column effluent (carrier gas + analyte in vapor form) enters. Before injection of a sample, both cells (reference and sample) receive the flow of carrier gas only. Due to the presence of pure carrier gas of same composition and thermal conductivity, temperature remains constant in the filament of both cells; the bridge is balanced giving a steady baseline on the recorder. However, after injecting sample in the column, the column effluent comprising of solute and mobile gas phase enters the sample cell, the temperature of filaments changes due to the different thermal conductivity of the gaseous mixture of the carrier gas and solute, leading to an abrupt change in the resistance. This causes the bridge to be unbalanced. The extent of imbalance is proportional to the concentration of the solute up to a certain limit and this gives a measure of concentration of the solute at that particular time. The signal of off-balance current is fed to a recorder to generate a chromatogram.

The best suited carrier gas in TCD is helium or hydrogen because their thermal conductivities are many folds greater than those of most organic compounds. As a result, the detector is able to operate in presence of small amount of solute in the carrier gas. This may not be possible if the carrier gas is nitrogen or argon because their thermal conductivities are nearly equal to many organic compounds. However, it is better not to use hydrogen; since it may reduce the bead material of the thermistor (thermistor is a semiconductor of fused metal oxides and is often used in place of metal filament).

8.12.3.4.2. Flame ionization detector (FID)

The principle of FID is based on the measurement of current which is produced by combusting the solute molecules in a hydrogen flame, generating ion fragments and free electrons. The amount of current is proportional to the amount of solute combusted. The detector consists of a capillary jet through which a mixture of column effluent and hydrogen gas comes out and through another inlet air/oxygen is introduced. A flame is maintained at the jet by igniting electrically. There is a platinum wire loop located at the tip of the flame, acting as an anode and the burner jet performs as a cathode. During combustion, if there is any ionization, the ions are collected in the electrodes producing an electric current. When in absence of solute the carrier gas alone is combusted, a negligible (due to production of few ions) but constant current is obtained. Impurities in the carrier gas and also column bleeding, if any, may cause a background current that can be suppressed by applying opposing bucking voltage so that a steady base line is obtained. Under this situation, if solute vapor along with the carrier gas enters into the flame, the solute molecules undergo ionization. Thus, ion concentration increases leading to an increase in current output. The resulting current is then amplified and recorded.

FID is sensitive only to oxidizable carbon atoms. It does not show any sensitivity toward functional groups, such as halogen, amine, thio-analogs, carboxyl, carbonyl, alcohol, etc. as their ionization in a flame is negligible or none. It also does not respond to CO, CO_2, SO_2, NO_2, H_2O, etc.; since they are non-combustible gases. Therefore, in terms of its application, TCD is more universal than FID. However, FID is more sensitive (10^{-13} g) than TCD. A major drawback of FID is its non-applicability to use the chromatogram in preparative scale. After passage through the detector the solutes cannot be trapped for further analysis because the detection of solutes by FID is a destructive process. However, this problem can be overcome by using a splitter before the column effluent enters the detector, so that a fraction of it is introduced into the detector chamber and the major fraction is trapped for further analysis.

8.12.3.4.3. Electron capture detector (ECD)

ECD is a selective detector that shows its selectivity to detect compounds that tend to capture electrons. Thus, organic compounds containing electronegative functional groups, such as halogens, quinines, nitro groups, etc. can be easily detected using ECD. It is insensitive to hydrocarbons.

In ECD, carrier gas (nitrogen) flows over a radioactive element, usually ^{63}Ni, that acts as a β-ray emitter. In the presence of β-ray, nitrogen ionizes with the formation of ions and low-energy 'slow' electrons. Under a fixed potential the electrons move toward anode generating a steady current that is translated by the recorder in the form of a steady base-line. When electron-capturing molecules come out from the column as effluent and enter into the detecting chamber, they scavenge the stream of electron reducing the current output. The amount of current is decreased in proportion with the number of electron-capturing molecules. Thus, the detector measures the decrease in current. Here lies its difference from the FID which measures the increase in current output, though both are ionization types of detectors. Methane (5%) is often mixed with carrier gas. Methane acts as a quench gas and reduces energy of the electrons so that more electrons are allowed to be captured. ECD is quite sensitive (10^{-14} g).

8.12.3.4.4. Flame photometric detector

FPD is also a selective type of detector, mainly used for detection of phosphorus and sulfur-containing compounds. This detecting system is based on the principle that when phosphorus and sulfur containing compounds are introduced into a hydrogen flame, they give rise chemiluminescent species such as HPO^* and S_2^*, respectively. In turn, these HPO^* and S_2^* species emit bands at 526 and 394 nm, respectively, that are monitored by photomultiplier tubes using filters transmitting at these wavelengths. The detector response is linear over the concentration in case of phosphorus containing compounds, whereas sulfur containing compounds receive a detector response that is proportional to the square of the concentration because two sulfur atoms constitute an excited S_2^* species. Unlike FID, in FPD the solute molecules along with carrier gas and air/oxygen come out through the jet of the burner and hydrogen through another inlet port around the jet.

It is possible to use FPD for detection of other elements like halogen, nitrogen, silicon, tin, etc. provided appropriate filters or grating monochromators are employed.

8.12.3.4.5. Thermionic detector (TID)

TID is a modified form of FID and is also known as alkali flame ionization detector (AFID). It is designed for selective determination of nitrogen and phosphorus containing compounds. TID is quite similar to the structure assembly of an FID. Only the difference is that it contains a sodium salt coated wire loop or rubidium silicate bead at the tip of the flame which is subjected to a potential of nearly 180V. The heated bead generates plasma with a temperature of 600-800 °C. This facilitates the production of large ion current. Unfortunately, the reasons for producing huge ions from nitrogen and phosphorus containing molecules by plasma remain unanswered.

8.12.3.4.6. Sulfur chemiluminescence detector (SCD)

This detector is selective and quite sensitive to sulfur containing compounds. Like FID, in SCD also the column effluent along with the carrier gas, hydrogen and air is burnt. But in SCD the burnt gas is additionally mixed with ozone. Sulfur compounds react with ozone and result in luminescence. The intensity of the emission is measured, which is proportional to the concentration of sulfur.

8.12.3.4.7. Microcoulometric detector (MCD)

Microcoulometric detection method is based on the coulometric methods of analysis which measures the quantity of electricity, or the number of coulombs required for a chemical reaction. According to Faraday's first law of electrolysis, the quantity of electricity passing through the electrode is directly proportional to the number of ions involved in the coulometric reaction. There are three main units in MCD: combustion chamber, titration cell and amplifier unit. In the combustion chamber, the effluents are converted into ionic species by oxidation or reduction. Few examples are given below:

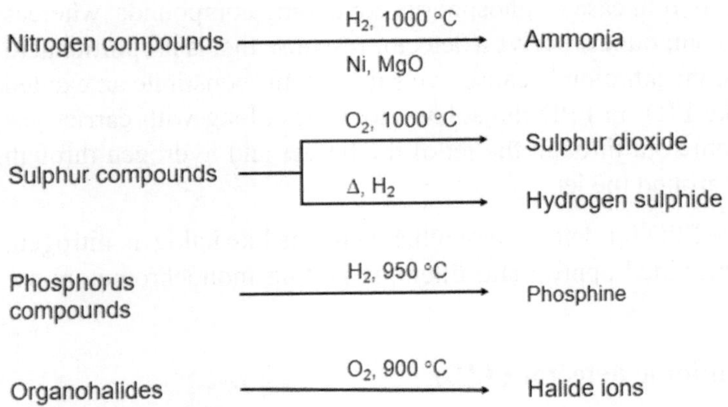

The ionic species then react with electrolyte ions in the titration cell which consists of a pair of reference and sensor electrodes and gives rise to a potential difference. The sensor electrode is located in the electrolyte solution. Initially when there is no reaction, the potential difference between the reference and sensor electrode is kept balanced sending zero input signal to the amplifier unit. But introduction of a solute into the detecting system reduces the titrating ion concentration in the electrolyte solution and as a result reduces the sensor electrode potential, creating an imbalance between the reference and sensor electrodes.

8.12.3.4.8. Atomic Emission detector (AED)

GLC column effluent is allowed to get an impact from microwave-energized helium plasma that atomizes all of the elements in a sample. The characteristic atomic emission spectrum is then monitored with a movable and flat-diode array working in the range of 170 – 780 nm.

8.12.3.4.9. Mass spectrometer detector

Gas chromatographic technique is often coupled with a mass spectrometer as a detecting system that is utilized to elucidate the structure of unknown compounds, getting a preliminary knowledge on the mass of the compound and its fragmentation pattern. Mass spectrum of each column effluent can be recorded. Mass spectrometer as a detecting system can be utilized to achieve quantitative estimation by monitoring selective ion in the mass spectrum. Before entering into the ionization chamber of the mass spectrometer, column effluent is made free from the carrier gas in a jet separator with the help of a skimmer under the influence of vacuum. Light helium atoms are pumped away, while a comparatively heavier solute molecule enters into the skimmer. This is followed by ionization of the molecule through electron impact ionization (MS-EI) or chemical ionization (MS-CI) and stored in a radio-frequency field. The trapped ions are then ejected from the storage field to an electron multiplier detector in such a manner that scanning is undergone based on the mass to charge ratio.

8.12.3.5. Data recorder and calculation

See section 8.11.3.10.

8.12.3.6. Fraction collector

The individual components emerging from the column can be trapped usually at low temperature for further analysis. For this purpose, a collector trap is fitted at the exit point at low temperature where it collects the individual component of interest by monitoring the peak during the time gap between its emergence from the base line and its dissipation to the base line. By repeating this process for a large number of times, individual components of sufficient amounts may be collected for further spectroscopic analysis to elucidate the components' chemical structures. Preparative instruments, using columns of larger diameter with provision of repeated automatic injection of sample and programmed sample collector, are available commercially. If at all, the individual components are not needed for further analysis, effluent's trapping is essential; otherwise, the components impose a health hazard after their diffusion into the laboratory atmosphere. Alternatively, the column effluent may be passed to a fume hood. A special care must be taken in case of radioactive samples for their proper disposal.

8.13. Model questions

i. Define chromatography. Outline the overall chromatographic technique. TLC is an adsorption type of chromatographic technique ---------- Justify.

ii. Explain why is available ion-exchange capacity of ion-exchanging resin more acceptable than total ion-exchange capacity. How does pH affect the ion-exchange capacity of ion-exchanger?

iii. How is a selection of ion-exchanger (cationic or anionic) made for an unknown sample? Why is paper chromatography considered as partition chromatography?

iv. What is an adsorption isotherm? Discuss various theories on adsorption isotherms. When do you expect 'tailing' and 'fronting' effects in the elution profile of the solute?

v. What are the forces responsible for adsorption processes in adsorption chromatography? Describe them in detail.

vi. Discuss the principle of paper chromatography. What are various modes of operation in paper chromatography? Why and how is lipid (if lipid is not a point of concern for its analysis) removed from a biological sample before performing paper chromatography?

vii. What is R_F? What does it signify when its value becomes zero and one? Why is knowledge on eluotropic solvent series essential for performing any type of liquid-liquid partition chromatography? Why is H_2SO_4 not used in paper chromatography during detection of spots?

viii. What are the differences between gel filtration chromatography and conventional chromatography? How is molecular separation accomplished in gel filtration chromatography? What are the other purposes of using gel filtration chromatography apart from molecular separation?

ix. When does K_D become zero and one in gel filtration chromatography? How is cross-linked dextran polymer synthesized for its use in gel filtration chromatography? How is molecular mass determined using gel filtration chromatography technique?

x. What is the basis of separation in affinity chromatography? How does it differ from other chromatographic techniques? How do the ligands bind to the cyanogen bromide activated agarose and epoxy activated agarose? Give a few reaction examples.

xi. Describe different types of affinity chromatography. What are the different ways to recover specific glycoprotein, bound to the specific lectin, after performing lectin affinity chromatography?

xii. How is the desired substance bound in an affinity chromatographic column released? Describe different methods.

xiii. Give a brief account on 2D-TLC. How do TLC and PC differ from each other? Describe it mentioning their merits and demerits.

xiv. List out the main differences between (a) normal- and reversed-phase HPLC, and (b) hydrophobic interaction and reversed-phase chromatography. What types of stationary phase matrices are used in size-exclusion HPLC? What is the significance of column selectivity (α) being unity?

xv. Define and describe capacity factor, column selectivity and column efficiency. Rs (resolution) = <0.8, 1.0 and 1.5. What do they signify?

xvi. What is the need of using guard columns in HPLC? Describe the common detectors used in HPLC.

xvii. What is the basis of separation in GLC and GSC? What is HETP? How is it related to column efficiency?

xviii. What are the different types of columns used in gas chromatography? Describe in detail. What are the main advantages of using capillary columns?

xiv. How are the particle size of the solid support and HETP related to each other in respect of column efficiency? What are pink and white supports used in GC? Why and how solid supports are deactivated?

xx. Why is temperature programming in GC columns more efficient than isothermal column temperature in solutes' separation process? Describe the principle of TCD, FID, ECD and MCD used in GC detection systems.

8.14. Suggested Readings

(i) Wilson, K.; Walker, J. (Eds) (2000) Practical Biochemistry – principles and techniques. 4th Edition, Cambridge Univ. Press.

(ii) Heftmann, E. (Ed) (1961), 3rd edition, Reinhold Publishing Corporation, New York.

(iii) Stahl, E. (Ed) (1990) Thin layer chromatography – A laboratory handbook. 2nd edition, New York Springer-Verlag.

(iv Renner, F.; Kanitz, R. D.(1997) Quantification of carbohydrate deficient transferrin by ion exchange chromatography with an enzymatically prepared calibrator. Clin. Chem. 43: 486 – 490.

(v) Cooper, T. G. (1977) Gel permeation chromatography. In : Cooper, T. G. (Ed) Tools of Biochemistry, New York John Willey, pp 169-192.

(vi) Gelotter, B.; Porath, J. (1967) Gel filtration. In: Heftmann, E. (Ed) Chromatography, 2nd edition, Reinhold.

(vii) Syska, K.; Perry, S. V.; Trayer, I. P. (1994) A new method of preparation of Troponin I (inhibitory protein) using affinity chromatography. Evidence for three different forms of troponin I in striated muscle. FEBS Lett. 40: 253-257.

(viii) Morvatr, G. (Ed) (1987) High performance liquid chromatography – Advances and perspectives. Vol. 4, New York Academic Press.

(ix) Niessen, W. M. A.; van der Greef, J. (1992) Liquid chromatography – mass spectrometry. Marcel Dekker Inc., New York.

(x) James, R. W. (1970) Undergraduate Instrumental analysis. Marcel Dekker Inc., New York.

9

Radiotracer Technique

9.1. Introduction

An atom consists of a positively charged nucleus surrounded by a cloud of negatively charged electrons. The positive nuclear charge is due to the presence of protons which carry a mass of 1 and a charge of +1. Besides protons in the nucleus, there are neutrons which are heavier than protons. They carry no charge. Comparatively electrons have no mass (1836 times lesser than that of a proton) but they carry a charge of –1. Hence, the whole mass of an atom is likely to be concentrated in the nucleus. All atomic nuclei contain protons and neutrons except the hydrogen nucleus which possesses only protons. Neutrons interact with protons to stabilize the nucleus, and overcome the mutual electrical repulsion of the positive protons. As a result, the neutrons and protons are held together very tightly. The sum of protons and neutrons in each nucleus is known as **mass number** (A). In the nucleus the total number of neutrons is the **neutron number** (N) and the total number of protons, which is obviously the same as the total number of orbital electrons in an electrically neutral atom, is referred to as **atomic number** (Z). Thus, A = Z + N.

Isotopes are defined as the atoms of the same element with the same number of protons but different numbers of neutrons in the nucleus. If we take hydrogen as an example, we find three of its isotopes, $^{1}H_{0}$, $^{2}H_{1}$, and $^{3}H_{2}$. All three isotopes have the same Z value, but they differ in A value due to different neutron numbers (N). Isotopes possess identical chemical properties. However, they exhibit different physical properties. This is due to the presence of the same atomic number in isotopes. The same atomic number confirms that they have the same number of protons, the same number of electrons, the same electronic configuration and more specifically, the same number of electrons in the valence shell. Identical chemical properties in isotopes are due to the same number of electrons in the valence shell, and their difference in physical properties is generated for their different mass number.

Isobars are the nuclides of different elements with the same mass number, whereas nuclides with the same neutron number but different atomic number and mass number are called **Isotones (Table 9.1)**.

Table 9.1: Isotope, Isobar and Isotone

Type	Characteristics	Example
Isotopes	Different mass number, Different neutron number, Same atomic number	$^{1}H_{0}$, $^{2}H_{1}$, and $^{3}H_{2}$
Isobars	Different proton number Different neutron number Same mass number	$^{3}H_{2}$ and $^{3}He_{1}$
Isotones	Different mass number Different proton number Same neutron number	$^{3}H_{2}$ and $^{4}He_{2}$

This chapter deals with only radioactive isotopes. Radioactive isotope is an isotope whose nucleus undergoes a spontaneous disintegration or transformation accompanied by the release of energy and the emission of one or more types of radiation and ultimately leads to the formation of stable isotope. Since a radioactive and a non-radioactive atom of the same element possess the same number of orbital electrons (as the same number of protons), their chemical and biological behavior are same, and this makes the tracer experiments with radioisotopes possible. Certain combinations of neutrons and protons produce nuclei of latent instability. In general, stable isotopes for elements with low atomic numbers tend to possess an equal number of neutrons and protons, whereas the stability is related to a neutron-proton ratio more than 1 for higher atomic numbered elements.

Majority of the nuclei, naturally found on the Earth, are stable. Approximately 270 stable isotopes and 50 naturally occurring radioactive isotopes are found on the Earth. However, innumerable man-made radioisotopes exist, that are produced artificially in the laboratory.

9.2. Types of particles and their properties

When different radioactive substances are put in the magnetic field, they deflect in different directions or not at all, showing that there are three classes of radioactivity: negative (β-particles), positive (α-particles), and electrically neutral (γ-rays). As positively charged alpha particles are more massive, and move slowly compared to beta and gamma particles; they interact much more easily with matter. Due to low penetration power, only a few centimetres of air were enough to stop the alpha radiation. Electrically charged beta particles are much less massive compared to alpha particles and move faster. An aluminium sheet of one millimetre thickness or several meters of air stops the electrons and positrons. As gamma rays carry no electric charge, they can penetrate large distances through materials. Several centimetres of lead or a meter of concrete is needed to stop most gamma rays.

9.2.1. Alpha particle (α-particle, $^4He^{2+}$)

An alpha (α) particle is a helium nucleus consisting of firmly bound two protons and two neutrons. It carries two units of positive charge and a mass of 4. Its penetration power is low (5-7 cm in air, 20-40 μ in water), but energy level is quite high, 3-8 MeV (million electron volt). These particles are mono-energetic (all α-particles from a given isotope possess the same amount of energy). The ionization power of α-particles is high due to its high energy level. Traversing a linear path of 3.5 cm in air, a 5 MeV α-particle may produce nearly 25000 ion pairs/cm. Alpha particles of considerable kinetic energy travel through a gas in a straight line with a limited range. Alpha-emitters (emitting α-particles while they decay) are generally isotopes of elements with high mass numbers (›140). These elements are seldom used in biochemical research. Ionization chamber is used to detect α-particles. Due to low penetration power, α-particles are relatively ineffective to produce artificial isotopes.

The initial velocity of an α-particle is dependent on the source, radio element from which it originates. It is observed that most of the α-particles of a given element possess the same velocity (1.4-2.0 × 10^9 cm/sec) and the initial velocity (V in cm/sec) of α-particles has a mathematical relationship (Equation 9.1) with their range (R in cm) in air.

$V^3 = k \times R$ Equation 9.1

Where, k is a constant with a value of about 1.0×10^{27}.

But this relationship does not exist when velocity falls below 40% of its initial value in air. 'Range' is defined as the travelled path length of the α-particle in air between the source and the point at which it has no appreciable ionization or scintillation power. The range of an α-particle in air is roughly related with its energy as follows:

$R \approx E$ Equation 9.2

Where,

R = range in air in cm, and

E = energy in MeV.

Therefore, it is observed that a range of 3-8 cm for α-particles corresponds to their 3-8 MeV energy.

9.2.2. Beta particle (β-particle)

β-particles are fast-moving electrons, existing in two forms; positron (β^+) & negatron (β^-), emitted during the decay of radioactive atoms possessing unstable nuclei (β-emitters). β-particles are considered as a stream of electrons shot out from the radioactive material with a very high velocity, almost approaching that of light. Like α–particles, the velocity of β-particles depends on the radioactive element from which they originate. But unlike α-particles, β-particles from a given source exhibit velocities varying over an appreciable range. The α-particles are mono-energetic, whereas β-particles show a continuous spectrum of energy up to a certain maximum value, E_{max}. When a β-particle from a given source is subjected to a magnetic field, it gives rise to a continuous β-ray spectrum which states that there is a continuous variation of velocity. But according to the law of conservation of energy, all electrons from a given substance possess the same kinetic energy leading to a constant velocity. To satisfy the basic concept of the law of conservation of energy, W. Pauli and E. Fermi (1934) suggested that β-emission is accompanied with emission of another particle, **neutrino**; so that the energy during radioactive decay is shared by both β-particle and neutrino. Therefore, a slow-moving β–particle is necessarily accompanied with a fast-moving neutrino. Neutrinos are neutral particles possessing a very small mass (less than that of an electron), but they are characterized with momentum and energy. β–particles show higher penetrating power, but due to small mass (nearly 1/7300 of α-particle) their ionization power is less compared to α-particles. Their small mass makes the β-particles not to move in a straight line but to follow a tortuous path through matter, causing a deflection from its source. The range or penetration of β-particles depends on the E_{max} (maximum energy) and is expressed as weight per unit area (mg/cm^2).

The range of β-particles can be approximately depicted by the following Glendenin's equation (Equation 9.3, applicable only to β-energy values which are above E_{max} of 0.8 MeV).

$$R_{max}\ (mg/cm^2) = 542\ E_{max} - 133 \qquad \text{Equation 9.3}$$

Most of the commonly used radioisotopes for biochemical studies are β-emitters such as ^{14}C, ^{35}S, ^{32}P, and ^{3}H. When the β-emitters disintegrate, β-particles with characteristic energies are emitted leading to the excitation and ionization of the molecules through which they traverse. The ionization properties are used for quantitative estimation of the radioisotopes. Though β-particles are emitted with a continuous range of energies from a given source; an average energy (E_{mean}) of β-particles, that is characteristic of that isotope, may be determined and is used to identify one β-emitter in the presence of the other.

9.2.3. Neutron

Neutrons possess a distinct mass which is nearly the same as a proton, but several hundreds of times more mass than an electron. Since they possess mass but devoid of charges, they exhibit a profound penetrating power. The main source of neutrons is fission type of nuclear reactions. However, they may be produced from the decay of radioactive nuclides.

9.2.4. Gamma (γ) ray

Gamma (γ) rays are electromagnetic radiations analogous to X-rays but are of shorter wavelength (10^{-8} and 10^{-11} cm) and more powerful. They are mono energetic high-energy photons. The spectrum of γ-ray consists of discrete lines. It cannot ionize atoms by direct interaction because it has no charge and mass. However, it has more penetrating power than α- and β-particles and it requires several inches of dense material (like lead) to shield them. It is observed that γ-ray from ^{60}Co penetrates 15 cm of steel to reduce its intensity to the level that cannot be detected. Gamma rays are considered to exhibit the least ionizing power but the greatest penetrating power. The emission of γ-ray takes place subsequently, not simultaneously with that of α- or β-particles. After the emission of α- or β-particles, the nucleus undergoes a rearrangement which causes liberation of energy in the form of γ-rays. Therefore, emission of γ-ray may be considered as a secondary process (Equation 9.4).

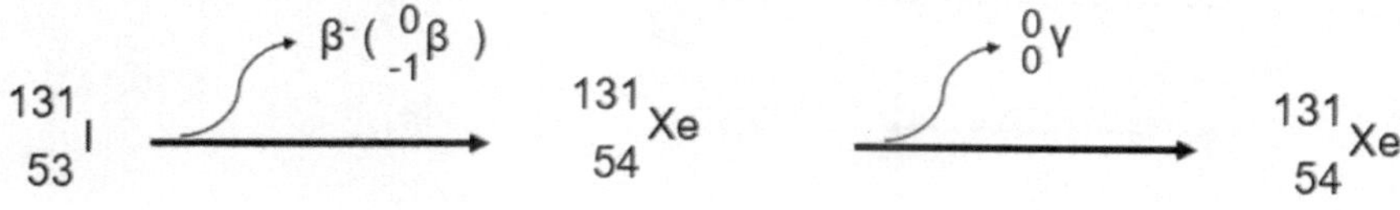

Equation 9.4

Emission of γ-ray does not change atomic number or mass. The γ-rays of distinct energy can be monitored for identification of the isotope emitting γ–rays. Source of γ-ray is inside the excited atomic nuclei, since γ-ray is produced by nuclear relaxation. Emission of initiating primary alpha or beta particles is often hard to detect, whereas fairly efficient systems exist to detect γ-rays. Gamma ray spectroscopy is used as one of the ways to detect γ-ray emitting radioisotopes.

9.3. Radioactive decay

Radioactive decay facilitates the changing of one unstable nucleus to a stable nucleus that possesses more nuclear binding energy than the unstable decaying nucleus. The difference in the nuclear binding energy between the unstable and stable state determines the type and nature of decay. The unstable nuclei, possessing excess of protons or neutrons compared to the stable nuclei, start decaying to be transformed to the stable nuclei by changing protons into neutrons or neutrons into protons either singly or in combination. When the nuclei remain in excited state, the decay may also occur by emitting gamma rays to release the excess energy and under this situation, the proton and neutron numbers remain unaltered. It should be remembered, whichever the case it may be, the decay process must follow various conservation laws. Taking into consideration all resultant decay products, the value of the conserved quantity after the decay must equal the same quantity evaluated for the nucleus before the decay.

9.3.1. Decay by alpha particle emission (alpha decay)

With an α-particle emission, a nucleus loses its atomic (Z) and neutron (N) number, both by 2 and thus, mass number (A) by 4 because an α-particle is a doubly charged helium atom. It may be expressed by the following way:

$$^{A}X_{N} \longrightarrow {}^{A-4}Y_{N-2} + {}^{4}He_{2} + \text{Energy}$$

Equation 9.5

e.g.,

$$^{222}Rn_{136} \longrightarrow {}^{218}Po_{134} + {}^{4}He_{2} + 5.58\ \text{MeV}$$

$$^{238}U \longrightarrow {}^{234}Th + {}^{4}He_{2} + 4.196\ \text{MeV}$$

Since the atomic number changes during alpha decay; the parent atom and the resultant daughter atom become different elements, differing in their chemical properties. Nuclei, possessing large proton to neutron ratio, exhibit alpha decay. Thus, the proton to neutron ratio in a massive nucleus is minimized by alpha decay to give rise to a more stable configuration. Many nuclei that are more massive than lead show alpha decay.

9.3.2. Decay by β-particle emission (Beta decay)

Beta decay is accomplished in three ways: a) Negatron (β^-) emission, b) Positron (β^+) emission, and c) Electron capture (EC). By these processes, protons are transformed into neutrons or vice-versa. In each case, the number *A* of nucleons remains the same, while both *Z* and *N* increase or decrease by 1. Since the atomic numbers of the parent and daughter atoms are different, consequently the parent and daughter atoms are different elements. Beta particles are electrons or positrons (electrons with positive electric charge, also known as antielectrons). Beta decay occurs in a nucleus containing too many protons or too many neutrons. In such cases, one of the protons or neutrons is transformed into the other. In beta negative decay, a neutron decays into a proton, an electron, and an antineutrino, while a beta positive decay corresponds to a proton-decay into a neutron, a positron, and a neutrino.

9.3.2.1. Negatron emission

With this, a nucleus having excess of neutrons relative to its more stable isobars achieves its stability by converting a neutron to a proton. During this conversion, ejection of a negatron is taken place. Conversion of neutron to proton leads to decreased N/Z ratio and atomic number is increased by one. As a result, the mass number remains the same. It may be expressed as follows:

$$^{A}X_{N} \longrightarrow {}^{A}Y_{N-1} + \beta^{-} + v^{-} \text{ (antineutrino)}$$

Equation 9.6

e.g.,

$$^{32}P_{17} \longrightarrow {}^{32}S_{16} + \beta^{-} + v^{-} \text{ (antineutrino)}$$

$$^{14}C_{8} \longrightarrow {}^{14}N_{7} + \beta^{-} + v^{-} \text{ (antineutrino)}$$

9.3.2.2. Positron emission

With this a nucleus having excess of protons relative to its more stable isobars achieves its stability by converting a proton to a neutron with positron emission. N/Z ratio is increased, and atomic number is decreased by one. The mass number does not change because decrease in atomic number is just compensated by increase in neutron number. It may be expressed as follows:

$$^{A}X_{N} \longrightarrow {}^{A}Y_{N+1} + \beta^{+} + v \text{ (neutrino)}$$

Equation 9.7

e.g.,

$$^{22}Na_{11} \longrightarrow {}^{22}Ne_{12} + \beta^+ + \nu \text{ (neutrino)}$$

$$^{65}Zn_{35} \longrightarrow {}^{65}Cu_{36} + \beta^+ + \nu \text{ (neutrino)}$$

The positron has a transitory existence, its ultimate fate being annihilation by reaction with an electron to yield two 0.511 MeV γ-ray photons.

9.3.2.3. Electron capture (EC)

Like positron emission, EC also occurs in the nucleus having excess protons relative to its more stable isobars. A deficiency of neutrons in the nucleus may alternatively be increased by capturing one of its own orbital electrons (mostly K-electrons, this is why EC is also known as 'K-capture') with the emission of a neutrino (Equation 9.8). When there is insufficient energy for positron emission, EC occurs as the only process to convert protons to neutrons. Like decay by positron emission, EC decay also leads to decrease of atomic number by one with concomitant increase of neutron number by one. Thus, the mass number remains constant. EC reaction is depicted in Equation 9.8 with examples.

$$^{A}X_{N} + e^- \longrightarrow {}^{A}Y_{N+1} + \nu \text{ (neutrino)}$$

Equation 9.8

e.g.,

$$^{65}Zn_{35} + e^- \longrightarrow {}^{65}Cu_{36} + \nu \text{ (neutrino)}$$

$$^{55}Fe_{29} + e^- \longrightarrow {}^{55}Mn_{30} + \nu \text{ (neutrino)}$$

$$^{48}Cr + e^- \longrightarrow {}^{48}V + \text{X-rays}$$

EC causes a loss of orbital electrons (K-, L- or M-shell) and generates a vacancy. Later, filling-up the vacancy by electrons from higher energy levels is carried out with emission of X-rays that are characteristic of daughter nuclide. As for example, when ^{55}Fe decays by electron capture, characteristic X-ray of manganese is emitted. But a gamma photon is also emitted in case of EC mode, as it is evidenced in the disintegration of zinc. It is important to note that photon emission is not a nuclear process, but electron capture by nucleus holds good as a nuclear process.

9.3.3. Decay by gamma ray emission (Gamma decay)

During γ-decay, a nucleus of a higher energy state is transformed into its lower energy counterpart without any change in proton and neutron number; but in the process, high energy electromagnetic radiations in the form of γ-ray photons are emitted. As a consequence, the parent and daughter atom remain as the same chemical element. Since γ-ray does not carry any charge and mass, γ-decay causes no alteration in the structure and composition of the atom. It only leads to the changing energy state of the atom. Gamma decay is represented by Equation 9.9. It should be remembered that unlike alpha and beta decay, no particles are ejected from the nucleus during gamma decay.

$$^{A}X^{*}_{N} \xrightarrow{\text{Relaxation}} {}^{A}X^{\circ}_{N} + {}^{0}_{0}\gamma$$

Parent atom (higher energy state) — Daughter atom (lower energy state) — γ-ray

Equation 9.9

e.g.,

$$^{152}_{66}Dy^{*} \longrightarrow {}^{152}_{66}Dy + {}^{0}_{0}\gamma$$

For gamma decay, a nucleus must achieve an excited energetic state. The nucleus in its excited state is obtained by jumping a proton or a neutron up to an excited state which is generally carried out by an alpha or beta decay. The excited proton or neutron of the daughter nucleus then emits energy in the form of gamma photons so that the nucleus goes down from a high to a low energy state. When the nucleon (collective term for protons and neutrons) makes this transition from a high to a low energy state, a gamma photon is emitted. This process may be illustrated with the decay scheme for Cobalt-60 (^{60}Co, **Figure 9.1**). In the first stage, ^{60}Ni in excited state is obtained following a beta decay of ^{60}Co where an electron of 0.31 MeV is emitted. In the next phase, the excited $^{60}Co^{*}$ comes down to its lower energy ground state by emitting γ-rays in succession of 1.17 MeV followed by 1.33 MeV.

$$\text{(i)} \quad {}^{60}_{27}Co \longrightarrow {}^{60}_{28}Ni^{*} + e^{-} + v^{-} + {}^{0}_{0}\gamma + 1.17\ MeV$$

$$\text{(ii)} \quad {}^{60}_{28}Ni^{*} \longrightarrow {}^{60}_{28}Ni + {}^{0}_{0}\gamma + 1.33\ MeV$$

Fig. 9.1: Radioactive decay scheme of cobalt-60

The frequency of γ-rays released by the process may be aptly determined by the Equation 9.10, if the initial and final states of the nucleon inside the nucleus are known.

$$Ei - Ef = h \cdot f \qquad \text{Equation 9.10}$$

Where,

Ei = the initial, higher energy state of the nucleon,

Ef = the final, lower energy state of the nucleon,

h = Planck's constant, and

f = the frequency of the emitted radiation.

9.4. Interactions of particle/radiation with matter

While interacting with matter, radiations collide with nuclei or orbital electrons or with the field surrounding a particle. As a resultant effect, different phenomena may occur. It may lead to conversion of energy to matter (pair production from photons), conversion of one type of radiation to another (photoelectrons from photons), scattering (change in direction) of photons or β-particles, ionization and excitation of atoms, complete absorption of photons, etc. Literally, 'collision' means the physical impact of one particle with another. But here, in true sense, the term 'collision' also refers to the interchange of momentum, energy and change within interacting particles without direct physical contact. Collision may be **elastic** (no change in internal energy or total kinetic energy of the colliding particles) and **inelastic** (a change in internal kinetic energy or total kinetic energy of the colliding particles).

The interaction of charged particles with matter involves the transfer of energy from the charged particles, α-particles (+2 charge) and β-particles (+ or -1 charge) or electrons, to the material through which they travel. Photons and neutrons, which have no charge, interact very differently. When charged particles pass through matter, they continuously interact with the electrons and nuclei of the surrounding atoms and as a consequence the velocity of charged particles is continually slowed down i.e., the alpha or beta particles lose kinetic energy. The interactions involve the electromagnetic forces of attraction or repulsion between the alpha or beta particles and the surrounding electrons and nuclei. The force associated with these interactions can be described by Coulomb's equation (Equation 9.11):

$$F = \frac{k q_1 q_2}{r^2}$$ Equation 9.11

Where,

F = the magnitude of the electrostatic force between two-point charges q_1 and q_2,

K = a Coulomb constant (8.988×10^9 N.m^2.C^{-2}),

q_1 = the charge on the incident particle in Coulombs,

q_2 = the charge on the "struck" particle, and

r = the distance between the particles in meters.

From the Equation 9.11, it is evident that the force increases as the charge increases, the force increases as the distance decreases and the force can be positive or negative (attractive or repulsive). If the product q_1q_2 is positive, the force between the two charges is repulsive; if the product is negative, the force between them is attractive.

It is essential to understand different types of interactions, since radioisotopes are detected by means of their radiations interacting with matter. But it is beyond the scope of this chapter to go in detail. Only a brief summary is presented here.

9.4.1. Interaction of matter with α-particles

When an alpha particle interacts with a matter, it strikes against the orbital electron and gives rise to two phenomena: electronic excitation and ionization. In the excitation process, energy from α-particle is transferred to the orbital electron resulting to its elevation to orbits of higher energy level. The charged particle (alpha or beta particle) exerts enough force to promote one of the atom's electrons to a higher energy state. The excited atom de-excites and emits a low energy ultraviolet photon. Each excitation event reduces the charged particle's velocity.

Ionization involves the complete removal of orbital electron, thus producing an ion-pair of a positively charged ion and an electron. Since α-particles are of heavy mass and low penetrating power, they collide frequently with atoms in their path and form a distinct cloud of ion-pairs as well as cause excitation of orbital electrons, resulting in a rapid dissipation of their own energies. Ionization turns a neutral atom into an ion pair. The electron stripped away from the atom is the negative member of the ion pair. It is known as a secondary electron. The atom, now with a vacancy in one of its electron shells, is the positive member of the ion pair. Energy of nearly 32.5 eV is required to remove an electron from an atom of gas. The number of ion-pairs generated by an α-particle is calculated as follows:

$$\text{Number of ion-pairs} = \frac{E_{Alpha}}{32.5} \qquad \text{Equation 9.12}$$

Where, E_{Alpha} denotes energy of α-particle.

The secondary electron has some, but not much, kinetic energy - usually less than 100 eV. Sometimes it has enough energy to ionize additional atoms. Then it is referred to as a delta ray (**Figure 9.2**). Formation of delta-rays may be considered as a secondary process of interaction between α-particles and matter.

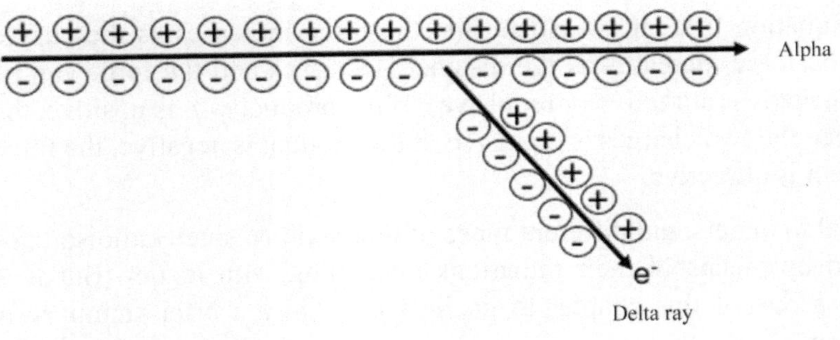

Fig. 9.2: Delta ray: a secondary electron, negative member of an ion pair, possessing enough kinetic energy for occurrence of additional ionization.

9.4.2. Interaction of matter with β-particles

9.4.2.1. Rutherford scattering

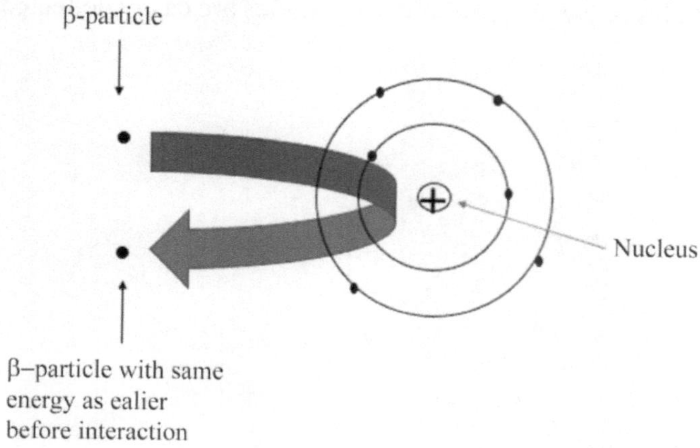

Fig. 9.3: Backscattering

Rutherford scattering is due to the elastic coulombic interaction between β-particle and nucleus, where β-particle changes its direction of travel sharply without losing its internal energy. Total kinetic energy of the colliding particles also remains the same. The backscattering (**Figure 9.3**) phenomenon of β-particles due to various backing materials is nothing but Rutherford scattering. Backscattering increases with increasing atomic number of the support.

9.4.2.2. Bremsstrahlung production

This is due to the inelastic interaction between a fast-moving β-particle and the nucleus of an atom. With this interaction, the β-particle is slowed down, loses energy and an electromagnetic radiation called Bremsstrahlung is emitted (**Figure 9.4**). The total kinetic energy of the colliding systems becomes less by an amount equal to the energy radiated as bremsstrahlung. This is increased with increase in atomic number of the absorbing material. Greater charge in the nucleus (atomic number) leads to greater deflection of the electrons and the greater the intensity of the bremsstrahlung. It is, therefore, advised to use low atomic number material like plastics as a shielding material for protection against β-particles. Bremsstrahlung photons may exhibit any energy up to the energy of the incident particle. It is evidenced that the bremsstrahlung photons produced by P-32 betas show a range of energies up to 1.7 MeV.

Bremsstrahlung production is not found during the interaction of matter with α-particles because α-particles are large particles. Due to their largeness, α-particles always prefer travelling in straight lines. Thus, they are not deflected to any significant level. On the other hand, bremsstrahlung production is found exclusively by β-particles or electrons because β-particles are easily deflected.

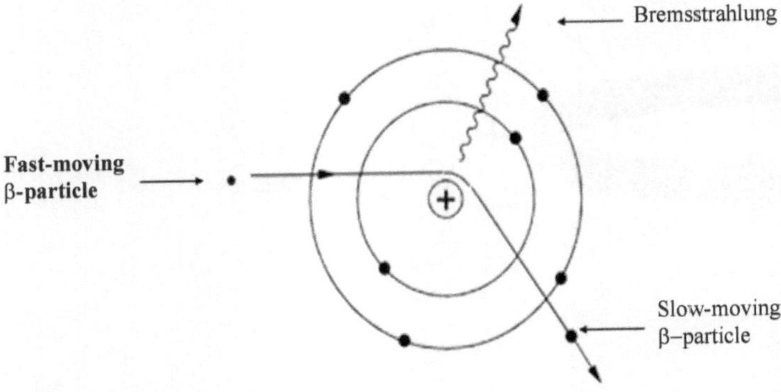

Fig. 9.4: Bremsstrahlung production

9.4.2.3. Ionization by β-particles

When a β-particle collides with an orbital electron (K-, L-, or M-electron), it undergoes an inelastic interaction expelling the electron completely from its atom and causes ionization (**Figure 9.5**).

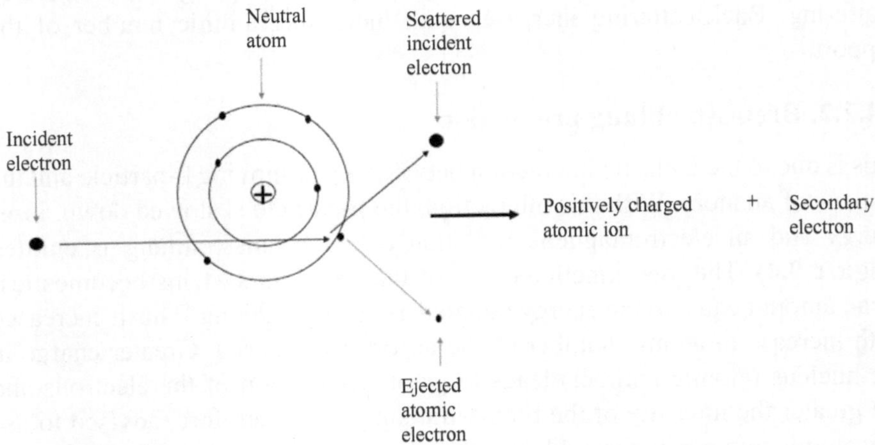

Fig. 9.5: Ionization process

9.4.3. Interactions of matter with γ-ray

9.4.3.1. Photoelectric effect

Photoelectric effect is an inelastic collision of a low energy level gamma photon with an orbital electron. An ion-pair is produced by complete removal of the electron. The ejected electron is now considered a photoelectron. Photoelectric effect arises from a photon-electron interaction. This is important for heavy absorbing elements (high atomic number) and for low γ-ray energies. However, photon energy must be higher than the binding energy of the electrons. The energy carried by γ-ray is completely transferred to the electron, with the resultant ejection of the electron from the atom (by overcoming binding energy of electron and providing kinetic energy to electron), after which the photon ceases to exist.The ejected electron retains whole energy of the incident photon except the energy spent to overcome the binding energy of the electron. As a consequence of the ejection of an inner shell electron, electron from outer shell drops down from outer to inner shells to fill the vacancy and while doing so, the electron releases its energy in the form of a characteristic photon (**Figure 9.6**). This process continues, with each electron emitting characteristic photons, until the atom is stable.

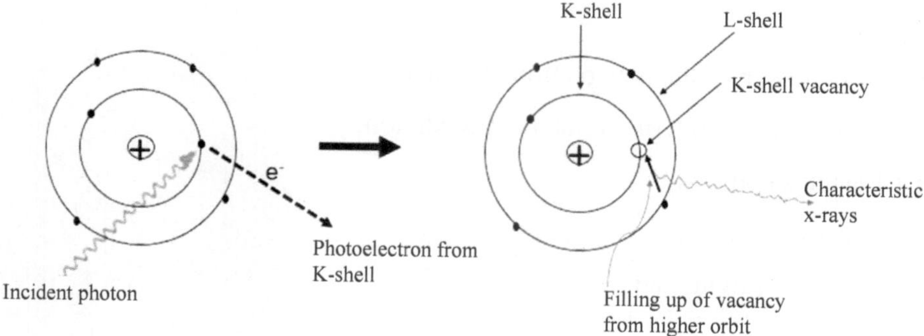

Fig. 9.6: Photoelectric effect

The energy of the ejected electron may be enumerated by Equation 9.13, using the energy of the incoming photon and the binding energy that held the electron to the atom.

$$E_e = E_\gamma - E_b \qquad \text{Equation 9.13}$$

Where,

E_e = the ejected electron (photoelectron) energy,

E_γ = the incident photon energy, and

E_b = the binding energy of orbital electrons.

For example, photoelectron energy becomes 50 keV, if incident photon energy and binding energy of orbital electrons are 70 keV and 20 keV, respectively.

9.4.3.2. Compton effect

This is an inelastic collision between a gamma photon of medium energy and an outer shell electron, so that a part of energy of the photon is transferred to the orbital electron. As a result, the electron is ejected from the atom and a new photon with reduced energy and increased wavelength is deflected in a new altered direction (**Figure 9.7**). The ejected electron is now referred to as a Compton electron. As the scattered photon possesses less energy, it exhibits a longer wavelength and less penetrating power than the incident photon. Compton scattered electrons possess kinetic energy and are capable of ionizing atoms. Finally, it recombines with an atom that has an electron deficiency. Compton effect may be expressed as follows:

$$\lambda_1 - \lambda = h\,(1\text{-}\cos\theta)\,/\,mc \qquad \text{Equation 9.14}$$

Where,

λ_1 = wavelength of the new photon after interaction,

λ = initial wavelength of the interacting photon,

h = Planck's constant,

m = mass of electron,

c = velocity of light, and

θ = scattering angle.

The term 'h/mc' is called Compton wavelength with a value of 2.43×10^{-10} cm. Compton scattering is important with light target elements (lower atomic weight) and with γ-rays of intermediate energy (< 3 MeV).

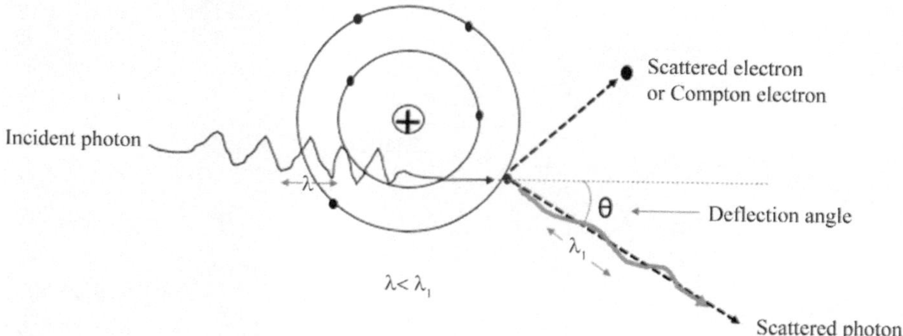

Fig. 9.7: Compton effect

9.4.3.3. Pair production

Pair production is the resultant effect of interaction between gamma photon of high energy (≥ 1.02 MeV) and high energy field in the vicinity of the nucleus where a photon is converted into two particles, one negatron and one positron (negatron-positron pair). As a result, the incoming photon disappears.With this, total energy of gamma photon is annihilated completely in creation of a negatron and positron (**Figure 9.8, Stage I**). Since photon has no rest mass, it may be said that energy is transformed into mass by pair production process according to $E = mc^2$. A minimum of 1.02 MeV energy is necessary for pair production, since one electron mass corresponds to 0.51 MeV energy. Photon energy in excess of 1.02 MeV appears as the kinetic energies of the negatron and positron produced. The interaction may be expressed as follows (Equation 9.15):

$$E = h\upsilon = 1.02 + Ke^- + Ke^+ \qquad \text{Equation 9.15}$$

Where, Ke^-& Ke^+ are kinetic energies of negatron and positron, respectively.

The pair production is most profound if the atomic weight of the absorbing material is high, and it increases with increasing photon energy. Pair production is used for positron emission tomography, a nuclear medicine imaging procedure. It is also used in radiation therapy.

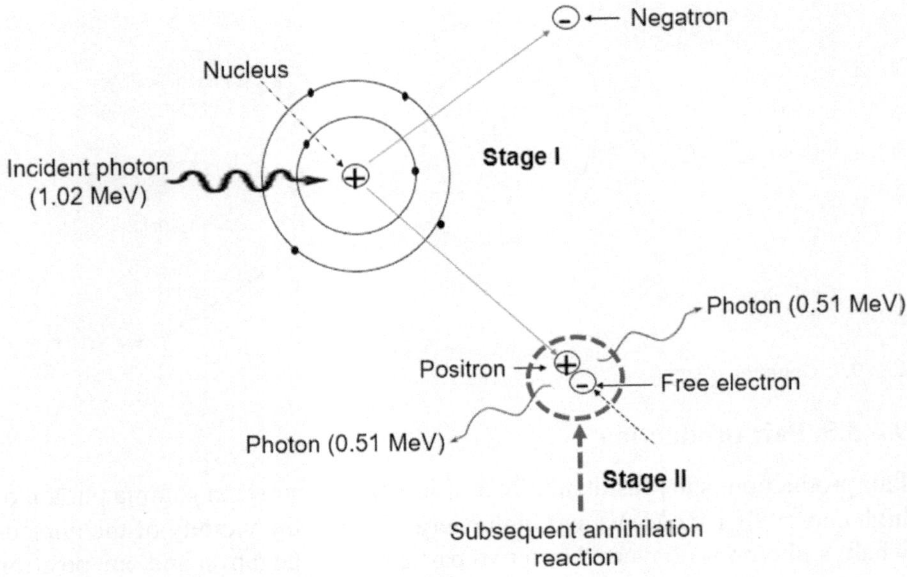

Fig. 9.8: Pair-production (negatron and positron) and subsequent annihilation reaction

Positron does not exist freely in nature. Thus, subsequently after its generation, it interacts with the first electron it encounters, leading to a process known as annihilation reaction where matter is converted back into energy. As a consequence, two gamma photons are released with energy of 0.51 MeV with total annihilation of positron and electron (**Figure 9.8, Stage II**). Looking at this point, it may be considered that pair production is the inverse process of the electron-positron annihilation.

9.4.3.4. Photodisintegration

Photodisintegration occurs on complete absorption of a high-energy gamma photon (7-15 MeV) by an atomic nucleus. The atomic nucleus enters an excited state, and consequently, ejects neutrons, protons or α-particles (one or more), forming lighter elements (**Figure 9.9**). Photodisintegration process is also referred to as phototransmutation or photonuclear reaction. When neutron is stripped off, the photodisintegration process is known as (γ, n) reaction; likewise, when proton or α-particle is knocked down from the nucleus, the process is referred as (γ,p), and (γ,α) reaction, respectively. Photodisintegration is an endothermic process for atomic nuclei lighter than iron, while for atomic nuclei heavier than iron it is exothermic.

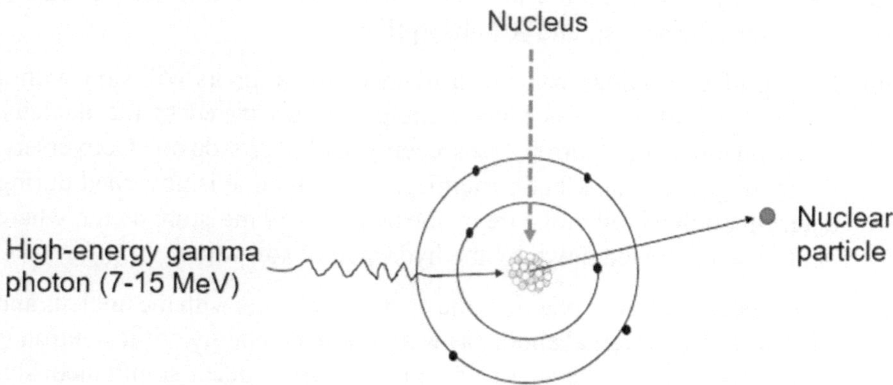

Fig. 9.9: Photodisintegration effect

During the core-collapse of a supernova, photodisintegration causes the nuclear fusion of the iron nuclei into helium nuclei and neutrons (Equation 9.16). Later, these helium nuclei are again transformed into protons and neutrons, the basic building blocks of elements, also through photodisintegration (Equation 9.17). [Source: https://astronomy.swin.edu.au/cosmos/P/Photodisintegration]

$\gamma + {}^{56}Fe \rightarrow 13\,{}^{4}He + 4n$ Equation 9.16

$\gamma + {}^{4}He \rightarrow 2p^{+} + 2n$ Equation 9.17

9.4.3.5. Rayleigh scattering

When a photon is unable to ionize or excite the atom while interacting with an orbital electron, it scatters through but not more than a few degrees. Rayleigh scattering is a scattering by the atom as a whole because it occurs without causing atomic excitation or ionization. Rayleigh scattering is a result of an elastic interaction since energy of the photon remains the same.

9.4.3.6. Thomson scattering

This is an interaction between electromagnetic radiation (photons) and orbital electrons. Electrons absorb the electromagnetic radiation, oscillate in an excited state and then re-radiate the electromagnetic radiation at the same frequency as the incident wave at random direction. As a consequence, the radiation is scattered.

9.4.4. Interactions of matter with neutron

Neutrons are devoid of any charge, but they exhibit large penetrating power. They are attenuated in respect of their energy and numbers when they interact with matter. Their interaction with matter involves three modes: (A) elastic scatter, (B) inelastic scatter, and (C) absorption.

(A) During elastic scatter, a neutron bounces off after its collision with a nucleus, a portion of its kinetic energy is transferred to the nucleus. Consequently, the neutron loses energy and slows down. Conversely, the atom gains some kinetic energies. This principle is governed during determination of soil moisture by the neutron soil moisture meter, where neutrons are slowed down by the hydrogen of soil water.

(B) In the inelastic scatter reaction, the neutron interacts with the nucleus and a portion of the internal energy as well as kinetic energy of the neutron is transferred to the nucleus. As a result, two events occur simultaneously: (i) slowing down of the neutron and (ii) attainment of the nucleus to its excited state. Later, the nucleus goes back to its original energy level from its excited state, emitting γ-ray.

(C) In the absorption mode of reaction, the nucleus of the atom in the matter absorbs the neutron. As a result, the atom reaches to its excited state and again comes to its ground state by emitting one or more gamma rays. This absorption type of interaction is more frequently observed in case of neutrons at lower energy levels, although it may occur at other energy levels of neutron too.

9.5. Disintegration rate

Radioactivity is a completely random and spontaneous process. Disintegration of radioactive substances is the characteristic of the particular nucleus involved and it remains unalterable by any chemical or physical process. The disintegration of radionuclide is a first-order process. The activity or the rate at which radioactive atoms disintegrate is proportional to the total number of radioactive atoms. If the activity of a radioactive substance is measured from time to time, the rate of decay is found to follow an exponential law. At any instant, the rate of decay is proportional to its activity. It is convenient to express this relationship mathematically as follows:

$$\frac{-dN}{dt} = \lambda N \qquad \text{Equation 9.18}$$

$$\frac{dN}{dt} = -\lambda N \qquad \text{Equation 9.19}$$

$$A = -\lambda N \qquad \text{Equation 9.20}$$

Where,

dN/dt = Number of atoms decaying per small increment of time. This is also known as **Activity** (A),

N = Total number of radioactive atoms present at any given time, and

λ = A decay constant, characteristic of the given radioisotope. Different radioisotopes have different values of λ.

The negative sign indicates that N decreases with time.

Decay constant (λ) may be enumerated from Equation 9.18 as follows:

$$\frac{-dN}{dt} = \lambda N$$ Equation 9.18

$$\lambda = \frac{-dN}{dt \times N}$$ Equation 9.21

Therefore, decay constant may be defined as the fraction of total radioactive atoms that decays per small increment of time.

Upon rearranging and integrating the differential decay equation (Equation 9.18), we may get:

$$\frac{dN}{N} = -\lambda \times dt$$

$$\int \frac{dN}{N} = -\lambda \int dt$$

Integrating over the interval between t = 0 and t = t, a given time during which number of radioactive nuclei in the sample decreased from N_0 to N, we may get

$$\int_{N_0}^{N} \frac{dN}{N} = -\lambda \int_{t_0}^{t} dt$$

$$\ln \frac{N}{N_0} = -\lambda t$$

$$\ln \frac{N_0}{N} = \lambda t$$ Equation 9.22

Or, in exponential form,

$$N = N_0 \times e^{\lambda t}$$ Equation 9.23

It should be remembered that the average lifetime of a radioactive sample (instead of the exact lifetime of one particular nucleus), being composed of the several nuclei of the same isotope, is predicted and measured. For this, the most suitable method to enumerate the lifetime of an isotope considers how long it takes for one-half of the nuclei in a sample to decay. The time required by a radioisotope to decay to half of its original activity is known as **half-life** (**$t_{1/2}$**). Thus, putting $N = N_0/2$, Equation 9.22 becomes

$$\ln\frac{N_0}{N_0/2} = \lambda t_{1/2}$$

$$\ln 2 = \lambda t_{1/2}$$

$$t_{1/2} = \frac{\ln 2}{\lambda}$$

$$t_{1/2} = \frac{0.693}{\lambda} \qquad \text{Equation 9.24}$$

Therefore, Equation 9.24 states that half-life is inversely proportional to the decay constant. A list of commonly used radioisotopes in biological research with their half-lives and emission characteristics is cited in **Table 9.2**. Depending on the nature of isotopes, the half-lives may vary from seconds to billions of years. Interestingly, Avogadro's number (N_A) can also be calculated from the study of radioactive decay. Let us take radium as an example. 1 g of radium emits α-particles at a rate of 11.6 x 10^{17} per annum. Since each α-particle is generated from the disintegration of one radium atom, we may consider $-dN/dt$ = 11.6 x 10^{17} with the time in year and 'N' is the number of atoms in 1g. Half-life of radium is 1590 years. Therefore, $\lambda_{Radium} = 0.693 / 1590$ and 'N' value may be calculated as follows:

$$\frac{-dN}{dt} = \lambda N$$

$$11.6\times10^{17} = \frac{0.693N}{1590}$$

$$N = \frac{11.6\times10^{17}\times1590}{0.693} \text{ atoms present in 1 g radium}$$

Therefore, Avogadro's number (N_A) = 226N (atomic weight of radium is 226)

$$N_A = \frac{226\times11.6\times10^{17}\times1590}{0.693}$$

$$N_A = 6.0 \times 10^{23}$$

Since detection efficiencies are hardly 100%, it is difficult to measure absolute activities in the laboratory. Therefore, one should be engaged in measuring the counting rate (R) that further adds up the effect of a constant, detection coefficient (C) to the absolute activity (A) [Equation 9.25].

$$R = C \times A \qquad \text{Equation 9.25}$$

$$R = C \times \lambda N \qquad \text{Equation 9.26}$$

Table 9.2: Useful isotopes in biological research with their half-lives and emission characteristics

Isotope	Half-life	Emission type
^{14}C	5730 years	Beta
^{32}P	14.3 days	Beta
^{3}H	12.3 years	Beta
^{35}S	87.2 days	Beta
^{42}K	12.4 hours	Beta, Gamma
^{22}Na	2.6 years	Beta, Gamma
^{24}Na	15 hours	Beta, Gamma
^{55}Fe	2.9 years	X-ray, Electron capture
^{59}Fe	45 days	Beta, Gamma
^{45}Ca	165 days	Beta
^{131}I	8 days	Beta, Gamma
^{60}Co	5.3 years	Beta, Gamma
^{36}Cl	3.1×10^5 years	Beta
^{65}Zn	245 days	Positron, Gamma, Electron capture
^{54}Mn	314 days	Beta, Gamma

9.6. Units used in radioactivity monitoring

Curie (Ci) is the basic unit of radioactivity. It is defined as the quantity of radioactivity material in which rate of disintegration per second or minute is same as that in 1 g radium (3.7×10^{10} disintegrations per second, dps or 2.2×10^{12} disintegrations per minute, dpm). Here, ^{226}Ra is used as a standard isotope to compare due its relatively long half-life (1590 years). For biological experiments, Curie is too large to use. It is convenient to use smaller units like millicurie (mCi, 3.7×10^7dps or 2.2×10^9 dpm) and microcurie (μCi, 3.7×10^4 dps or 2.2×10^6 dpm). It is well known that there is a distinction between DPM (disintegrations per minute) and CPM (counts per minute). DPM refers to actual disintegrations per minute and CPM refers to counts per minute that is actually detected by a radiation counter, since efficiency of radiation detection devices is practically less than 100%. Curie refers to the number of disintegrations actually occurring in the sample (dps or dpm; not cps or cpm). Becquerel (Bq) is another unit of radioactivity used in place of Curie and is defined as that quantity of a radioactive isotope that decays at the rate of one disintegration per second. So, by definition Curie and Becquerel are related as follows:

1 Curie = 3.7×10^{10} Bq (Becquerel)

= 3.7×10^{7} kBq (kilo becquerel)

= 3.7×10^{4} MBq (Mega becquerel)

The relationships among Curie, Becquerel and dpm are further stated in **Table 9.3**.

Table 9.3: Conversions among dpm, Curie and Becquerel

dpm	Curie (Ci)	Becquerel (Bq)
	(a)	
2.22×10^{12}	1 Ci	3.7×10^{10} (37 GBq)
2.22×10^{9}	1 mCi	3.7×10^{7} (37 MBq)
2.22×10^{6}	1 µCi	3.7×10^{4} (37 kBq)
2.22×10^{3}	1 nCi	3.7×10^{1} (37 Bq)
	(b)	
6×10^{10}	27 mCi	1000 MBq
6×10^{10}	27 µCi	1 MBq
6×10^{10}	27 nCi	1 kBq
6×10^{10}	27 pCi	1 Bq

Radioactivity is also expressed in terms of specific activity that refers to the amount of radioactivity per unit amount of substance. The concept of specific activity arises due to the fact that sometimes radioisotopes are mixed with stable isotopes of the element, and it is then inevitable to express the amount of radioisotope present per unit mass. Specific activity is generally expressed as Ci/g, mCi/mg, dpm/mmole, cpm/µmole, etc.

9.7. Radioactivity detection

Radioactivity detection systems vary with the nature and types of particles or electromagnetic waves to be detected. Most commonly three types of radioactivity detectors are used: Gas ionization, Scintillation and Semiconductive (**Figure 9.10**).

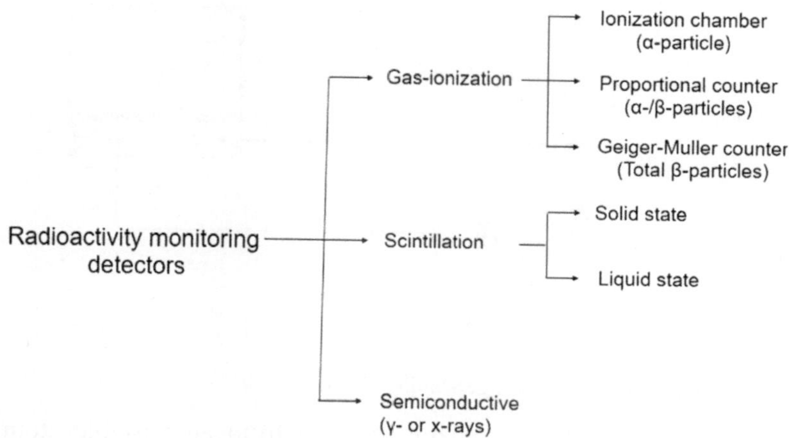

Fig. 9.10: Different types of radioactivity monitoring detectors

9.7.1. Gas Ionization / Geiger-Muller Counter (GM Counter)

The principle of radioactive detection and measurement is based on that of the gas ionization chamber, where a gas atom or molecule when ionized by radiation produces a small pulse of current which is amplified and measured by electronic device. A GM Counter consists of (i) a GM tube, a hollow metallic chamber, which contains a sealed window of very thin material for the entry of radiation, and a thin wire made of tungsten and mounted along its axis. The whole metal case is enclosed in a thin glass tube and it acts as a cathode, while the tungsten wire acts as an anode. The chamber is filled with a special gas mixture that may be helium or argon, mixed with a bit of another gas, butane or alcohol, at low pressure. Usually, a mixture of 90% argon and 10% ethyl alcohol vapor is used. (ii) Scaler, a combination of voltage source that activates the GM tube, and electronic adding machine that counts each ray as it is detected, (iii) a timer that records time spent during measurement, and (iv) a planchet containing radioactive sample that is placed below the thin mica window of GM tube. The central fine wire anode, suspended inside the tube, is attached to the positive end of a battery through a resistance and the cylindrical cathode is attached to the negative end (**Figure 9.11**).

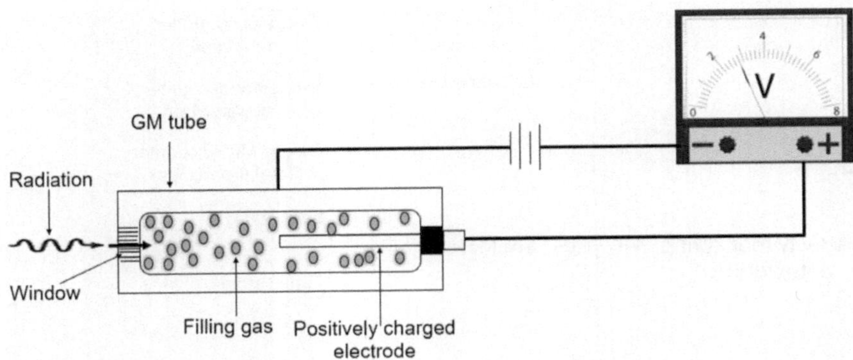

Fig. 9.11: Schematic representation of a Geiger-Muller Counter

Entering through the window an ionizing particle from an unstable atom collides with filling gas atoms and ejects electrons. As it takes place near two electrodes, the positively charged gas atoms move to the cathode and negatively charged electrons to the anode under a voltage gradient producing a current. Current generated is amplified, measured and converted to counts per minute. β-particles of high energy (e.g., ^{32}P) face no problem to penetrate the mica window of the GM tube. But this is not suitable for weak β-emitters (e.g., ^{14}C, ^{3}H). Generally, Mylar Window, Flow Window are used for weak β-particles.

The different stages of ion-pair formation (**Figure 9.12**) are utilized to detect radiation in the form of different counting methods. Since specific ionization for particles is in the order of $\alpha > \beta > \gamma$ (1000, 100 and 1, respectively), one might be able to detect α- and β-particles by gas ionization method. However, it is quite insensitive to detect γ-radiation.

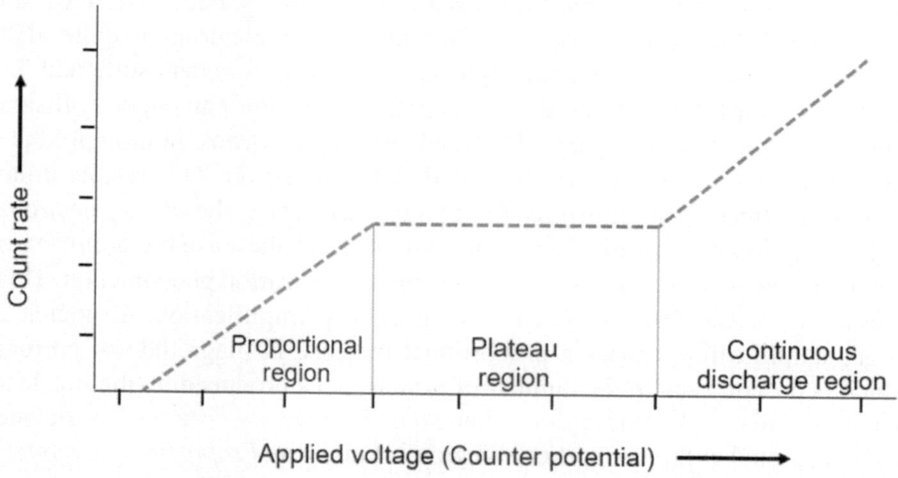

Fig. 9.12: Various phases of ion-pair formation: Proportional region, Plateau region and Continuous discharge region

When α- or β-particles produce ion-pairs and under this situation, if there is no potential applied in the electrodes, the ion-pairs recombine again to form neutral atoms resulting in no pulse in the electrical circuit. Under a small potential applied, the charged ions and electrons move toward their corresponding electrode initiating a current flow, obviously with recombination of few charged ions and electrons forming neutral atoms. But the numbers of ions and electrons reaching the electrodes go on increasing proportionately with increasing applied voltage, gradually minimizing the recombination phenomenon and a point is achieved when at a particular applied voltage all ions and electrons reach the electrodes without any recombination. Under this situation, further increase in voltage does not create any difference in the magnitude of current. *Simple ionization chamber* operates usually at this region of saturation current. Therefore, it is useful to mention that in an ionization chamber the potential across the electrodes is adjusted to minimize recombination of ion-pairs without causing any gas amplification. Electroscopes, pocket dosimeters, etc. are unique examples of ionization chambers. An electroscope, one of the earliest devices for detecting and measuring radioactivity, consists of a rod with two aluminum or gold foils as positive electrodes and a metal case, the wall of which acts as negative electrode. When the electroscope is kept near any radioactive source, air inside the electroscope becomes ionized and deflection of the metal foils occurs. The amount of deflection (angle) is a function of charge accumulated.

Further increase in potential between the electrodes, after reaching the region of saturation current, causes an increase in current flow due to secondary ion-pair production. It has been observed that usually an energy of about 30-35 electron volts is needed to produce ion-pairs. Mobility of few electrons is quite high at higher voltage difference, attaining energy which is more than sufficient for further ion-pair formation and thus, they produce new ion-pairs upon collision with another atom of filling gas. The newly formed electrons, in turn, produce more electrons which produce more collisions and so on. This results in an avalanche effect (also known as Townsend Avalanche), shower of electrons originating from the original ionizing particle from the unstable atom. As a result, current flow is more compared to primary ionization phenomenon. This is easier to detect. This process is known as Gas Amplification. An increase in gas amplification causes a proportional increase in magnitude of current which is proportional to the number of primary ions produced by the incident radiation. This is why this region is known as *Proportional region*. The device that is operated at this proportional region is known as *Proportional Counter*. Proportional counter is able to differentiate between α- and β-radiations due to their pronounced difference in ionizing power. Proportional counter requires a very stable voltage supply; otherwise, a small fluctuation in voltage may lead to a significant change in gas amplification.

The proportional increase in gas amplification with gradual increment in applied potential (found in proportional region) operates up to a limit, and after that, counter does not respond in a proportional manner. One point is reached when there is no further increase in current magnitude. This region is known as *Limited Proportional Region* which is not often used in detection and measurement of radioactivity. The existence of a limited proportional region in the current-voltage characteristic curve for gas ionization is due to the physical dimension of the counter that dictates the total number of ion-pairs possibly produced.

After reaching a limited proportional region, further increase in applied voltage does not result in any change in the magnitude of current. Current flow remains the same due to the attainment of maximum value of gas amplification. This region is called the *Geiger Region* and the device operated in this region is known as the *Geiger Counter*. Unlike the proportional counter, the Geiger counter cannot differentiate between α- and β-radiations. This is the major disadvantage of the Geiger counter. Another major limitation of the Geiger Muller counter is that it cannot measure high radiation rates. This is due to the occurrence of a 'dead time', after each ionization. As positive ions are heavier than electrons, they travel a short distance toward cathode when the electron avalanche is collected on anode. The positive ions usually take 200 μs to reach completely to the cathode. Before the positive ions reach cathode, the slow-moving positive ions make a sheath around anode and lower down the potential gradient in a manner that the counter becomes insensitive to the entry of fresh ionizing particles. This is referred to as *dead time* of the counter. Or in other words, the dead time of a Geiger Muller counter refers to the time during which the counter fails to record any other ionizing particle entering the counter.

A small quantity of alcohol or ethyl formate or the halogens are added to the gas filling as common quenching agents to control the continuous discharge. These common quenching agents are easily dissociated by the energy contained in the avalanche discharge. The dissociation absorbs the energy and stops the discharge, preparing the counter for the next particle.

Unfortunately, GM Counter does not count every atomic disintegration that occurs in the sample. The counting yield of one instrument may differ with another due to some factors. It is better to stick to the same counter for comparing activities in different radioactive samples, keeping conditions constant. The factors may be listed as follows:

(a) ***Geometry***: Expelled β-particles may go out in any direction during the disintegration of a radioactive nucleus. Some particles will go sideways or down, avoiding the GM tube.

(b) ***Back scattering***: Few particles that proceed down away from the GM tube hit the planchet and bounce back up. This effect is observed more if the planchet is made of atoms of higher atomic number.

(c) ***Self-scattering***: Like back scattering, some particles are reflected back by the sample itself. The self-scattering is pronounced by the presence of heavy atoms in the sample and also by the shape and size of the sample crystals.

Few β-particles are absorbed by air above the sample, the window of the tube and the sample itself. This self-absorption increases with greater sample thickness, and lesser β-particle emission energy. If the sample is very thin (infinitely thin sample), self-absorption is negligible. In the opposite case, self-absorption is more. A correction for self-absorption is made by preparing a self-absorption curve for individual isotopes to present a meaningful interpretation and comparison of data from different samples.

9.7.1.1. GM counter efficiency

The counter efficiency refers to the counting ability of the GM counter. It is a ratio of the observed counts per second to the number of ionizing particles entering the counter per second. It is also expressed as follows:

Counting efficiency (η) = 1 – exp(spl)

Where,

s = specific ionization at one atmosphere,

p = atmospheric pressure, and

l = path length of the ionization particle in the counter.

9.7.2. Scintillation Counter

Scintillation counting is based on the principle of interacting radiation emitted from radioactive material with scintillators that are responsible to convert radiation energy to light energy. Generated light energy is then detected and amplified by a photomultiplier that converts light energy to electrical energy. As the photoelectron number is directly proportional to γ-ray energy lost in the crystal; the electrical pulse, thus produced, is directly proportional to the original amount of radioactivity. Two forms of scintillation counter exist in detection and measurement of radioactivity: solid scintillation and liquid scintillation counter. They differ with each other mainly on the nature of scintillators used. A scintillation counter usually consists of scintillators, photomultiplier (a photocell system causing multiplication of electron with the help of dynodes), high voltage supply unit (capable of providing up to about 1200 volts which is extremely necessary for the operation of photomultiplier), an amplifier, a pulse-height analyzer and a scalar or recorder. A block diagram on the mode of working of a scintillation counter is depicted in **Figure 9.13**.

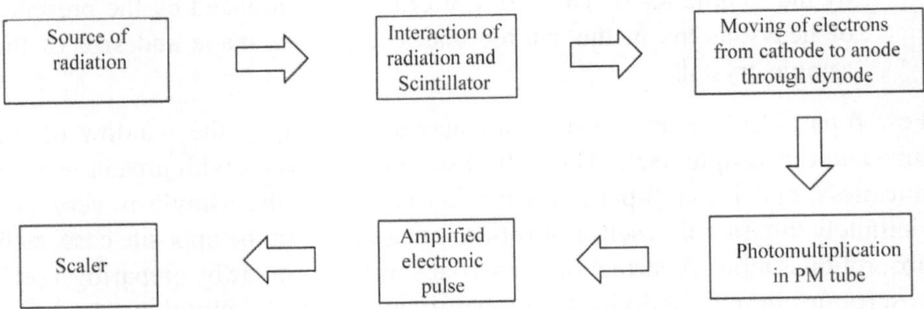

Fig. 9.13: Mode of working of a scintillation counter

9.7.2.1. Solid scintillation counter

Different kinds of crystalline scintillators (Fluors) are used for different kinds of radiation particles. For alpha emitters, zinc sulfide crystals activated by silver are used as alpha fluors, while large crystals of anthracene or naphthalene containing a small amount of anthracene or plastic fluors are used for beta emitters. Large NaI crystals with thallium iodide (acting as an activator) in trace amounts are commonly used for gamma emitters. Gamma rays lack both charge and mass, but they have more penetrating power. For this reason, they require dense material for their absorption and efficient recording. Solid scintillation counting is the most efficient detecting method for γ-rays and is frequently used in biological work like radio-immunoassay.

Gamma ray falls on NaI (TI) crystals and excites iodine atoms which, in turn, returns back to its ground state by emitting a light pulse in the ultraviolet region. This light pulse is then spontaneously absorbed by thallium atom which re-emits a fluorescent light at 4100Å. NaI (TI) crystal is enveloped by an aluminum foil that serves dual functions: acting as internal reflector and protecting the crystal from atmospheric moisture and extraneous light. In order to get more counting efficiency, a hole is drilled at the center of the crystal to accommodate the sample; so that the radioactive source is completely surrounded by the crystal (known as *Well counter*).

9.7.2.2. Liquid scintillation counter

Crystal scintillators are not efficient enough to detect low energy beta particles. This is due to their low penetrating power, and possibility of self-absorption and absorption by the surroundings too, that hinder the interaction of radiation with the scintillator. This problem can be overcome by directly mixing radioactive materials with the scintillator and counting efficiency is thus increased. A liquid scintillator in which radioactive material is either dissolved or suspended as a fine dispersion, thereby, replaces solid scintillator particularly for weak beta particles. This method is quite efficient to quantify ^{3}H, ^{14}C and ^{35}S which are weak β-emitters and frequently used in biological research. Counting high energy β-emission (^{32}P) does not require a fluor because β-particles can be detected directly by the PMT. A liquid scintillator, sometimes also known as 'Scintillation Cocktail', contains an excitable aromatic solvent in which radioactive material is dissolved or suspended with one or more organic fluorescent substances. Aromatic solvents, such as toluene, xylene, pseudocumene or other benzene derivatives (**Figure 9.14**) for organic samples, are preferred because of their high electron density and the electrons are easily elevated to an excited state. The selected solvent must exhibit the capacity of solubilizing the scintillators, show the feature of absorbing the energy efficiently, impart good energy transfer efficiency to the scintillators and should possess no quenching properties.

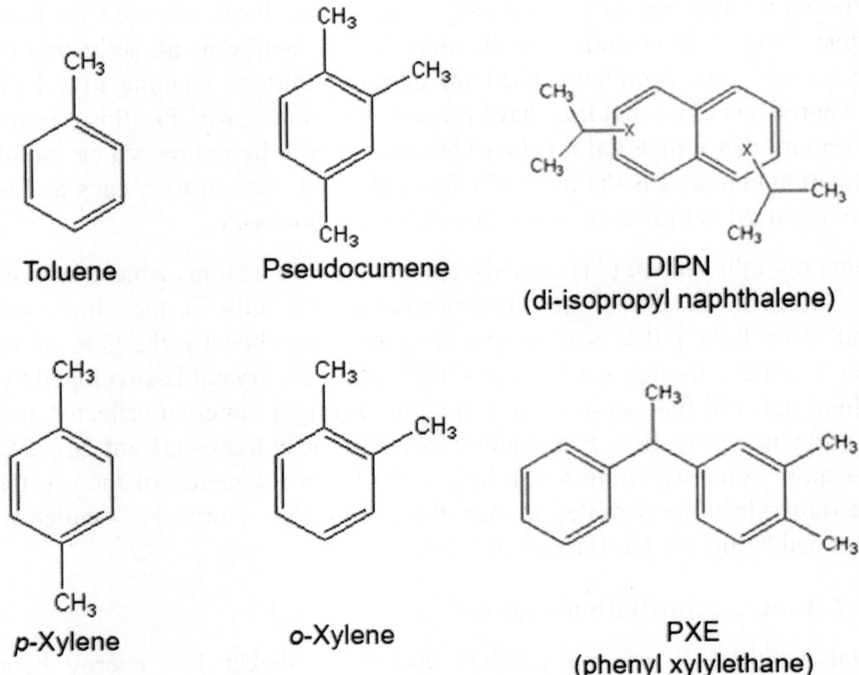

Fig. 9.14: Classical aromatic organic solvents for liquid scintillation counting

PPO (2, 5-diphenyloxazole)

P-terphenyl

Anthracene

Butyl-PBD [2(4-Biphenyl)-5-(4-tert-butylphenyl)-1,3,4-oxadiazole]

(a) Primary scintillators

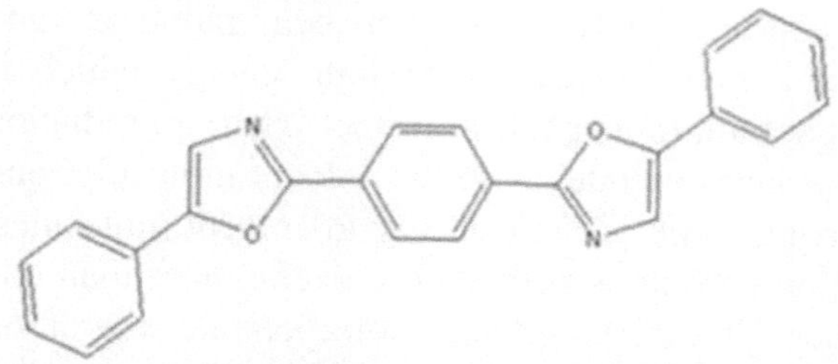

POPOP [1-4,bis-2-(5-phenyloxazolyl)-benzene]

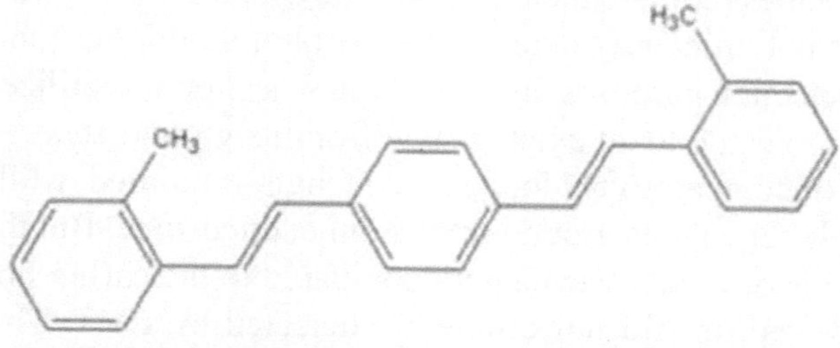

Bis-MSB [*p*-bis-(*o*-methyl styryl)-benzene]

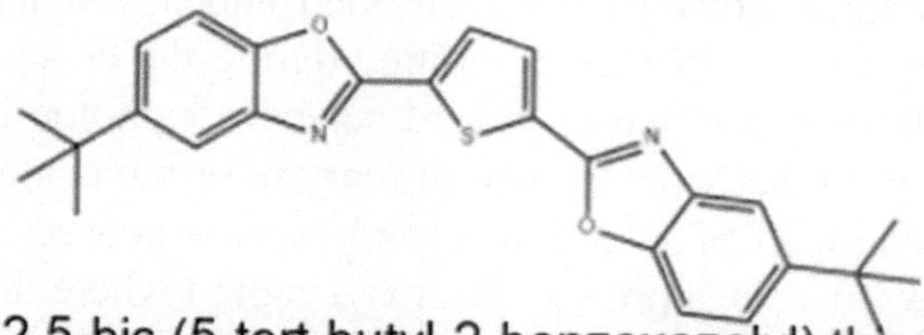

2,5-bis (5-tert-butyl-2-benzoxazolyl) thiophene

(b) Secondary scintillators

Fig. 9.15: Common primary (a) and secondary (b) scintillators used in liquid scintillation counting

Sometimes complicated samples, such as tissue slices containing ^{14}C are processed to produce $^{14}CO_2$ with the help of biological oxidizer. $^{14}CO_2$ is then absorbed in a scintillator solution like hydroxide of hyamine 10-X [p-(diisobutyl–cresoxy ethoxy ethyl)-dimethyl-benzyl-ammonium chloride]. Ethanolamine may also be used for the same purpose. Typical primary scintillators [**Figure 9.15 (a)**] are 2, 5-diphenyloxazole (PPO) and Butyl PBD [2(4-Biphenyl)-5-(4-tert-butylphenyl)-1,3,4-oxadiazole], while 1,4-bis (5-phenyloxazol-2-yl)-benzene (POPOP) and 1, 4-bis-(4-methyl-5-phenyloxazol-2-yl)-benzene (DM-POPOP) are preferred as secondary scintillators [**Figure 9.15 (b)**]. A new type of sulfur analogue fluor, 2,5-bis (5-tert-butyl-2-benzoxazolyl) thiophene (BBOT) is also used with toluene and does not require any secondary fluor.

Figure 9.16 shows the step-wise interaction between radiation and scintillation-cocktail converting radiation energy to light energy which is then detected efficiently by photomultiplier tubes. In the first step, radiation emitted from radioactive substances collides with the solvent molecules and makes them excited, transferring some of their energy to solvent molecules. Energy remaining in a radiation particle is sufficient to excite more than one solvent molecule. The numbers of excited solvent molecules are, therefore, proportional to the energy of the radiation particle in question. In the second step, excited solvent molecules return to their ground state by emitting a photon of short wavelength. It is not efficiently detected by the photomultiplier tube. To make it practical for detection purposes, in the third step, a photon is utilized to make a primary fluor molecule to an excited state from its ground state. In the fourth step, a photon of comparatively longer wavelength is emitted, while the primary fluor molecule regains its ground state from excited state. But the emitted photon, though it is of longer wavelength compared to the earlier one evidenced in the second step, is still not efficiently detected by PMT. This photon is then, in the fifth step, absorbed by the secondary fluor molecule and as a result the secondary fluor molecule is transformed into its excited state. In the sixth and last step, again a photon is emitted during the change in secondary fluor molecule from its excited to ground state. This photon is of longer wavelength compared to that from excited primary fluor molecules and is more efficiently detected by a PMT. Thus, a secondary fluor is used to shift the wavelength of light emitted by a primary fluor to a region where the photomultiplier is more sensitive, and thereby, count efficiency is increased.

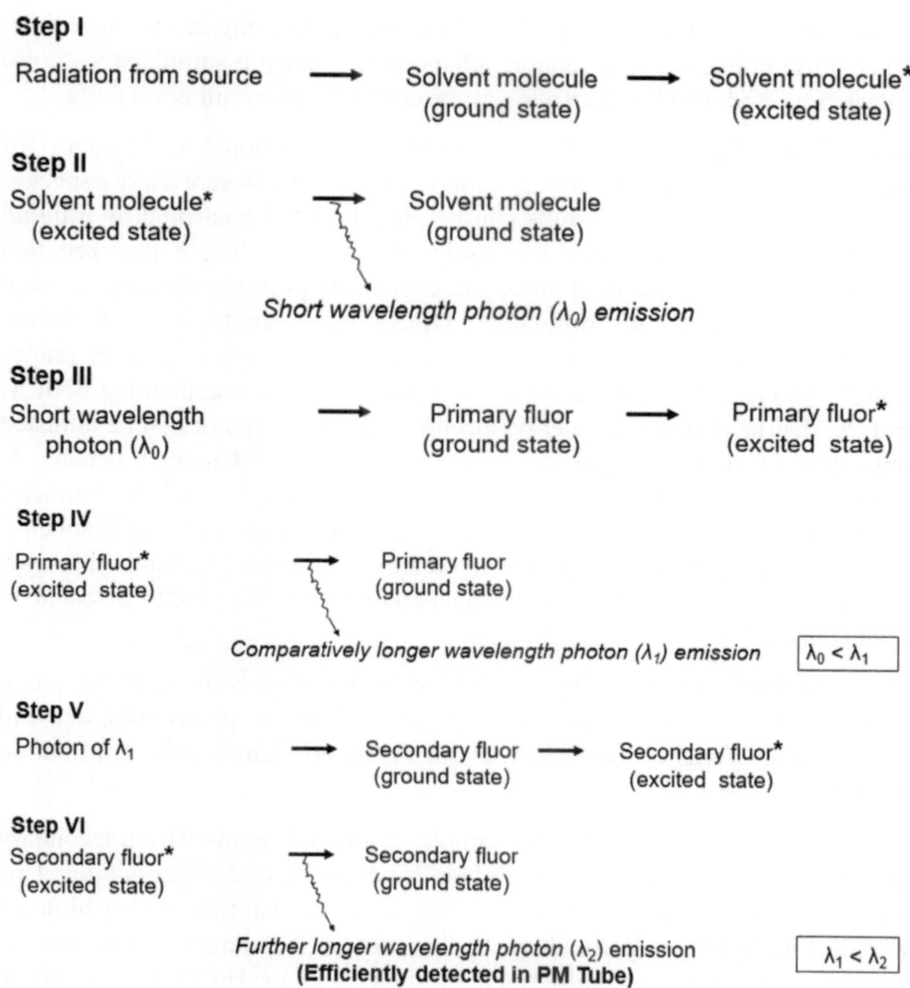

Fig. 9.16: Working principle of liquid scintillation counting

Emitted light then falls on the photomultiplier tube. As the PMT is operated at high voltage with occurrence of electronic events, it produces a thermal background noise that can be directly minimized, though not completely, by reducing operating temperature with an inclusion of a freezing system. Usually -8 °C is chosen as optimum temperature, since too low temperature may freeze the liquid scintillator. Alternatively, thermal noise may also be overcome by introduction of a coincidence circuit along with two photomultiplier tubes. The flash of light will be registered at both photo-cathodes at the same time. But the coincidence circuit registers pulse only when it receives pulses simultaneously coming from both PMT and when pulses are in 'coincidence'.

A coincidence circuit rejects pulses which are not arriving at the same time. Since noise pulses occur at random, chances of occurring simultaneous noise pulses from both PMT and arriving at the coincidence circuit are remote.

Since electrical pulse generated in the PMT is proportional to the energy of β-particles and different β-emitters emit β-particles of different energy spectra, two isotopes in the same sample can be identified and measured by a liquid scintillation counter, provided it is equipped with pulse height analyzers that measure the size of electrical pulse and count only pulses within pre-selected energy limits set by discriminators. Discriminator contains an electronic gate that allows counting β-particles within certain energy or voltage ranges called *Channels*. This may be best explained by a sample containing both ^{3}H and ^{14}C that have different energy spectra. The discriminator can be adjusted separately in channel-1 and in channel-2. Therefore, channel-1 detects ^{3}H along with a small number of counts originating from ^{14}C, while channel-2 detects only counts for ^{14}C, but less than a number that interferes in channel-1. This overlapping phenomenon is corrected by counting a ^{14}C standard in both channel-1 and –2 to determine percentages of total ^{14}C counts detected in channel-1 and –2.

The main disadvantage of liquid scintillation counting is the quenching that causes a decrease in amount of light received at the photomultiplier and affects count efficiency. Quenching is not a great problem in solid scintillation counting.

Colored substances, if present in the sample, absorb light emitted from secondary fluor before it is detected by PMT. This is known as *Color Quenching*. This can be reduced by removal of colors using activated charcoal or bleaching the sample with hydrogen peroxide or by combustion techniques where samples containing ^{14}C or ^{3}H are combusted to produce $^{14}CO_2$ or $^{3}H_2O$ before counting.

Scintillation efficiency is reduced when any chemical substance, present either in the sample or in the scintillation cocktail and interacting with excited solvent or fluor molecules, interferes in transferring energy. This process is known as *Chemical Quenching*. Dissolved O_2 from air and water may chemically quench the scintillation counting to some extent. To minimize chemical quenching effect, samples should be purified and fluor concentration should be increased.

The insolubility of a radioactive sample to the scintillation solvent results in a *point quenching* problem to a greater extent because emitted β-particles may be absorbed by the sample itself without interacting with the solvent molecule. The extent of the point quenching phenomenon is reduced when a solubilizing agent like Thixin is added to the sample making liquid scintillator to a gel.

Dirty scintillation vials also result in quenching (*Optical quenching*). In this quenching problem, emitted light is being absorbed before reaching a photomultiplier. Besides quenching effect, liquid scintillation counting is also affected by *chemiluminescence*, chemical nature of the scintillation vials, etc.

Chemiluminescence occurs when a sample reacts chemically with the scintillation cocktail resulting in light emission. *Chemiluminescence* is quite different from original light flash coming from interaction between radiation and the fluor molecules, and is thus considered as photomultiplier noise. This problem can be minimized if the radioactive sample, after mixing with a scintillation cocktail and before counting, should be stored for sufficient time to allow the *chemiluminescence* to decay.

The vials, in which a radioactive sample is mixed with a scintillation cocktail, should not contain radioactive ^{40}K; otherwise, the background count will be too high. Normal glass contains natural potassium with a higher amount of ^{40}K. Therefore, vials should be made of special glass containing negligible amounts of potassium. Now-a-days vials of quartz and polyethylene are used to avoid interference of ^{40}K.

9.7.2.2.1. Quenching correction by determining counting efficiency

(a) **Internal standard method**: After getting sample counting (X), a small known amount of standard possessing the same radionuclide as of the sample with known disintegration per minute (Z) is added and the activity (Y) is measured once again. The counting efficiency will be calculated as follows:

Counting Efficiency (%) = $100\,(Y - X) / Z$

(b) **External standard method**: This is a fast and precise method of quenching correction, based on an in-built external γ-radiation source in a lead-shielded chamber that is kept before the sample vial. The advantage of this method is that the sample is not contaminated with the standard. The relative decrease in the standard count for each sample counting vial is proportional to the amount of quenching material and this relationship provides quench correction for that sample.

9.7.3. Semiconductive detector

This is used to monitor γ-or x-rays. Lithium-drifted silicon [Si(Li)], Lithium-drifted germanium [Ge(Li)], intrinsic Ge detectors are a few examples of semiconductive detectors. Each incident photon leads to a current pulse. The output of the detector is directly proportional to the energy of the incident photon.

9.8. Importance and application of radioisotopes

Radioisotopes offer few major advantages over conventional analytical techniques: sensitivity, specificity, and accuracy in measurement. Radioisotopes are also unique in nature because they offer the ability to differentiate between substances that are chemically indistinguishable. Thus, they avoid chemical separations that are mostly required in other analytical methods. These characteristics enable radioisotopes to be used in all branches of science, such as chemistry, biochemistry, forensic chemistry, archeology, agriculture, medical science, industry, ecological science, molecular biology, etc. Ecologists use radioisotopes to study migration behavior patterns of many animals as well as to examine the food chain, making primary producers radioactive. Molecular biologists use radioisotopes in DNA and RNA sequencing, DNA replication, transcription, recombinant DNA technology, synthesis of cDNA and many others.

Due to their high energy and significant penetrating power, gamma rays are most useful radiation for medical purposes, especially in radiotherapy for cancer treatment (Cesium-137). Cobalt-60 and cesium-137 are examples of few most widely used gamma emitters. Cesium-137 is used for soil density measurement at construction sites and also for investigating the subterranean layers of the Earth in oil wells. Cobalt-60 is used to sterilize medical equipment and irradiate food, killing bacteria and pasteurizing the food. In addition, radioisotopes (uranium-238 and uranium-235) are used for nuclear power plants for power generation.

Radioisotopes are used as tracers or indicators in various processes, since they behave chemically and biologically like other atoms of the same element. The characteristic property of isotope, its radioactivity or its mass, enables it to be used as 'tag' or 'label'; so that the fate of the element in various physico-chemical processes may be known. Tracers may be of isotopic or non-isotopic. When a stable atom of a molecule is replaced by a radioactive atom of the same element, it is called *isotopic tracer*. But use of an isotopic tag is not always necessary. Non-isotopic tracer is used to study flowing liquid through a pipe, circulating currency, insects and many others by touching them with a drop of radioactive solution. Radioactive tracers, in a true sense, help to get information related to the location and quantity of an element even after they are intermixed with a large number of stable atoms of the same element. When radio-labeled compound is injected or fed to an animal or plant, it quickly becomes a part of the total metabolic pool and in this form, it may undergo many reactions, or it may remain unchanged, or it may be degraded to simpler compounds. The tissues and excretions can be examined for the presence of the original compound or its metabolic transformation products. From this, attempts can be made to prepare a balance sheet of radioactivity stating the amount of compounds utilized by the body system, amount of degradation products formed and their chemical identity (after going through isolation processes). This is made possible by taking samples from different tissues or organs at different time intervals followed by extraction and separation (by different chromatographic, electrophoresis or other techniques) processes. Parent compounds and their metabolites can be monitored by GLC or HPLC, equipped with radioactivity detectors. This may also be accomplished by paper or thin-layer chromatography, coupled with autoradiography. Isotopes are indispensable for establishing metabolic origins of complex metabolites, such as heme, cholesterol, and phospholipids. It is done by administering isotopically labeled starting materials to animals and isolating the resultant products. It has been proved that nitrogen atoms of heme are derived from glycine rather than from ammonia, proline, glutamic acid, or leucine when ^{15}N-glycine are fed to rats. Heme from their blood was isolated and analyzed for ^{15}N. An affirmative result on the presence of ^{15}N in heme confirms glycine as its source of nitrogen. Likewise, it has also been established that cholesterol's carbon atoms are derived from acetyl-CoA.

Tagging must be done at an appropriate position to gather specific information. Tagging depends on the kind of information to be obtained. Physical and chemical nature of compounds must be known for proper tagging. For instance, individual tagging may be done in an amino acid with carbon, nitrogen, hydrogen and oxygen depending on the nature of information required. Only ^{14}C-labeled amino acid may be used to know whether amino acid becomes a 'glycogen former' or 'lipid former', ^{15}N-labeled amino acid to know the contribution of the amino acid to amide pool and ^{3}H- or ^{18}O-amino acid for kinetic study of peptide bond formation. Radioisotopes are conveniently used to study metabolic turnover, pathway of metabolism, absorption, route of translocation, and sites of accumulation of a molecule of biological interest.

9.8.1. Neutron activation

Transformation of a non-radioactive sample into a radioactive substance by using particle or gamma radiation bombardment is, in general, known as activation analysis. When the technique utilizes gamma rays for bombardment, it is known as Photon Activation Analysis (PAA). Similarly, Neutron Activation Analysis (NAA) refers to the bombardment of the nucleus by neutrons. More elaborately, neutron activation is defined as a process where neutrons from other sources induce radioactivity in substances on capturing free neutrons by the atomic nuclei. This is the only common process of transforming a stable material into intrinsically radioactive. While interacting with matter, it may be either scattered (elastic or inelastic scattering) or captured by nucleus. Consequently, the interacting substance becomes heavier and transforms into excited state. Excited nucleus is then immediately disintegrated along with emission of gamma rays [$^{23}_{11}Na$ (n, γ) → $^{24}_{11}Na$], or α-/β-particles. Neutron-capturing by nucleus also causes fission leading to its fragmentation. However, an ultimately unstable activation product is formed due to the neutron activation process through neutron capturing, even after the occurrence of any intermediate decay. Such radioactive nuclei show half-lives ranging from fractions of a second to several years. A schematic diagram of Neutron activation is depicted in **Figure 9.17.**

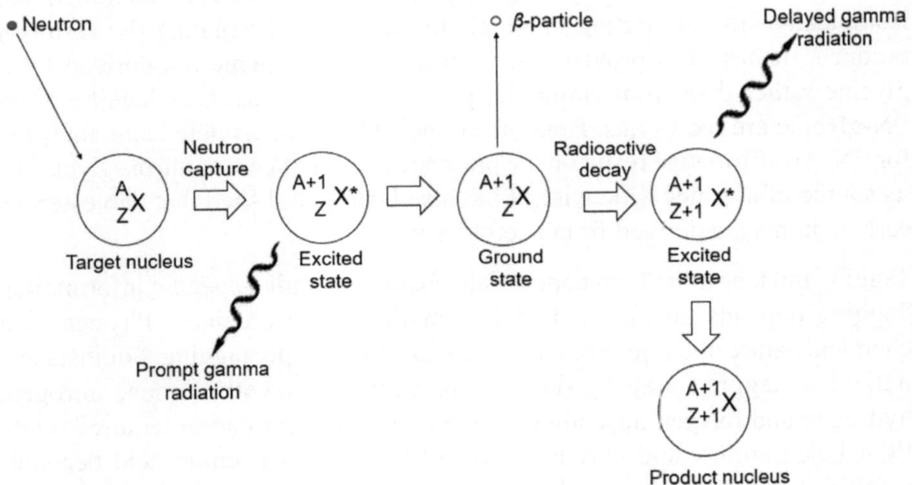

Fig. 9.17: A schematic diagram of neutron activation

Neutron has a mass of 1.68 x 10^{-24} g, slightly heavier than a proton. Since it does not carry charge, it shows high penetrating power. Based on energy content there are three types of neutrons: thermal (0.04 eV), intermediate (0.5 to 10 KeV) and fast (10 KeV to 20 MeV) neutrons. The free proton is stable, while the free neutron is not stable ($t_{1/2}$ ≈ 11 min) and it decays as follows:

$^{1}_{0}n \rightarrow p^{+} + e^{-}$ + antineutrino

Irrespective of the nature of the neutron sources employed to generate neutrons, neutrons are initially at high energy levels. But usually, their energy level is reduced using moderators of low molecular weight substances, such as paraffin, water and deuterium oxide. There are mainly three sources of neutrons used in neutron activation methods: reactors, radio nuclides and accelerators. Thermal and fast neutrons are produced from nuclear reactors and accelerators, respectively. A nuclear reactor is an assembly of fissionable material (viz., ^{235}U or ^{239}Pu), offering a self-sustaining reaction that results in neutron emission. In contrast, in a typical accelerator, deuterium ions from a source are accelerated through a potential of about 150 KV and interacted with tritium which is absorbed on titanium or zirconium. As a result, neutrons are emitted as follows:

$$^{2}_{1}H + ^{3}_{1}H \rightarrow ^{4}_{2}He + ^{1}_{0}n$$

Neutrons are also produced by mixing intimately an α- or γ-emitter with a light element (beryllium) as per the following nuclear reactions:

(i) $^{9}_{4}Be + ^{4}_{2}He \rightarrow ^{12}_{6}C + ^{1}_{0}n + 5.7\ MeV$

(It is a sort of nuclear fusion reaction)

(ii) $^{9}_{4}Be + \gamma \rightarrow ^{1}_{0}n + ^{8}_{4}Be - 1.66\ MeV$

($^{8}_{4}Be$ is immediately decomposed to 2 α particles)

In NAA, rate of formation of induced radioactivity is expressed as follows (Equation 9.27).

$$\frac{dN'}{dt} = N\varphi\sigma \qquad \text{Equation 9.27}$$

Where,

dN'/dt = the formation rate of induced radioactivity in a neutron per second,

N = the number of stable target atoms capable of forming radioisotopes,

ϕ = average neutron flux (n cm^{-2} s^{-1}), and

σ = neutron capture cross section in cm^2 target $atom^{-1}$ (This is a measure of interaction probability between target atom and neutron).

Along with the formation of radioactive nuclei, they start decaying simultaneously and rate of decay is expressed as

$$\frac{dN'}{dt} = \lambda N' \qquad \text{Equation 9.28}$$

Therefore, from Equation 9.27 and Equation 9.28 one may get a net rate of formation of active particles (Equation 9.29).

$$\frac{dN'}{dt} = N\varphi\sigma - \lambda N' \qquad \text{Equation 9.29}$$

Integrating Equation 9.29 from time *0* to *t*, the activity *A* produced (expressed in disintegrations per second) in a specific nuclear species in an irradiation time *t* is given by

$$A = N\varphi\sigma f(1 - e^{-0.693t/t_{1/2}})$$

$$A = N\varphi\sigma fs \qquad \text{Equation 9.30}$$

Where,

s = the saturation factor which represents the ratio of the amount of the activity produced during irradiation period to that produced in infinite time, and

f = fractional abundance of the target nuclide.

Equation 9.30 may be further extended to obtain weight of the element

$$W = \frac{AM}{N_A\varphi\sigma fs} \qquad \text{Equation 9.31}$$

Where,

W = weight of the element,

N_A = the Avogadro's number, and

M = atomic weight of the element.

It is observed that after exposing the target atoms in any neutron-flux (high, medium and low) for a period of 4-6 half-lives, the count rate remains constant because at that time period the rate of formation and disintegration of the isotope are the same. The nuclide decays as fast as it is being produced.

Neutron activation analysis may be performed by both destructive and nondestructive methods. In both cases the standard (with a known elemental weight) and sample (with a known gross weight) should be irradiated together with the same neutron flux for a period of nearly 4-6 half-life time of the element. Sample and standard are cooled after irradiation, and directly and selectively measured for their radioactivity using a γ-ray spectrometer that can select the right radioactive species and discriminate it from other interfering radiation energies. Cooling helps to reduce short-lived interferences. But along with the desired radioactive species there are possibilities to get other radioactive species that pose problems in measuring the desired radioisotope produced by the element. This problem may be overcome by using a destructive method where unknown and standard samples are treated identically to isolate the desired element by extraction, precipitation, or chromatography after adding a known amount of the stable element as carrier. Because of intermediate treatments for isolation of the desired radioisotope species, it is difficult to get 100% quantitative yield, which may be rectified by monitoring recovery of fortified known amounts of the carrier. Radioactivity of the isolated material is then measured and the amount of the element in the sample is computed from the Equation 9.32.

$$W_{Sample} = \frac{C_{Sample} \times W_{S\tan dard}}{C_{S\tan dard}} \qquad \text{Equation 9.32}$$

Where,

C = count rate, and

W = weight of the element.

NAA is used for qualitative and quantitative multi-elemental analysis based on the measurement of characteristic radiation from radionuclides generated through neutron irradiation of the material or measuring the decay rate of the radioactive element, which is specific to a specific radioactive element. The radioactive emissions and radioactive decay paths for each element are well known. Using this information, it is possible to study spectra of the emissions of the radioactive sample and determine the concentrations of the elements within it. NAA has its wide applications in chemistry, geology, archaeology, medicine, environmental monitoring, forensic science and in many other domains. NAA is a sensitive and non-destructive method of analysing major, minor, and trace elements, present in very low concentrations in soil, water, plant and other biological material, minerals, and other environmental matrices. NAA is especially utilized to measure rare earth elements due to their very high thermal neutron cross sections.

9.8.2. Autoradiography

It is also known as *radioautography*. Becquerel's findings, that atomic radiations show pronounced effect on photographic film, scintillated the idea of autoradiography. It is a technique that involves sensitization of silver grains present in the photographic emulsion by radiations coming from a radioactive source when photographic emulsion and specimen containing radioactive atoms or molecules are kept in close contact. After a suitable exposure time, development of the plate results in production of an image stating location and amount of radioactivity in the specimen. Thus, location of a drug accumulation can be determined by injecting a radioactive drug into an animal. The animal is sacrificed after a suitable time. Thin slices of various organs are prepared and placed on sensitive photographic emulsions. The exposure time is selected on the basis of the nature of radiation and concentration of radioactivity. It may vary from an hour to a month. After development, darkened areas on the film show the locations where labeled drug and its degradation products are accumulated. The intensity of blackening at a given region is the function of the amount of radioactivity at that place and the exposure time. Autoradiography is also applicable to study the distribution pattern of pesticides or nutrients in plants using radioactive compounds. A quantitative estimation of radioactivity may also be possible by scanning an autoradiograph with a densitometer. Autoradiographic resolution depends on the gap width between sample and photographic emulsion, and their thickness. A minimum value of these factors provides a maximum resolution. It also depends on specific ionization of radiation. α- and soft β-particles are highly effective in autoradiography due to their higher specific ionization. Gamma ray hardly produces any effect due to its low specific ionization. Normally, a plastic film is placed in between the specimen and emulsion to avoid artifacts. Artifacts are the result of chemical reactions between specimen and emulsion. The plastic film prevents any direct contact between specimen and emulsion and thus, avoids any artifactual image.

9.8.3. Radiometric titration

It refers to quantitative estimation of a substance involving a reaction between a substance and a radioactive reagent. The reaction results in stoichiometric production of a radioactive compound. Basic requirement of this method is to standardize radioactivity in terms of chemical equivalence. After completion of the reaction, radioactivity of the product or radioactivity of unconsumed reagent is measured to determine the number of equivalents of the reagent consumed. A radiometric titration curve may be prepared by plotting radioactivity as a function of titrant volume. This is most commonly used for precipitation, complexometric and redox titration. As for instance, amount of chromate may be determined by measuring the activity of $*Ag_2CrO_4$ which is formed by precipitating chromate with radioactive *Ag. Three types of radiometric titration may be possible based on the fact whether the reagent and /or titrant are radioactive or non-radioactive (**Figure 9.18**).

Type I

Where a non-radioactive substance (NaCl) is titrated with radioactive titrant ($*AgNO_3$) and the activity of solution is measured, solution activity is found low due to the precipitation of *AgCl. A plateau is maintained up to the equivalence point and after that a sudden upsurge in the titration curve is observed due to gradual increase in $*AgNO_3$ content in solution.

Type II

When a radioactive substance (*$AgNO_3$) is titrated with a non-radioactive titrant (NaCl), the solution activity is initially found more and then gradually decreases due to the decreasing quantity of radioactive *Ag in solution for precipitation of *AgCl. The titration curve finally maintains a plateau after reaching the equivalence point, exhibiting minimum radioactivity in solution.

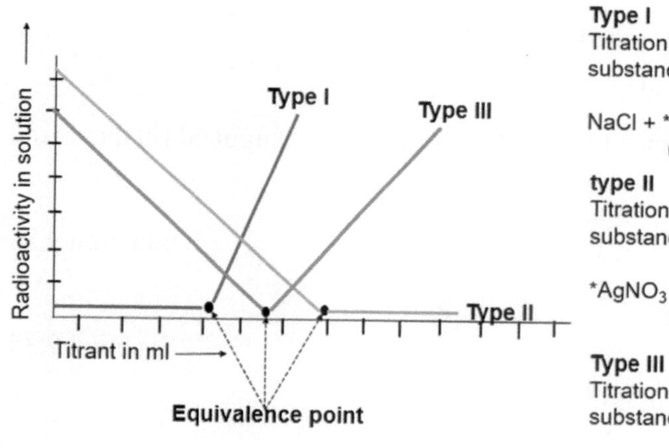

Type I
Titration between non-radioactive substance and radioactive titrant

NaCl + *$AgNO_3$ → *AgCl ↓+ $NaNO_3$
(Titrant)

type II
Titration between radioactive substance and non-radioactive titrant

*$AgNO_3$ + NaCl → *AgCl ↓+ $NaNO_3$
(Titrant)

Type III
Titration between radioactive substance and radioactive titrant

Na*I + *$AgNO_3$ → *Ag*I ↓+ $NaNO_3$
(Titrant)

Fig. 9.18: Radiometric titration curve

Type III

When both titrant (*$AgNO_3$) and reagent (Na*I) are radioactive, a 'V-shaped titration curve is obtained. Initially solution activity is high due to the presence of more amount of radioactive *I. Once the titration is started with *$AgNO_3$, radioactivity of solution decreases due to precipitation of *Ag*I. When the equivalence point is reached, the activity of the solution becomes minimum and immediately there is an upsurge of the titration curve after further addition of *$AgNO_3$.

9.8.4. Isotope dilution analysis

It is an analytical method for quantitative estimation of a compound, containing traces of natural isotopic composition by adding a known amount of tracer (carrier-free or with a known amount of carrier). After equilibration, a small amount of compound is re-isolated and specific activity of the re-isolated compound is measured. This is more specifically known as *Direct Isotopic Dilution*. Quantitative recovery of the species is not important, provided it is sufficient to weigh and to count. The amount of inactive element or compound may be obtained as follows:

$$S_0 = \frac{C_0}{M_0} \quad \text{Equation 9.33}$$

Where,

S_o = specific activity of tracer added,

C_o = counts per minute for the tracer added, and

M_o = amount of tracer added.

After adding tracer, the specific activity of re-isolated compound (in the mixed form) is obtained by

$$\overline{} \quad \text{Equation 9.34}$$

$$M \quad M$$

$$S^{/} = \frac{C_r}{M_r} \quad \text{Equation 9.35}$$

Where,

$S^{/}$ = specific activity of re-isolated compound,

$M^{/}$ = unknown amount of non-radioactive compound in the mixture,

C_r = counts per minute for the re-isolated compound, and

M_r = amount of re-isolated compound.

Dividing Equation 9.33 by Equation 9.34, unknown amount of non-radioactive compound / element, $M^{/}$ is obtained (Equation 9.36).

$$\frac{S_0}{S^{/}} = \frac{\dfrac{C_0}{M_0}}{\dfrac{C_0}{M_0 + M^{/}}}$$

$$\frac{S_0}{S^{/}} = \frac{M_0 + M^{/}}{M_0}$$

$$\frac{S_0}{S^{/}} = 1 + \frac{M^{/}}{M_0}$$

$$\frac{M^{/}}{M_0} = \frac{S_0}{S^{/}} - 1$$

$$M' = M_0(\frac{S_0}{S'} - 1)$$ Equation 9.36

The same principle may be applied to measure an unknown amount of a tagged substance of known specific activity. At that time a known quantity of unlabeled inactive substance is added, and active and inactive species are equilibrated. This is known as *Inverse Isotope Dilution*.

Isotopic dilution method is widely used in organic chemistry, biochemistry and other branches of chemistry, and in radiocarbon dating of archeological and anthropological specimens. This is successfully used for substances that are present in low concentration and cannot be assayed by conventional methods. The positive side of this method is that quantitative recovery of the substance is not required. Both stable and radioactive isotopes are employed in isotope dilution techniques.

9.8.5. Radiocarbon dating

It is a method of dating objects whose life terminated hundreds or thousands of years ago. All living matters contain a proportion of radiocarbon-^{14}C, since in the atmosphere ^{14}C is produced by a reaction between atmospheric nitrogen and the neutrons from cosmic rays [^{14}N (n, p) ^{14}C]. As a result, $^{14}CO_2$ also shares a definite proportion to the total pool of CO_2 in the atmosphere. $^{14}CO_2$ is assimilated in plants through photosynthetic processes, via plants in the animals and again returned to the atmosphere by respiration and other processes (so called carbon cycle). Thus, an equilibrium is established and living matter contains a constant equilibrium concentration of ^{14}C-isotope which is nearly equivalent to 15.3 dpm g^{-1} C. When the living matter is dead, it ceases to be a part of the carbon cycle. Hence, ^{14}C of the dead material starts decaying with a half-life of 5730 years without any further assimilation of new-^{14}C. The age of the material can then be readily calculated using decay equation (Equation 9.23)

$$N_t = N_o \times e^{\lambda t}$$ Equation 9.23

Where,

N_o = 15.3 dpm g^{-1} C,

N_t = number of dpm g^{-1} C at time *t*, and

λ = decay constant for ^{14}C.

The procedure is dependent on the assumption that intensity of cosmic radiation and ^{14}C-specific activity in a biological cycle has been reasonably constant over past thousands of years and that a dead organism ceases to exchange carbon with its surroundings.

9.8.6. Use of radioisotopes to study reaction mechanism

Radioisotopes are considered as a powerful tool to monitor reaction processes. Radioisotopes are widely used as markers to obtain knowledge on the reaction mechanisms in biological reactions. Carrying out experiments with biomolecules, xenobiotics or inorganics tagged with radioisotopes at their various positions help to deduce their metabolic pathways under different conditions. Several research publications are available, that are dealing with reaction mechanisms obtained with the help of radio elements. For instance, when we study the photosynthesis process using CO_2 labelled with radioactive oxygen isotope and after analyzing the nature of free oxygen obtained during the photosynthesis process, we find that the oxygen produced in the reaction is non-radioactive. Thus, we conclude that the production of oxygen in the photosynthesis process is not contributed by carbon dioxide but by water. Similarly, during ester hydrolysis (**Figure 9.19**), an alcohol and an acid are produced. If water is tagged with radioactive oxygen, produced acid is found to contain radioactivity. This observation proves the attainment of hydroxyl groups from water in the produced acid during ester hydrolysis.

$$R{-}C(=O){-}OR + H\overset{*}{O}H \xrightarrow{\text{Hydrolysis}} R{-}C(=O){-}\overset{*}{O}H + ROH$$

Radioactive

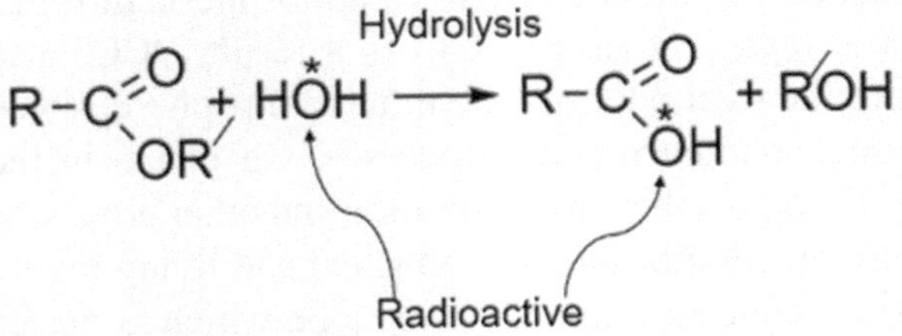

Fig. 9.19: Ester hydrolysis using water containing radioactive oxygen

Radioisotopes also find an important place in the study of the mechanistic ways of enzymes and special protein molecules with catalytic powers.

9.8.7. Use of radioisotope in medicine

Diagnostic and therapeutic uses of radioisotopes are now common practices in medicine. Radioactivity is efficiently employed to diagnose hyperthyroidism, malfunctioning of lungs, kidney infections or imbalance in function of two kidneys, bone fracture, to detect regions of poor circulation inside the heart, etc. High energy radiations (Radiation therapy), in therapeutic uses, are used successfully against cancerous tissues to damage the DNA of cancer cells. A cancer patient may receive external beam radiation therapy, or internal radiation therapy (brachytherapy). Problem of hyperthyroidism had previously been solved surgically, but now the same is treated using radioactive ^{131}I. The patient is offered a dose of radio-iodide, which is preferentially taken up by thyroid. The intense β-radiation is absorbed largely in thyroid cells and kills them off. Charged particles from accelerators in a form of beam are used to treat brain tumors. Narrow wires of ^{60}Co are used by pushing them into the tumors for internal irradiation.

9.8.8. Use of radioisotope in agriculture

Radioisotopes in agricultural research are used mainly in mutation breeding, food preservation, soil moisture measurement, insect control, assessing available soil nutrient to plan optimum fertilizer use, rate of decomposition of organic matter in soil, elucidation of mechanisms involved in ion absorption in soil-plant system, metabolism study and in many other purposes. Rice seeds' γ-irradiation is frequently applied to produce mutant varieties which are heavy-yielding and disease resistant. In the food industry, γ-irradiation (^{60}Co) causes sterilization of packed foods, such as milk, meat, etc. Ionizing radiations are used to protect food from harmful and disease-causing organisms, such as *Salmonella* sp., *E. coli,* etc. Reproductive sterilization of insects (reproductive male sterile technique) with a view to reduce reproduction rate is also possible using electromagnetic radiations. Subsequent release of sterile males into a native population reduces the native population to a level that can be very well controlled by conventional methods.

It is possible to gather information on the distribution pattern, translocation behavior and accumulation site of nutrient elements in the plants, using fertilizers containing radioisotope of that element. Radioisotopes also enable us to study the fate of any compound, once it is applied to soil, water and plant, and to prepare accurately a complete balance-sheet of the radioactivity in respect of their initially applied concentration and concentration remaining in its parent form and other transformed forms at a specific time period after its application.

9.9. Precautions

Following precautions and safety measures should be strictly considered while working with the radioactive substances.

(i) One should use disposable latex or nitrile gloves, close-toed shoes, full-length laboratory coats and safety glasses for working with any radioactive sources. Besides, the sources should only be handled by the forceps and should never be touched by hand.

(ii) One should avoid eating, drinking, and smoking in any room designated for storing or working with any radioactive materials.

(iii) Storing food, beverages, or medicines in refrigerators or cold rooms where radioactive materials are used or stored is strictly prohibited.

(iv) All containers and equipments used for radioactive substances must be labelled properly. There should be a set of dedicated equipment that is only used for radioisotopic work.

(v) It is better to cover the working bench completely with absorbent paper to soak or trap any radioactive contamination while working. Thus, it is easier to remove the radioactive spillage from the working area.

(vi) Wearing of radiation monitoring badges is essential, so that one could be able to know the level of one's exposure with radioactivity.

(vii) Mouth pipetting is strictly avoided while working with radioactive substances.

(viii) One should use fume hood designated for radioactive materials while working with volatile radiolabeled compounds. Biological safety cabinets, especially with laminar flow hoods must be avoided for working with volatile radioactive materials.

9.10. Model questions

(i) What are radioisotopes? Briefly discuss 'isotope', 'isobar' and 'isotone', giving appropriate examples.

(ii) Law of conservation of energy states that all electrons from a given substance possess the same kinetic energy leading to a constant velocity. However, when β-particles from a given source are subjected to a magnetic field, a continuous β-ray spectrum is achieved confirming a continuous variation of velocity. Why?

(iii) What happens to the atomic number and mass of the elements when they emit an α-particle, a β-particle and γ-radiation separately?

(iv) What happens when (a) a radioactive nucleus ($^{A}_{Z}X$) emits an alpha particle followed by a beta particle (b) a beta particle followed by a photon and (c) an alpha particle followed by a photon?

(v) Briefly discuss the various types of radiation from radioactive atoms.

(vi) What happens to a nucleus when β-decay occurs with emission of negatron, positron and electron capturing? Describe in detail.

(vii) Illustrate the use of radioisotopes in the field of agriculture and biochemistry.

(viii) Complete the following reactions:

A. $^{32}P_{17}$ → $^{32}S_{16}$ + ? + antineutrino

B. $^{222}Rn_{136}$ → ? + $^{4}He_{2}$ + 5.58MeV

C. $^{14}C_{8}$ → ? + β^{-} + antineutrino

D. ? → $^{65}Cu_{36}$ + β^{+} neutrino

E. $^{55}Fe_{29}$+ e^{-} → ? + neutrino

(ix) What is bremsstrahlung? Why does it not occur during α-particles' interaction with matter?

(x) Describe the photoelectric effect when photons (γ-rays) interact with matter. How does it differ from the Compton effect?

(xi) 'Pair production is the inverse process of the electron-positron annihilation'------Explain.

(xii) Define half-life of a radioisotope? How can it be deduced?

(xiii) Explain the principle of Geiger Muller counter and liquid scintillation counter. What are the main drawbacks of using a GM counter?

(xiv) Write short notes on the followings:

(a) Radiocarbon dating

(b) Isotope dilution analysis

(C) Radiometric analysis

(d) Autoradiography

(e) Neutron activation analysis

9.11. Suggested Readings

(i) Mathews, C.; van Holde, K.; Ahern, K. (2000) Biochemistry. 3rd Edition, Benjamin/ Cummings, San Francisco, pp 440 – 444.

(ii) Holme, D.; Peck, H. (1998) Analytical Biochemistry. 3rd Edition, Addison Wesley Longman, Essex, U.K., pp 196 – 209.

(iii) Skoog, D. A.; Holler, F. J.; Nieman, A. T. (1998) Principles of Instrumental Analysis. 5th Edition, Harcourt Brace and company, USA.

(iv) Chase G. D.; Rabinowitz, J. L. (1967) Principles of radioisotope Methodology, 3rd Edition, Burgess Publishing Company, Minneapolis.

(v) Clark, J. M.; Switzer, R. L. (1977) Experimental Biochemistry. 2nd Edition, Watt-Freeman and Company, San Francisco, USA, pp 29-42.

(vi) Goulding, K. H.; Slater, R. J. (1994) Radioisotope techniques. In: Keith Wilson and John Walker (Eds) Practical Biochemistry – principles and techniques. 4th Edition, Cambridge Univ. Press, pp 227-274.

(vii) Slater, R. J. (1991) Radioisotopes in Biology – A Practical Approach. IRL Press, Oxford.

(viii) James, T. (1995) Atoms, Radiation, and Radiation Protection, 2nd edition. John Wiley and Sons.